Quantum Computing for Multimedia Processing, Teleportation, and Secure Networks

Quantum Computing for Multimedia Processing, Teleportation, and Secure Networks offers a comprehensive exploration of quantum-enhanced technologies across multimedia, communication, and cryptography. Bridging foundational theory with cutting-edge applications, this book systematically addresses how quantum computing transforms classical approaches to data representation, transmission, and security.

Structured in two parts, the book first delves into Quantum Computing for Multimedia Processing, covering quantum image and video representation, arithmetic and geometric operations, compression, encryption, and data hiding. The second part, Quantum Communication and Cryptography, examines quantum teleportation protocols, secure network design, quantum state dynamics, and the principles of quantum cryptography.

By leveraging the unique properties of quantum mechanics – such as superposition, entanglement, and no-cloning – this work provides scalable solutions to challenges in computational complexity, secure transmission, and robust multimedia systems. It serves as an essential reference for graduate students, researchers, and professionals in quantum computing, multimedia processing, telecommunications, and cybersecurity.

Key features:

- Integrates quantum computing principles with real-world applications in multimedia, teleportation, and secure networking.
- Covers the full spectrum of quantum multimedia processing – from representation and compression to encryption and steganography.
- Explores quantum teleportation as a core mechanism for achieving secure and high-fidelity information transfer.
- Introduces geometric and dynamic perspectives on quantum systems for enhanced modeling and transmission reliability.
- Provides a forward-looking perspective on quantum cryptography and its role in reinventing secure digital communication.

CRC Press Series on Quantum Information Processing and Computing

This interdisciplinary series delves into the multifaceted world of quantum information processing and quantum computing. It covers applications, challenges, and implications for next-generation communications, cybersecurity, and smart emerging technologies. Focusing on cutting-edge research in quantum computing and quantum technologies, this series is aimed at researchers, professionals in applied fields, and graduate-level students with a strong interest in quantum technologies, communications, or cybersecurity.

Authors are also welcome to contact the Editor-in-Chief Ahmed A. Abd El-Latif (ahmedabdellatif@ieee.org) or publisher (Physics Editor, Carolina Antunes: carolina.antunes@tandf.co.uk) to discuss new title ideas.

Quantum Computing for Multimedia Processing, Teleportation, and Secure Networks
Ahmed A. Abd El-Latif, Abdallah Slaoui, and Xianhua Song

Quantum Computing for Multimedia Processing, Teleportation, and Secure Networks

Edited by

Ahmed A. Abd El-Latif, Abdallah Slaoui, and Xianhua Song

CRC Press
Taylor & Francis Group
Boca Raton London New York

CRC Press is an imprint of the
Taylor & Francis Group, an **informa** business

Designed cover image: Shutterstock

First edition published 2026
by CRC Press
2385 NW Executive Center Drive, Suite 320, Boca Raton, FL 33431

and by CRC Press
4 Park Square, Milton Park, Abingdon, Oxon, OX14 4RN

CRC Press is an imprint of Taylor & Francis Group, LLC

ISBN: 9781041006954 (hbk)
ISBN: 9781041250982 (pbk)
ISBN: 9781003621423 (ebk)

DOI:10.1201/9781003621423

Typeset in Times
by Newgen Publishing UK

Contents

PART II Quantum Communication and Cryptography

Preface

The dawn of the quantum computing era has ushered in transformative possibilities across science and technology, promising to redefine how we process information, secure communications, and interact with multimedia data. *Quantum Computing for Multimedia Processing, Teleportation, and Secure Networks* emerges at this pivotal juncture, offering a comprehensive exploration of quantum technologies at the intersection of visual data processing and secure communication.

MOTIVATION AND VISION

This book was born from a recognition of two converging trends: the growing sophistication of quantum algorithms and the urgent need for secure, high-efficiency systems in our multimedia-driven world. While quantum computing has often been discussed in abstract or theoretical terms, our goal is to ground its potential in tangible applications – from quantum-enhanced image compression to teleportation-based video encryption. We bridge the gap between quantum theory and practical implementation, equipping readers to harness quantum advantages in real-world scenarios.

JOURNEY THROUGH THE BOOK

The text is structured to guide readers from foundational principles to cutting-edge innovations:

- Part I delves into quantum multimedia processing, introducing novel representation models (e.g., QIRHSI, NCQI) and operations (arithmetic, geometric, compression) that redefine how images and videos are stored and manipulated on quantum hardware. It culminates in secure multimedia systems, with dedicated chapters on encryption and steganography tailored for quantum video.
- Part II shifts focus to quantum communication, where teleportation and cryptography take center stage. Here, we explore how quantum mechanics enables unhackable networks, linking protocols like QKD to the multimedia security frameworks of Part I.

FOR WHOM THIS BOOK IS WRITTEN

We designed this work to serve a diverse audience:

- Researchers in quantum computing seeking applications in multimedia and cybersecurity.
- Engineers developing quantum algorithms or secure communication infrastructure.
- Students in physics, computer science, or engineering who aspire to work at the quantum frontier.
- Industry professionals evaluating quantum technologies for future deployment.

No prior expertise in quantum multimedia is assumed, though familiarity with classical computing concepts will aid comprehension.

Ahmed A. Abd El-Latif, Abdallah Slaoui, and Xianhua Song

Acknowledgments

This work is derived from a research grant funded by the Research, Development, and Innovation Authority (RDIA) – Kingdom of Saudi Arabia – with grant number (13325-psu-2023-PSNU-R-3-1-EF). Also, the authors would like to thank Prince Sultan University for their support.

About the Editors

Ahmed A. Abd El-Latif is Professor of Quantum Cybersecurity, with over 18 years of professional experience. He earned his PhD with honors from Harbin Institute of Technology, China, in 2013, and has since led numerous successful research projects and grants across Egypt, the Russian Federation, Saudi Arabia, and Tunisia. He has held the position of Full Professor at Menoufia University since July 2024. His extensive research portfolio includes over 330 publications in high-impact journals and conferences, 20 books, and more than 15,000 citations. Since 2022, he has served as the Head of the MEGANET 6G Lab in the Russian Federation and is the Vice Chair of the EIAS Research Lab. Additionally, he is the Founder and Deputy Director of the Centre of Excellence in Quantum & Intelligent Computing. Prof. Abd El-Latif has held prestigious academic and research roles globally, including Head of the MEGANET6G Lab in Russia, Postdoctoral Fellow at Harbin Institute of Technology, China, and Visiting Professorships at the Polish Academy of Sciences and Warsaw University of Technology, Poland, producing over 36 high-impact publications during these engagements. His research focuses on pioneering advancements in quantum computing, chaotic dynamical systems, and artificial intelligence, with applications in cybersecurity and the Artificial Intelligence of Things (AIoT). His work in quantum computing emphasizes the development of novel algorithms for data representation, cryptography, privacy, authentication, and information hiding in quantum scenarios. He is currently leading groundbreaking research in quantum information processing and quantum-inspired algorithms to address the challenges of secure data transmission and storage in the quantum computing era. His contributions extend to applications in wireless sensor networks, IoT, healthcare, and smart cities. A recognized leader in his field, Prof. Abd El-Latif has received several prestigious awards, including the State Encouragement Award in Engineering Sciences (2016, Egypt), the Best PhD Student Award from Harbin Institute of Technology (2013, China), and the Young Scientific Award from Menoufia University (2014, Egypt). He serves as Editor-in-Chief of the *International Journal of Information Security and Privacy* and as a series editor for Quantum Information Processing and Computing and Advances in Cybersecurity Management. He also holds editorial roles in numerous indexed journals (WoS and Scopus quartile-ranked). Prof. Abd El-Latif's vision is to lead the transition to quantum-safe cybersecurity by developing innovative cryptographic protocols and educating the next generation of researchers and professionals in post-quantum cryptography and quantum-inspired security solutions. His work continues to shape the future of secure data transmission and storage in the digital and quantum eras.

Abdallah Slaoui is Professor of Physics at the Faculty of Sciences, Mohammed V University in Rabat, Morocco. He is a theoretical physicist specializing in quantum technologies and mathematical physics. Slaoui earned his PhD degree from Mohammed V University with a thesis titled "Intrinsic relationships of quantum resource theories and their roles in quantum metrology." Following that, he served as a postdoctoral researcher at the Centre of Physics and Mathematics (CPM). Slaoui's research focuses primarily on the fundamental concept of quantumness, studying its manifestations through nonclassicality, quantum coherence, and entanglement. A major thrust of his work is understanding the role these resources play in the dynamics of quantum many-body systems and in the burgeoning field of quantum metrology. More broadly, his interests lie in the dynamics of quantum resources within open quantum systems, which naturally incorporates an interest in quantum metrology and measurement theory. Furthermore, he explores the critical interplay between information and thermodynamics in the quantum regime, alongside the development and implementation of quantum communication protocols for quantum computers. Prof. Slaoui has authored and co-authored several scientific papers, actively participating in numerous national and international congresses where he presented his research. Additionally, he has been invited to act as a reviewer for many scientific manuscripts from prestigious journals. Beyond his academic pursuits,

Prof. Slaoui is passionate about engaging the public in Quantum Physics activities. He is a member of the National Network of Young Researchers in Sciences, and he is involved in the organization of national and international events.

Xianhua Song received her BS and MS degrees from Jilin University in 2005 and 2007, respectively, and the PhD degree from Harbin Institute of Technology in 2015. She is currently an associate professor with the School of Science, Harbin University of Science and Technology, China. She is a permanent member of Heilongjiang Mathematical Society and member of China Computer Society. Her research interests include multimedia security, machine learning, and quantum computation.

Contributors

Bassem Abd-El-Atty
Luxor University
Egypt

Ahmed A. Abd El-Latif
Prince Sultan University
Saudia Arabia

Basma Abd El-Rahiem
Menoufia University
Egypt

Rachid Ahl Laamara
Mohammed V University in Rabat
Morocco

Brahim Amghar
Chouaïb Doukkali University in El Jadida
Morocco

Lalla Btissam Drissi
Mohammed V University in Rabat
Morocco

Yassine Dakir
Mohammed V University in Rabat
Morocco

Bouchra El Alaoui
Mohammed V University in Rabat
Morocco

Mohammed El Falaki
Mohammed V University in Rabat
Morocco

Hanane El Hadfi
Mohammed V University in Rabat
Morocco

Meryem El Kirdi
Mohammed V University in Rabat
Morocco

Safae Gaidi
Mohammed V University in Rabat
Morocco

Nada Ikken
Mohammed V University in Rabat
Morocco

Shaocong Lan
Harbin University of Science and Technology
China

Oussama Latifi
Mohammed V University in Rabat
Morocco

Youssef Oulouda
Mohammed V University in Rabat
Morocco

Abdallah Slaoui
Mohammed V University in Rabat
Morocco

Xianhua Song
Harbin University of Science and Technology
China

Fuzhou Wang
Harbin University of Science and Technology
China

Mouhcine Yachi
Chouaïb Doukkali University in El Jadida
Morocco

Zhixin Zhang
Harbin University of Science and Technology
China

Zhiheng Zhao
Harbin University of Science and Technology
China

Part I

Quantum Computing for Multimedia Processing

1 Fundamentals of Quantum Representations and Processing for Visual Data

Fuzhou Wang, Xianhua Song, and Basma Abd El-Rahiem

1.1 RESEARCH PURPOSE AND SIGNIFICANCE

5G/6G can meet the transmission of high-definition video, virtual reality, and other large amounts of data in the era of big data due to its high data rate, low latency, large capacity, and large-scale equipment connection. The data information (especially the image and video media information) will be more and more huge, and the data to be analyzed will be more and more complex. In addition, the requirements of specific fields for the calculation accuracy, the calculation resources will increasingly become a bottleneck in the development of big data technology. In order to improve computing efficiency, researchers proposed many supercomputing models while developing supercomputing equipment. Among all supercomputing models, quantum computer is expected to solve the computational efficiency problem faced by Turing computing model due to its natural and powerful parallel computing ability. In particular, quantum states have the properties of coherent superposition and quantum entanglement, which are not found in classical physics, making the computing power of quantum computers much higher than that of classical Turing computers. It can be seen that quantum information processing technology will have important research value in the era of big data.

Media (text, audio, image, and video) is an important source of human information and an important means of using information. Compared with text and audio media, image and video (i.e., image sequence), as the information body closest to human senses, contain more information and are more intuitive, and play a greater role in information dissemination. With the rapid development of Internet technology, the security of image and video information has become increasingly prominent. How to effectively keep information confidential for massive image and video big data at the same time, quantum computing technology is an effective way to solve this problem.

Generally, the security issues of image and video mainly include the secure transmission of image and video and content authentication.

Secure transmission technology mainly includes:

- *Encryption:* achieve the purpose of content confidentiality by modifying or scrambling the image and video content to make it look meaningless.
- *Secret Sharing:* divide the secret image or video into some shadow images or videos and distribute them to different participants. Only a sufficient number of participants can restore or approximately restore the original secret information together.

Content authentication is to confirm the authenticity of the statement through a trusted way. The statement generally refers to the source, integrity, and authenticity of images and videos. Content authentication can be divided into active authentication and passive authentication.

DOI: 10.1201/9781003621423-2

The active authentication technology mainly includes:

- *Hashing:* mapping image content to a hash sequence, also known as a digital signature based on image content.
- *Watermark:* the watermark information is pre-embedded to realize the content authentication and copyright protection of the image.

The most representative method of passive authentication is image forensics. Passive image blind forensics tracks and authenticates the source of the image through analysis and detection, and distinguishes and identifies the authenticity of the image.

Quantum image and video security method is based on the representation and processing of quantum images and quantum videos. Unlike classical media processing, quantum image and video processing algorithms first need to encode the image and video as quantum states, then perform evolutionary operations according to the laws of quantum mechanics, and finally extract the results of quantum image and video processing according to the theory of quantum measurement. Therefore, the representation and processing technology of quantum images and videos is an inevitable key issue in the research of quantum media security technology.

This book first offers a general framework for the representation of quantum image and video, and under this framework, gives several specific representation models and preparation methods of quantum image and video. Then the basic processing methods of quantum image and video are introduced, including algebraic operation, geometric operation, compression, and frame motion detection. Finally, the security methods of quantum image and video are introduced, including quantum image and video watermarking, steganography, encryption and secret sharing, etc.

To sum up, the research on quantum image and video security has the following significance:

1. Promote the research of quantum multimedia technology.
2. Promote the development of the multimedia security field.

1.2 QUANTUM IMAGE AND VIDEO FUNDAMENTALS

Quantum image and video are stored and represented on quantum computers, and should follow the basic theories and methods of quantum computing, which is supported by the principles of quantum mechanics. Because the world we live in is a classic world, we cannot display and evaluate images and videos without the assistance of classic computers. The basic theories of quantum mechanics, the general processes of quantum computing theory, and the basic concepts of quantum image and video are introduced below.

1.2.1 Quantum Mechanics-Related Concepts

The logic system of quantum computing is based on quantum mechanics. Quantum physical processes determine the results of quantum computation. The remarkable characteristics of quantum computing are the coherent superposition and the unitary transformation of operations. Here, we briefly introduce the principles of quantum mechanics related to quantum computing.

1. *Superposition:* In quantum mechanics, the quantum state of microscopic particles is described by the wave function, which can be completely described by a unit vector in Hilbert space. Schrödinger equation in quantum mechanics Equation 1.1 is as follows:

$$i\hbar\frac{d}{dt}\psi = H\psi \tag{1.1}$$

It is shown that if $\psi_1, \psi_2, \ldots, \psi_n$ describes the possible quantum states of the system, their linear superposition

$$\psi = c_1\psi_1 + c_2\psi_2 + \cdots + c_n\psi_n \tag{1.2}$$

It is also a possible quantum state of the quantum system, where $c_i\,(i = 1,2,\ldots,n)$ is a constant. Corresponding to quantum computing, it shows that quantum computers can represent multiple states in classical computers at the same time, which makes large-scale parallel storage and computing possible.

2. *Coherence*: The superposition of quantum state superposition principle refers to coherent superposition, that is, the amplitude of superposition interferes with each other, resulting in mutual growth or elimination. In quantum computation, the interference of quantum states can be represented by a unitary transformation (unitary transformation, unitary matrix) of quantum gates.
3. *Unitary Transformation:* The process of quantum computation corresponds to the evolution of wave functions. The quantum state evolves according to the unitary transformation law of the quantum gate, and each state in the system evolves simultaneously according to the unitary transformation, so a quantum calculation can act on multiple data. In the quantum register, the quantum state evolves through the unitary transformation of the quantum gate. When the ground state is known, the quantum gate can be directly described by the vector matrix in Hilbert space. Due to the linear constraint of quantum gates, quantum gates transform all ground states of quantum states in Hilbert space.
4. *Entanglement:* Quantum entanglement refers to the non-localized and non-classical correlation between composite quantum systems, which is the mechanical property of the correlation between various subsystems or degrees of freedom in a quantum system, and is a pure quantum effect. For example, in two interacting subsystems, if some quantum states cannot be represented by the tensor product of the ground state in the two subsystems, these states are called quantum entangled states. The entanglement of quantum states is an important theoretical basis for high-speed and indecipherable information communication.
5. *No-Cloning Theorem:* Cloning is the creation of an identical quantum state in another system without changing the original quantum state. An unknown quantum state cannot be observed accurately without interference. It makes quantum communication safe from eavesdropping or eavesdropping information that cannot be interpreted, and it is a perfect anti-theft method of secrecy. However, this also limits the precise "copy" operation in quantum computers. Although it cannot be accurately copied, the "probabilistic quantum cloning" is possible.

1.2.2 Quantum Computing-Related Concepts

A basic quantum system is the integration of qubits, quantum gates, and output reading. The operation process is performed by preparation and storage technology, manipulation technology, and measurement technology, respectively. The following are the basic principles of quantum mechanics related to quantum computing and the three main parts of quantum systems: quantum bits, quantum circuits in quantum evolution, and quantum measurement used for output.

Quantum algorithm determines the process of quantum computing, and different quantum computing processes correspond to different unitary operation sequences. As shown in Figure 1.1, the quantum computing process is actually the selective operation of different qubits in the quantum state to make it evolve according to certain requirements. These operations must be unitary and reversible. The simulation process of quantum algorithm mainly includes:

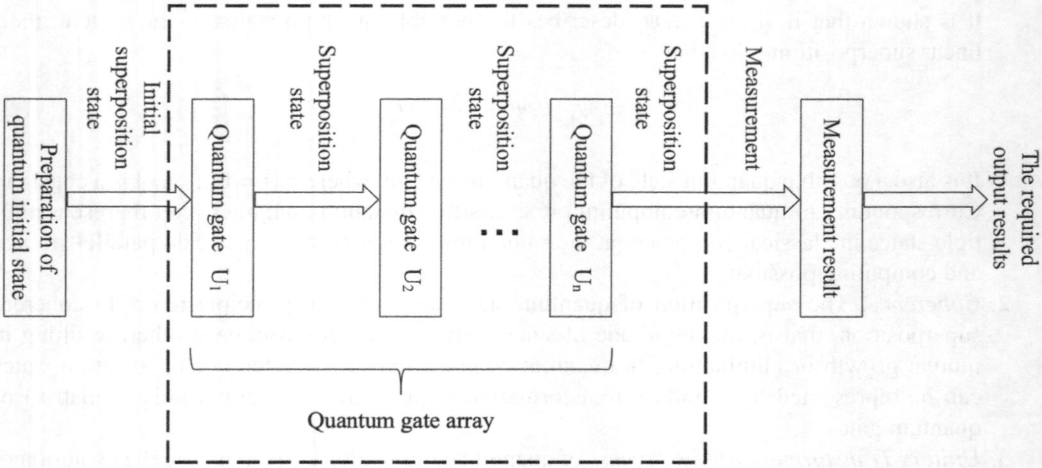

FIGURE 1.1 General process of quantum computing.

1. *Preparation of Quantum Initial State:* The input data is prepared into the initial superposition state and stored in the quantum register.
2. *Quantum Gate Operation:* After the input data quantum state undergoes unitary transformation U_i, $i=1,2,...,n$, the processed superposition state is obtained.
3. *Result Measurement:* Finally, the quantum measurement operation is carried out to obtain the required output results.

1.2.2.1 Quantum Bit

Bit (bit, represented by 0 and 1) is a basic concept of classical computing. Quantum information and quantum computing are based on the concept of quantum bit (quantum bit, abbreviated as qubit)[1]. One possible state of a qubit is $|0\rangle$, that of another qubit $|1\rangle$, The Dirac symbol is "$|\cdot\rangle$", and these two states are common quantum ground states, which constitute a set of orthogonal bases of a two-dimensional Hilbert space.

$$|0\rangle=\begin{pmatrix}1\\0\end{pmatrix},\ |1\rangle=\begin{pmatrix}0\\1\end{pmatrix} \tag{1.3}$$

The value of qubits can be $|0\rangle$, or can be $|1\rangle$ or can even be taken $|0\rangle$ and $|1\rangle$ as the superposition state of the base, as shown in Equation 1.4,

$$|\varphi\rangle=\alpha|0\rangle+\beta|1\rangle\equiv\begin{pmatrix}\alpha\\\beta\end{pmatrix} \tag{1.4}$$

where the quantum bit is expressed as $|\varphi\rangle$ and $\alpha,\beta\in\mathbb{C}$ is the probability amplitude and meets the normalization condition, that is, $|\alpha|^2+|\beta|^2=1$. When measuring the state of $|\varphi\rangle$, $|\varphi\rangle$ will collapse to the ground state $|0\rangle$ with the probability of $|\alpha|^2$, and collapse to the ground state $|1\rangle$ with the probability of $|\beta|^2$. This collapse requires measurement. A multiqubit quantum system can be described as a vector in a complex Hilbert space, $|\varphi\rangle$ called blade arrow or right arrow (ket). The duality of $\langle\varphi|$ is called diao arrow or left arrow (bra), the quantum state $|\varphi\rangle$ represented by a complex vector in a 2^n-dimensional Hilbert space is composed of n bit qubits; see Equation 1.5:

$$|\varphi\rangle = \begin{pmatrix} \varphi_0 \\ \varphi_1 \\ \vdots \\ \varphi_{2^n-1} \end{pmatrix}, \quad \varphi_i \in \mathbb{C}\left(i = 0,1,\ldots,2^n - 1\right) \tag{1.5}$$

Apply the conjugate transpose operation represented by the symbol "†" to the $|\varphi\rangle$, and the duality of $|\varphi\rangle$ can be obtained as follows:

$$\langle\varphi| = |\varphi\rangle^{\dagger} = \left(\varphi_0^{\dagger}\ \varphi_1^{\dagger} \cdots \varphi_{2^n-1}^{\dagger}\right) \tag{1.6}$$

For example, $\langle 0| = |0\rangle^{\dagger} = (1\ \ 0)$, $\langle 1| = |1\rangle^{\dagger} = (0\ \ 1)$.

Because direct product and direct sum operations are involved in the process of quantum circuit design. The two most commonly used operations are described below.

The direct product, also known as the Kronecker product, is represented by the symbol "⊗". Let A be a complex matrix of order $m \times n$, and B be a complex matrix of order $p \times q$:

$$A = \begin{pmatrix} a_{00} & a_{01} & \cdots & a_{0,n-1} \\ a_{10} & a_{11} & \cdots & a_{1,n-1} \\ \vdots & \vdots & \ddots & \vdots \\ a_{m-1,0} & a_{m-1,1} & \cdots & a_{m-1,n-1} \end{pmatrix}, \quad B = \begin{pmatrix} b_{00} & b_{01} & \cdots & b_{0,q-1} \\ b_{10} & b_{11} & \cdots & b_{1,q-1} \\ \vdots & \vdots & \ddots & \vdots \\ b_{p-1,0} & b_{p-1,1} & \cdots & b_{p-1,q-1} \end{pmatrix} \tag{1.7}$$

Then the tensor product of matrices A and B is:

$$A \otimes B = \begin{pmatrix} a_{00}B & a_{01}B & \cdots & a_{0,n-1}B \\ a_{10}B & a_{11}B & \cdots & a_{1,n-1}B \\ \vdots & \vdots & \ddots & \vdots \\ a_{m-1,0}B & a_{m-1,1}B & \cdots & a_{m-1,n-1}B \end{pmatrix} \tag{1.8}$$

From Equation 1.8, $A \otimes B$ is a matrix of order $mp \times nq$. For example:

$$A \otimes B = \begin{pmatrix} 1 & 2 \\ 3 & 4 \end{pmatrix} \otimes \begin{pmatrix} 1 & 2 \\ 2 & 3 \\ 3 & 4 \end{pmatrix} = \begin{pmatrix} 1B & 2B \\ 3B & 4B \end{pmatrix} = \begin{pmatrix} 1 & 2 & 2 & 4 \\ 2 & 3 & 4 & 6 \\ 3 & 4 & 6 & 8 \\ 3 & 6 & 4 & 8 \\ 6 & 9 & 8 & 12 \\ 6 & 12 & 12 & 16 \end{pmatrix}$$

The tensor product of a quantum state $|\varphi\rangle$ and a quantum state $|\psi\rangle$ is $|\varphi\rangle \otimes |\psi\rangle$, sometimes abbreviated to $|\varphi\rangle|\psi\rangle$ or $|\varphi\psi\rangle$. The nth tensor product of $|\varphi\rangle$ the secondary tensor is denoted as $|\varphi\rangle^{\otimes n} = |\varphi\rangle \otimes |\varphi\rangle \otimes \cdots \otimes |\varphi\rangle$. Tensor product is the operation that two vector spaces form a larger vector space. In quantum mechanics, the state of quantum is described by the unit vector in Hilbert space. Therefore, the following form of tensor product operation is often used:

$$|00\rangle = |0\rangle \otimes |0\rangle = \begin{pmatrix} 1 \\ 0 \\ 0 \\ 0 \end{pmatrix} . |0\varphi\rangle \otimes |\varphi\rangle = \begin{pmatrix} 1 \\ 0 \end{pmatrix} \otimes \begin{pmatrix} \alpha \\ \beta \end{pmatrix} = \begin{pmatrix} \alpha \\ \beta \\ 0 \\ 0 \end{pmatrix}$$

The symbol "⊕" is used to represent the direct sum, given the matrix A and matrix B in Equation 1.7, which defines the direct sum as

$$A \oplus B = \begin{pmatrix} A & 0_{m\times q} \\ 0_{p\times n} & B \end{pmatrix} \tag{1.9}$$

Obviously, $A \oplus B$ is a matrix of order $(m+p)\times(n+q)$. For example,

$$A \oplus B = \begin{pmatrix} 1 & 2 \\ 3 & 4 \end{pmatrix} \oplus \begin{pmatrix} 1 & 2 \\ 2 & 3 \\ 3 & 4 \end{pmatrix} = \begin{pmatrix} 1 & 2 & 0 & 0 \\ 3 & 4 & 0 & 0 \\ 0 & 0 & 1 & 2 \\ 0 & 0 & 2 & 3 \\ 0 & 0 & 3 & 4 \end{pmatrix}$$

In addition to direct product and direct sum, inner product and outer product are commonly used operations. $\langle x | y \rangle = \langle x || y \rangle$ is called the inner product of $|x\rangle$ and $|y\rangle$. It is a scalar and has $\langle 0|0 \rangle = \langle 1|1 \rangle = 1$, $\langle 0|1 \rangle = \langle 1|0 \rangle = 0$. If they are multiplied in the reverse order, that is, a column vector is multiplied by a row vector to obtain a matrix, $|x\rangle\langle y|$ is called the outer product of $|x\rangle$ and $|y\rangle$, and is a matrix operator, for example

$$|0\rangle\langle 0| = \begin{pmatrix} 1 & 0 \\ 0 & 0 \end{pmatrix}, |1\rangle\langle 0| = \begin{pmatrix} 0 & 0 \\ 1 & 0 \end{pmatrix}$$

If $|0\rangle\langle 0|$ will act on $|\varphi\rangle$, there are

$$|0\rangle\langle 0|(|\varphi\rangle) = \begin{pmatrix} 1 & 0 \\ 0 & 0 \end{pmatrix}(\alpha|0\rangle + \beta|0\rangle) = \alpha|0\rangle$$

It can be seen that $|0\rangle\langle 0|$ puts forward the component belonging to $|\varphi\rangle$ in $|0\rangle$, or $|0\rangle\langle 0|$ makes $|\varphi\rangle$ project on $|0\rangle$, that is, $|\varphi\rangle$ measures in the $|0\rangle$ direction. Similarly $|1\rangle\langle 1|(|\varphi\rangle) = \beta|1\rangle$, $|\varphi\rangle$ measures in the $|1\rangle$ direction.

1.2.2.2 Quantum Logic Gate

Classical computers use electronic circuits to calculate, while quantum computers are implemented through quantum logic gates. Unlike most traditional logic gates, quantum logic gates are reversible[2]. Unitarity is the only constraint for the establishment of quantum gates, and each unitary matrix corresponds to an effective quantum gate.

1.2.2.2.1 Single-Qubit Gate

In quantum computing, there are many commonly used 1-qubit gates, and the quantum circuits and corresponding unitary matrices (unitary matrices) of some commonly used 1-qubit gates are shown in Table 1.1.

TABLE 1.1
Commonly Used 1-Qubit Gates

Quantum Gate	Line Symbol	Unitary Matrix
Identity Gate	I ≡ —	$\begin{bmatrix} 1 & 0 \\ 0 & 1 \end{bmatrix}$
Hadamard Gate	H	$\frac{1}{\sqrt{2}}\begin{bmatrix} 1 & 1 \\ 1 & -1 \end{bmatrix}$
Pauli-X Gate	X ≡ ⊕	$\begin{bmatrix} 0 & 1 \\ 1 & 0 \end{bmatrix}$
Pauli-Y Gate	Y	$\begin{bmatrix} 0 & -i \\ i & 0 \end{bmatrix}$
Pauli-Z Gate	Z	$\begin{bmatrix} 1 & 0 \\ 0 & -1 \end{bmatrix}$
Phase Gate	S	$\begin{bmatrix} 1 & 0 \\ 0 & i \end{bmatrix}$
π/8 Gate	T	$\begin{bmatrix} 1 & 0 \\ 0 & e^{i\pi/4} \end{bmatrix}$
Rotate α along the x-axis	$R_x(\alpha)$	$\begin{bmatrix} \cos\frac{\alpha}{2} & -i\sin\frac{\alpha}{2} \\ -i\sin\frac{\alpha}{2} & \cos\frac{\alpha}{2} \end{bmatrix}$
Rotate β along the y-axis	$R_y(\beta)$	$\begin{bmatrix} \cos\frac{\beta}{2} & -\sin\frac{\beta}{2} \\ \sin\frac{\beta}{2} & \cos\frac{\beta}{2} \end{bmatrix}$
Rotate γ along the z-axis	$R_z(\gamma)$	$\begin{bmatrix} e^{-i\gamma/2} & 0 \\ 0 & e^{-i\gamma/2} \end{bmatrix}$

Taking Hadamard gate (H gate) as an example, when H gate acts on $|0\rangle$ and $|1\rangle$, we can get

$$H|0\rangle=\frac{1}{\sqrt{2}}\left(|0\rangle+|1\rangle\right),\ H|1\rangle=\frac{1}{\sqrt{2}}\left(|0\rangle-|1\rangle\right)$$

If $H\otimes H$ is applied to $|00\rangle$, then

$$\left(H\otimes H\right)|00\rangle=\frac{1}{2}\left(|00\rangle+|01\rangle+|10\rangle+|11\rangle\right)$$

It can be obtained that the H-transform of each bit of $|00\rangle$ can produce the superposition of 2^2 basic states of 2 qubits, that is, $|00\rangle$, $|01\rangle$, $|10\rangle$, and $|11\rangle$ exist at the same time, and the probability of existence is $\frac{1}{4}$. When all n states in the quantum register are $|0\rangle$, the application of H-transform to each bit can produce a superposition of 2^n basic states that is, all binary numbers from 0 to 2^n-1 are generated, and the probability of simultaneous existence is $\frac{1}{2^n}$.

TABLE 1.2
Commonly Used 2-Qubit Gates

Quantum Gate	Line Symbol	Unitary Matrix
Controlled NOT Gate	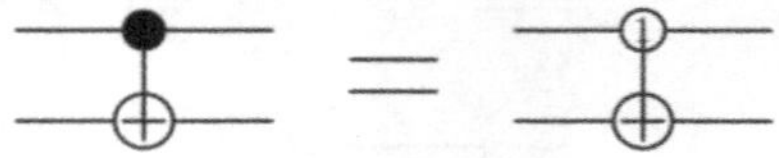	$\begin{bmatrix}1&0&0&0\\0&1&0&0\\0&0&0&1\\0&0&1&0\end{bmatrix}$
Zero-Control NAND Gate		$\begin{bmatrix}0&1&0&0\\1&0&0&0\\0&0&1&0\\0&0&0&1\end{bmatrix}$
Swap Gate	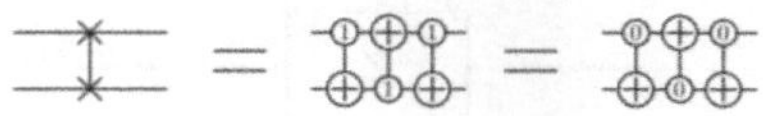	$\begin{bmatrix}1&0&0&0\\0&0&1&0\\0&1&0&0\\0&0&0&1\end{bmatrix}$
Controlled Phase Gate	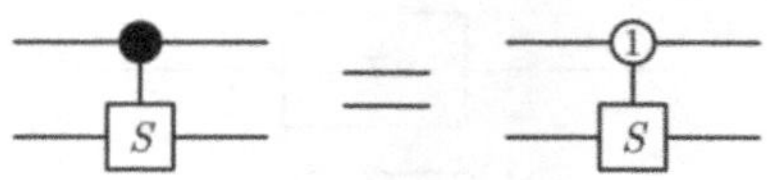	$\begin{bmatrix}1&0&0&0\\0&1&0&0\\0&0&1&0\\0&0&0&i\end{bmatrix}$

TABLE 1.3
Commonly Used 3-Qubit Gates

Quantum Gate	Line Symbol	Unitary Matrix
Toffoli Gate	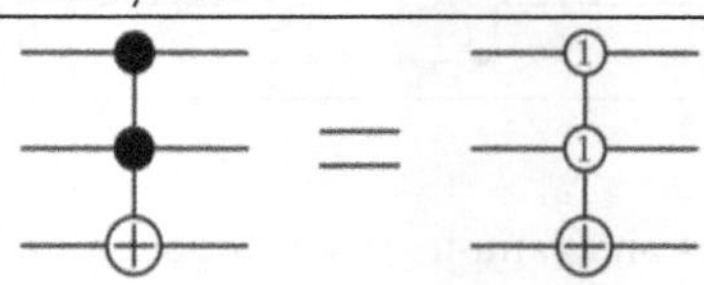	$\begin{bmatrix}1&0&0&0&0&0&0&0\\0&1&0&0&0&0&0&0\\0&0&1&0&0&0&0&0\\0&0&0&1&0&0&0&0\\0&0&0&0&1&0&0&0\\0&0&0&0&0&1&0&0\\0&0&0&0&0&0&0&1\\0&0&0&0&0&0&1&0\end{bmatrix}$
Fredkin Gate	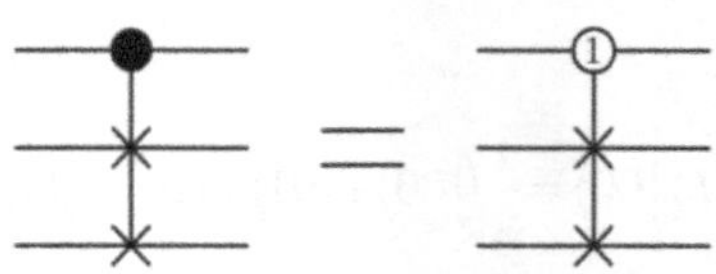	$\begin{bmatrix}1&0&0&0&0&0&0&0\\0&1&0&0&0&0&0&0\\0&0&1&0&0&0&0&0\\0&0&0&1&0&0&0&0\\0&0&0&0&1&0&0&0\\0&0&0&0&0&0&1&0\\0&0&0&0&0&1&0&0\\0&0&0&0&0&0&0&1\end{bmatrix}$

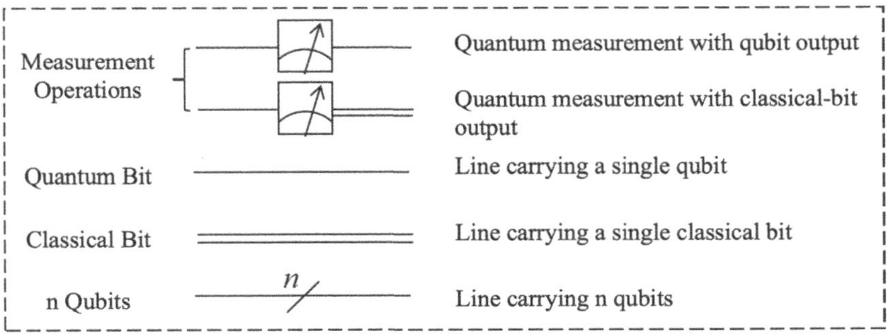

FIGURE 1.2 Commonly used quantum circuit symbols.

1.2.2.2.2 Multiqubit Gate

Table 1.2 shows four commonly used 2-qubit gates, and Table 1.3 shows two types of 3-qubit gates.

1.2.2.3 Quantum Measurement

The last operation used in quantum circuits is quantum measurement, which is implicit in some cases[1]. A set of measurement operators $\{M_m\}$ is used to describe the quantum measurement, and these operators are executed on the state space of the system under test, and the subscript m represents the possible measurement results in the experiment. Figure 1.2 shows the commonly used quantum circuit symbols.

Assuming $|\psi\rangle$ is the quantum state to be measured, and the probability of occurrence of the measurement result m is determined by

$$p(m) = \langle\psi| M_m^\dagger M_m |\psi\rangle \tag{1.10}$$

Given the status of the system after measurement is:

$$\frac{M_m|\psi\rangle}{\sqrt{\langle\psi| M_m^\dagger M_m |\psi\rangle}} \tag{1.11}$$

The measurement operator M_m must satisfy the completeness equation:

$$\sum_m M_m^\dagger M_m = 1 \tag{1.12}$$

The completeness equation states the fact that the sum of probabilities is 1:

$$1=\sum_m p(m)=\sum_m \langle\psi| M_m^\dagger M_m |\psi\rangle \tag{1.13}$$

Here, we take the measurement results of a 1 qubit on a computational basis as an example. The measurement operator $\{M_m\} = \{M_0, M_1\}$ is used to measure the quantum state $|\psi\rangle=\alpha|0\rangle+\beta|1\rangle$, it should be noted that $M_0 = |0\rangle\langle 0|$, $M_1 = |1\rangle\langle 1|$ and

$$I = M_0^\dagger M_0 + M_1^\dagger M_1=|0\rangle\langle 0|+|1\rangle\langle 1| = M_0 + M_1 \tag{1.14}$$

Then the probability of obtaining the measurement results 0 and 1 is:

$$p(0)=\langle\psi|M_0^{\dagger}M_0|\psi\rangle=\langle\psi|M_0|\psi\rangle=|\alpha|^2 \tag{1.15}$$

$$p(1)=\langle\psi|M_1^{\dagger}M_1|\psi\rangle=\langle\psi|M_1|\psi\rangle=|\beta|^2 \tag{1.16}$$

Therefore, the measured states are:

$$\frac{M_0|\psi\rangle}{|\alpha|}=\frac{\alpha}{|\alpha|}|0\rangle \tag{1.17}$$

$$\frac{M_1|\psi\rangle}{|\beta|}=\frac{\beta}{|\beta|}|1\rangle \tag{1.18}$$

1.2.2.4 Quantum Circuit

According to the related concepts of quantum computing, the key to implementing a quantum algorithm is to design a quantum circuit (corresponding to a unitary matrix) that meets the functional requirements. By using the matrix product, direct product, and direct sum operations, a quantum circuit diagram can be transformed into its implicit unitary matrix. Therefore, the entire circuit corresponding to the quantum circuit can be realized. Figure 1.3 shows a circuit of Toffoli gate composed of Hadamard gate, phase gate, CNOT gate, and π/8 gate[1].

In general, given a complex multi-qubit gate, there are different but equivalent decomposition methods, and the use form of the basic quantum gate determines how to decompose. The constructed arbitrary quantum circuit is based on the general quantum gate of quantum computing[3]. Here are three lemmas that will be used later.

Lemma 1.1 For any second-order unitary matrix U, a $\Lambda_1(U)$ gate is equivalent to at most six basic gates, namely four 1-bit Λ_0 gates and two CNOT gates.

Lemma 1.2 For any second-order unitary matrix U, a $\Lambda_2(U)$gate is at most equivalent to sixteen fundamental gates, namely eight 1-bit Λ_0 and eight CNOT gates.

Lemma 1.3 For any $n \geq 3$ and second-order unitary matrix U, a $\Lambda_{n-1}(U)$gate is equivalent to $2^{n-1}-1\Lambda_1(V)$ gates and $2^{n-1}-2$ CNOT gate, where V is a unitary matrix and satisfies $V^{2^{n-2}}=U$.

A key factor to evaluate a quantum algorithm is the complexity of the quantum circuit, which can be discussed from three aspects: width, length, and size. The width is the total number of qubits

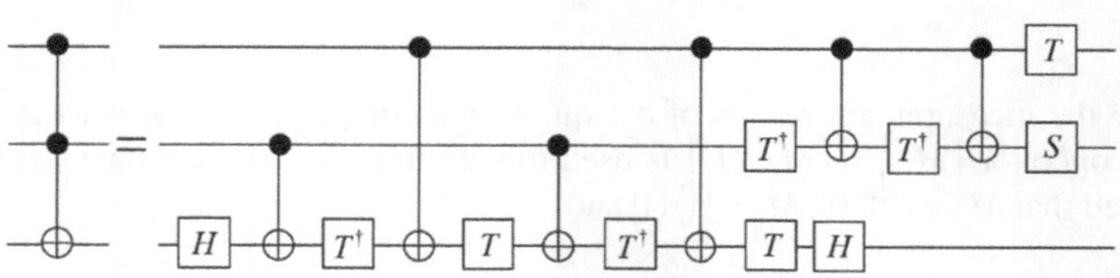

FIGURE 1.3 Toffoli gate realized by Hadamard gate, S gate, CNOT gate, and T gate.

acted by the quantum circuit (including the auxiliary qubits required for calculation), the length is the number of serial gate operations after the quantum circuit is processed in parallel as far as possible, and the size is the total number of quantum gates used by the whole circuit. If the size (or other complexity measures) of a quantum circuit increases polynomial with the number of qubits n, that is, it has the form $n^k, k > 0$. Then the circuit has "polynomial-size," and the quantum computation is said to be effective at this time. However, if the size increases exponentially with the number of qubits n, that is, it has the form 2^n or e^n, then the circuit is called "exponential-size," which is an inefficient way to achieve the purpose of calculation. For the calculation of the complexity of a quantum circuit, it is mainly measured by the size of the circuit, that is, the number of basic quantum gates required to realize the circuit, that is, the number of universal 1-qubit gates, 2-qubit gates, and 3-qubit gates[4].

Quantum adder and modulo-N adder are two quantum circuits to be used in this book.

1.2.2.4.1 General Adder

To calculate the sum of two numbers stored in two registers $|a\rangle$ and $|b\rangle$, it needs to be realized with the help of an ordinary adder. Addition is one of the most important operations, as shown in Equation 1.19:

$$|a,b,0\rangle \rightarrow |a,b,a+b\rangle \tag{1.19}$$

A more complex operation is to write the calculation result to the input register again, that is

$$|a,b\rangle \rightarrow |a,a+b\rangle \tag{1.20}$$

As the input (a,b) can be reconstructed from the output $(a,a+b)$, the information will not be lost, and the calculation can be performed reversibly. The circuit of the full adder network is shown in Figure 1.4, and the basic carry and sum operations of the ordinary adder network are shown in Figure 1.5.

It is important to note the position of the thick black bars to the right or left of the basic carry and sum operations. The line with a bar on the left represents an inverse sequence of basic gates embedded in the same network as the bar on the right.

If the operation of the line is opposite to the input (a,b), when $b \geq a$, the output is $(a,b-a)$; when $b < a$, the output is $(a,2^n-(a-b))$.

1.2.2.4.2 Modulo-N Adder

The Modulo-N adder[5] is a complex quantum network structure, which is used to calculate the modulo sum of two numbers. The specific form is shown in Equation 1.21:

$$|a,b\rangle \rightarrow |a,(a+b) \bmod N\rangle \tag{1.21}$$

where $a,b \in [0,N)$. Figure 1.6 shows the quantum network of modulo-N adders. Vedral et al.[5] used the method based on the output of the ordinary adder network to determine the subtraction of N, according to the relationship between the value of $a+b$ and the size of N.

1.3 QUANTUM IMAGE AND VIDEO-RELATED CONCEPTS

Quantum image and video are images and videos that can be recognized and processed by quantum computers in the quantum computer environment. The research field of quantum image and video

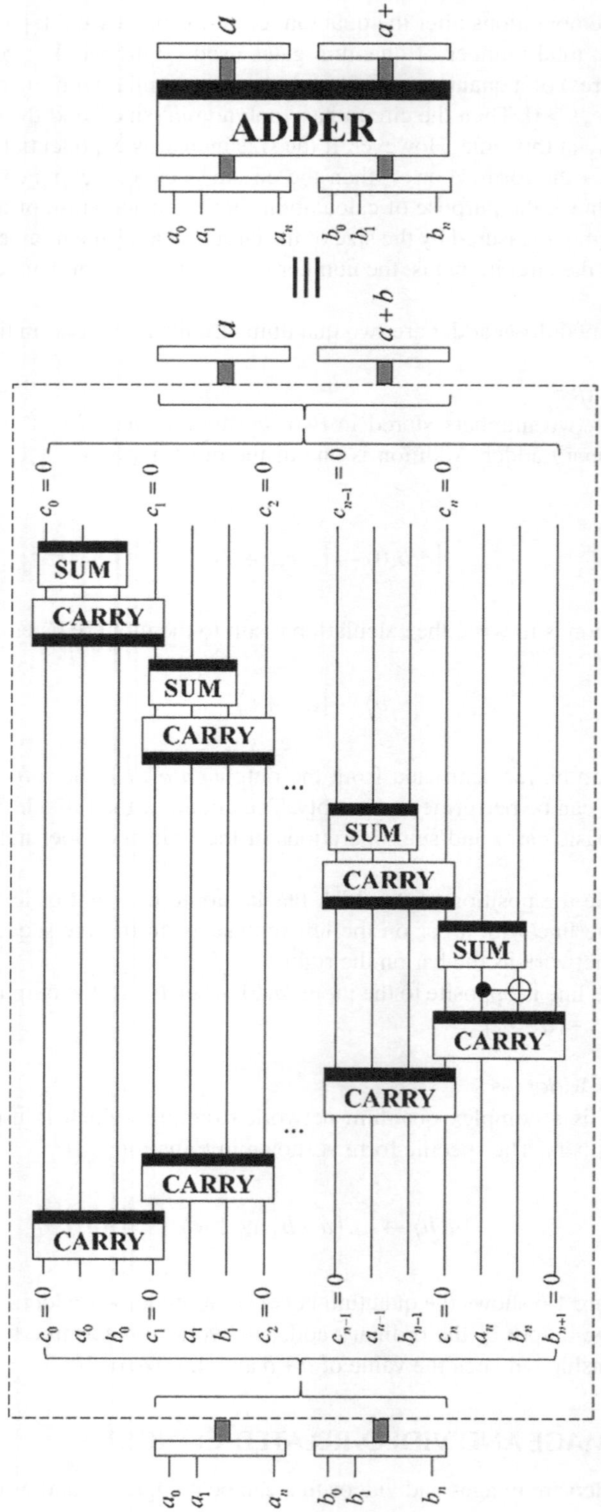

FIGURE 1.4 General adder circuit.

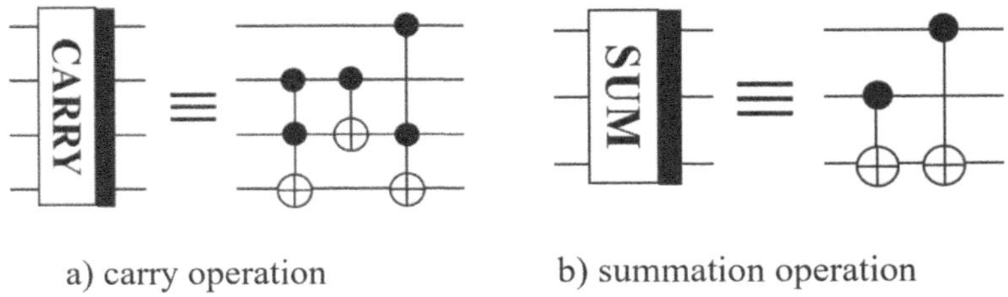

FIGURE 1.5 Basic carry and sum operations.

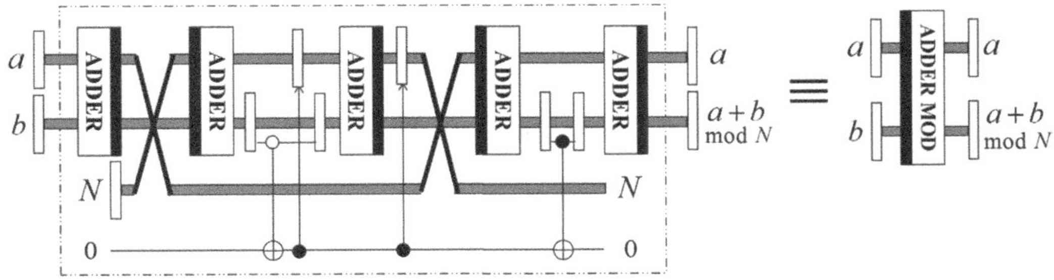

FIGURE 1.6 Modulo-N adder.

processing mainly includes the construction of quantum image and video representation methods, the research of quantum image and video processing and application under each representation method, and the design of quantum image and video security algorithms. This section mainly describes the current research status of quantum image and video representation, processing, and security.

1.3.1 Quantum Image and Video Representation Model

Quantum image and video representation has the characteristics of flexibility and diversity. The design of representation methods needs to be beneficial for storage and various quantum processing operations. According to quantum computer theory, quantum images and videos should be stored in a quantum computer in the form of a quantum superposition state or entangled state. At present, quantum image representation methods mainly focus on the quantization coding of different classical images. Classical image representation is mainly composed of pixel position and color. Therefore, the difference between the current main quantum image representation methods is in how to encode these two elements.

Typical quantum image representation methods are introduced below.

1.3.1.1 Quantum Image Representation Model Qubit Lattice

Venegas-Andraca et al.[6] proposed the quantum image representation method Qubit Lattice Model (QLM) in 2003. Similar to the matrix representation of classical images, the qubits need to be arranged orderly one by one into a second-order quantum matrix. The quantum state $|\Psi\rangle$ of each pixel can be expressed in the form of Equation 1.22:

$$|\Psi\rangle = \cos\frac{\theta}{2}|0\rangle + e^{i\gamma}\sin\frac{\theta}{2}|1\rangle \tag{1.22}$$

The color information of pixels is encoded by θ. Because the superposition and entanglement properties of quantum states are not considered in QLM representation method, a large number of quantum bits are required to store a quantum image.

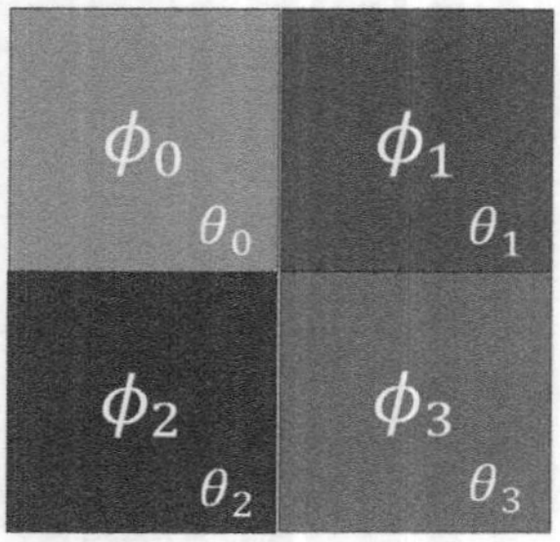

$$
\begin{aligned}
|I\rangle = \frac{1}{2}\Big[& \left(\cos\phi_0|0\rangle + \sin\phi_0|1\rangle\right)\otimes\left(\cos\theta_0|0\rangle + \sin\theta_0|1\rangle\right) \\
& + \left(\cos\phi_1|0\rangle + \sin\phi_1|1\rangle\right)\otimes\left(\cos\theta_1|0\rangle + \sin\theta_1|1\rangle\right) \\
& + \left(\cos\phi_2|0\rangle + \sin\phi_2|1\rangle\right)\otimes\left(\cos\theta_2|0\rangle + \sin\theta_2|1\rangle\right) \\
& + \left(\cos\phi_3|0\rangle + \sin\phi_3|1\rangle\right)\otimes\left(\cos\theta_3|0\rangle + \sin\theta_3|1\rangle\right)\Big]
\end{aligned}
$$

FIGURE 1.7 A 2 × 2 image and QSMC&QSNC representation.

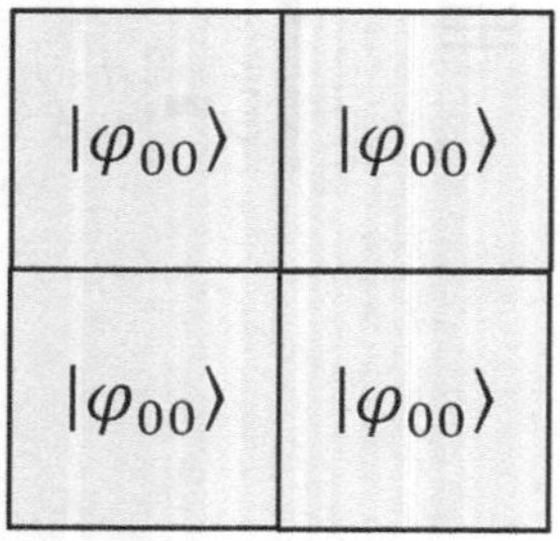

$$
\begin{aligned}
&|Q\rangle = \{|\varphi_{00}\rangle|\varphi_{01}\rangle|\varphi_{10}\rangle|\varphi_{11}\rangle\} \\
&|\varphi_{00}\rangle = \cos\frac{\theta_{00}}{2}|0\rangle + \sin\frac{\theta_{00}}{2}|1\rangle, |\varphi_{01}\rangle = \cos\frac{\theta_{01}}{2}|0\rangle + \sin\frac{\theta_{01}}{2}|1\rangle \\
&|\varphi_{10}\rangle = \cos\frac{\theta_{10}}{2}|0\rangle + \sin\frac{\theta_{10}}{2}|1\rangle, |\varphi_{11}\rangle = \cos\frac{\theta_{11}}{2}|0\rangle + \sin\frac{\theta_{11}}{2}|1\rangle
\end{aligned}
$$

FIGURE 1.8 A 2 × 2 image and SQR representation.

Based on QLM presentation, with the help of two sets of quantum state sets, Li et al.[7] proposed the quantum image representation method Quantum States for M Colors & Quantum States for N Coordinates (QSMC&QSNC), which contains *m* colors and n coordinates, as shown in Figure 1.7; Yuan et al.[8] proposed the Simple Quantum Representation (SQR) method for focusing infrared quantum images, as shown in Figure 1.8.

1.3.1.2 Quantum Image Representation Model Real Ket

The image is continuously quartered and stored in the real state vector. This is the quantum image representation Real Ket[9] proposed by Latorre in 2005. A Real Ket quantum image of size $2^n \times 2^n$ is represented by the Equation 1.23:

$$
\left|\psi_{2^n\times 2^n}\right\rangle = \sum_{i_1,\dots,i_n=1,\dots,4} c_{i_n,\dots,i_1}\left|i_n,\dots,i_1\right\rangle \tag{1.23}
$$

where the pixel value is recorded with $c_{i_n,\dots,i_1}$, and the position information of the image after successive quartering is recorded as $\left|i_n,\dots,i_1\right\rangle$. See Figure 1.9 for a $2^2 \times 2^2$ image and the expression of Real Kets shown in Equation 1.24:

$$
\begin{aligned}
\left|\psi_{2^2\times 2^2}\right\rangle &= \sum_{i_1,i_2=1,2,3,4} c_{i_2,i_1}\left|i_2,i_1\right\rangle \\
&= c_{11}|11\rangle + c_{12}|12\rangle + c_{13}|13\rangle + c_{14}|14\rangle + \\
&\quad c_{21}|21\rangle + c_{22}|22\rangle + c_{23}|23\rangle + c_{24}|24\rangle + \\
&\quad c_{31}|31\rangle + c_{32}|32\rangle + c_{33}|33\rangle + c_{34}|34\rangle + \\
&\quad c_{41}|41\rangle + c_{42}|42\rangle + c_{43}|43\rangle + c_{44}|44\rangle
\end{aligned} \tag{1.24}
$$

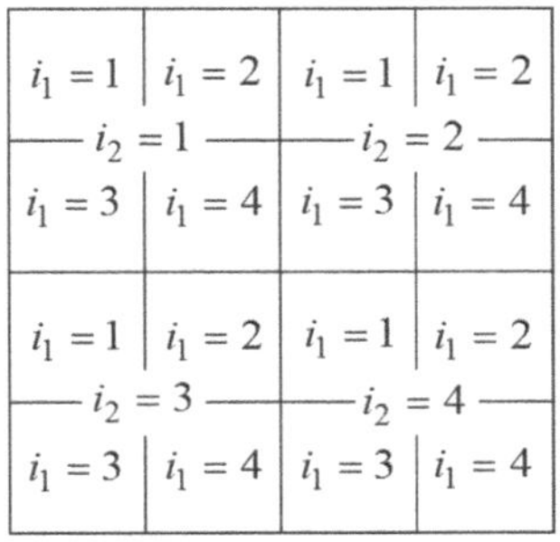

FIGURE 1.9 A real ket quantum image of $2^2 \times 2^2$.

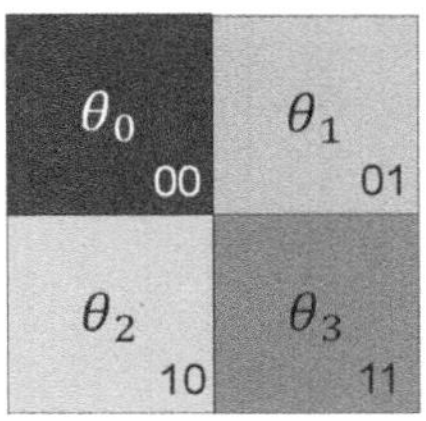

$$
\begin{aligned}
|I(\theta)\rangle = \frac{1}{2}\big[&(\cos\theta_0|0\rangle + \sin\theta_0|1\rangle)\otimes|00\rangle + (\cos\theta_1|0\rangle + \sin\theta_1|1\rangle)\otimes|01\rangle \\
&+ (\cos\theta_2|0\rangle + \sin\theta_2|1\rangle)\otimes|10\rangle + (\cos\theta_3|0\rangle + \sin\theta_3|1\rangle)\otimes|11\rangle \big]
\end{aligned}
$$

FIGURE 1.10 A 2 × 2 image and FRQI representation.

The Real Ket model takes into account the superposition property of quantum states, and only n qubits are required for the storage of $2^n \times 2^n$ size images.

1.3.1.3 Quantum Image Representation Model Flexible Representation of Quantum Images

Le et al.[10] gave Flexible Representation of Quantum Images (FRQI) in 2011. The FRQI method represents the image as a normalized quantum superposition state. The FRQI of a grayscale image of size $2^n \times 2^n$ is expressed as

$$
\begin{gathered}
|I(\theta)\rangle = \frac{1}{2^n}\sum_{k=0}^{2^{2n}-1}(\cos\theta_k|0\rangle + \sin\theta_k|1\rangle)\otimes|k\rangle \\
\theta_k \in \left[0, \frac{\pi}{2}\right], k = 0, 1, \ldots, 2^{2n} - 1
\end{gathered}
\tag{1.25}
$$

where $\cos\theta_k|0\rangle + \sin\theta_k|1\rangle$ encodes the gray information of the pixel position $|k\rangle$ of the image. A simple 2 × 2 image and FRQI representation are shown in Figure 1.10.

FRQI indicates that $2n+1$ qubits are required for a grayscale image of size $2^n \times 2^n$ 1 qubit, which is required for encoding color information, and $2n$ qubits are required for encoding pixel positions. FRQI stores the gray level of the image with 1 qubit, so FRQI can only obtain the gray-level information of the original image in a probabilistic way.

Based on FRQI representation, Yang et al.[11] presented the Flexible Quantum Representation for Color Images (FQRCI) representation method, as shown in Figure 1.11; Lit et al.[12] proposed the quantum color image model Flexible Representation of Quantum Color Images (FRQCI).

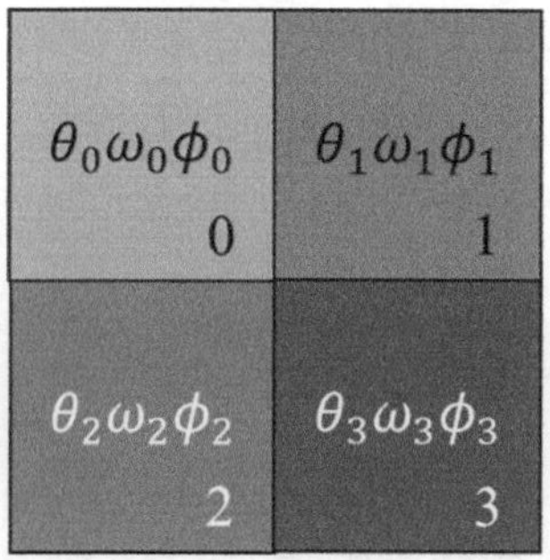

$$
\begin{aligned}
\left|I(\theta,\omega,\phi)\right\rangle = \frac{1}{2}\Big[& \left(\cos\theta_0|0\rangle+\sin\theta_0|1\rangle\right)_r \left(\cos\omega_0|0\rangle+\sin\omega_0|1\rangle\right)_g \left(\cos\phi_0|0\rangle+\sin\phi_0|1\rangle\right)_b \otimes|0\rangle+ \\
& \left(\cos\theta_1|0\rangle+\sin\theta_1|1\rangle\right)_r \left(\cos\omega_1|0\rangle+\sin\omega_1|1\rangle\right)_g \left(\cos\phi_1|0\rangle+\sin\phi_1|1\rangle\right)_b \otimes|1\rangle+ \\
& \left(\cos\theta_2|0\rangle+\sin\theta_2|1\rangle\right)_r \left(\cos\omega_2|0\rangle+\sin\omega_2|1\rangle\right)_g \left(\cos\phi_2|0\rangle+\sin\phi_2|1\rangle\right)_b \otimes|2\rangle+ \\
& \left(\cos\theta_3|0\rangle+\sin\theta_3|1\rangle\right)_r \left(\cos\omega_3|0\rangle+\sin\omega_3|1\rangle\right)_g \left(\cos\phi_3|0\rangle+\sin\phi_3|1\rangle\right)_b \otimes|3\rangle\Big]
\end{aligned}
$$

FIGURE 1.11 A 2 × 2 image and FQRCI representation.

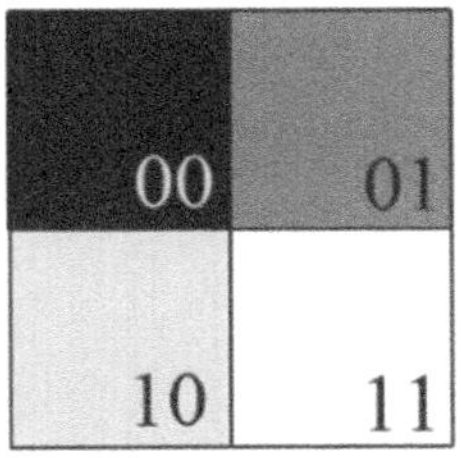

$$
\begin{aligned}
|I\rangle &= \frac{1}{2}\big(|0\rangle \otimes |00\rangle + |100\rangle \otimes |01\rangle + |200\rangle \otimes |10\rangle + |255\rangle \otimes |11\rangle\big) \\
&= \frac{1}{2}\big(|00000000\rangle \otimes |00\rangle + |01100100\rangle \otimes |01\rangle + \\
&\qquad |11001000\rangle \otimes |10\rangle + |11111111\rangle \otimes |11\rangle\big)
\end{aligned}
$$

FIGURE 1.12 A 2 × 2 image and NEQR representation.

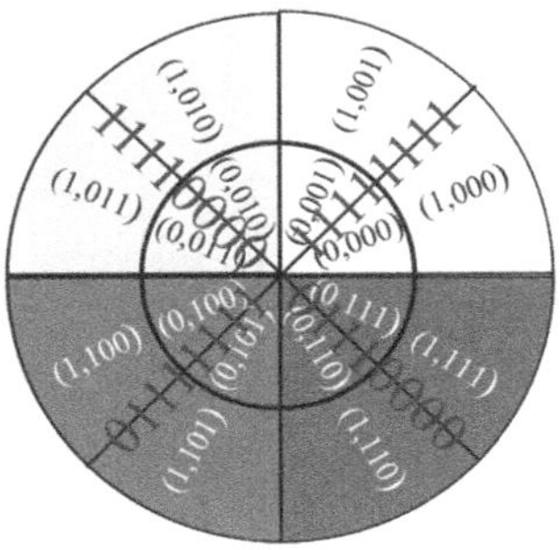

$$
\begin{aligned}
|I\rangle = \frac{1}{4}\Big[&|11111111\rangle \otimes \big(|0000\rangle + |0001\rangle + |1000\rangle + |1001\rangle\big) \\
&+ |11110000\rangle \otimes \big(|0010\rangle + |0011\rangle + |1010\rangle + |1011\rangle\big) \\
&+ |01111111\rangle \otimes \big(|0100\rangle + |0101\rangle + |1100\rangle + |1101\rangle\big) \\
&+ |01110000\rangle \otimes \big(|0110\rangle + |0111\rangle + |1110\rangle + |1111\rangle\big)\Big]
\end{aligned}
$$

FIGURE 1.13 A 16 × 8 image and QUALPI representation.

1.3.1.4 Quantum Image Representation Model NEQR

In 2013, Zhang et al.[13] proposed the Novel Enhanced Quantum Representation (NEQR) for digital images representation method for quantum grayscale images. The NEQR model uses q qubits to store the gray information of the image in the form of an entangled state. Therefore, the NEQR of a grayscale image with size $2^n \times 2^n$ is expressed as

$$
\begin{aligned}
&|I\rangle = \frac{1}{2^n}\sum_{y=0}^{2^n-1}\sum_{x=0}^{2^n-1}|f(y,x)\rangle|yx\rangle = \frac{1}{2^n}\sum_{y=0}^{2^n-1}\sum_{x=0}^{2^n-1}\bigotimes_{l=0}^{q-1}\left|C_{yx}^{l}\right\rangle|yx\rangle \\
&f(y,x) = C_{yx}^{0}C_{yx}^{1}\cdots C_{yx}^{q-1},\ C_{yx}^{l} \in \{0,1\},\ f(y,x) \in \left[0, 2^q - 1\right]
\end{aligned}
\tag{1.26}
$$

It is more convenient to design more complex image processing operations based on NEQR representation. See Figure 1.12 for a simple 2 × 2 image and NEQR representation.

Based on the NEQR model, Zhang et al.[14] proposed the Quantum Log-Polar Image (QUALPI) model in polar coordinates, as shown in Figure 1.13; Jiang et al.[15] proposed the Generalized Quantum Image Representation (GQIR) method with size $H \times W$, as shown in Figure 1.14. Li et al.[16] proposed a bit plane representation of quantum images Bit Plane Representation of Quantum Images (BRQI), as shown in Figure 1.15.

1.3.1.5 Quantum Image Representation Model Mult- Channel Representation for Quantum Images

In 2013, Sun et al.[17] proposed the color quantum image representation method Multi-Channel representation for Quantum Images (MCQI). MCQI method uses 3 qubits to store the color of color image; see Equation 1.27:

$$
\left|I(\theta)_{mc}\right\rangle = \frac{1}{2^{n+1}}\sum_{i=0}^{2^{2n}-1}\left|C_{RGB}^{i}\right\rangle \otimes |i\rangle
\tag{1.27}
$$

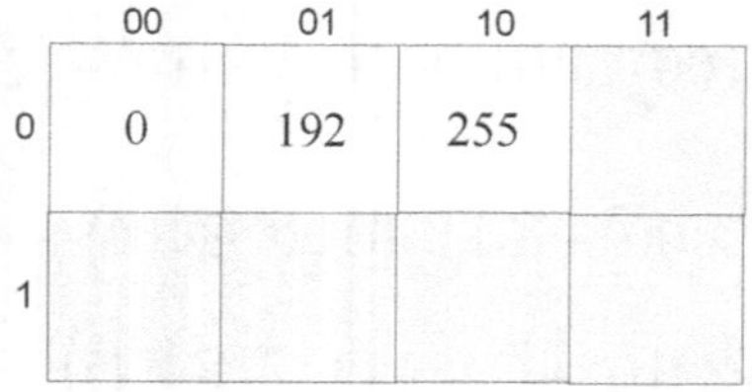

$$|I\rangle = \frac{1}{\sqrt{2}^{3}}\left(|00000000\rangle \otimes |000\rangle + |10000000\rangle \otimes |001\rangle + |11111111\rangle \otimes |010\rangle\right)$$

FIGURE 1.14 A 1 × 3 image and GQIR representation.

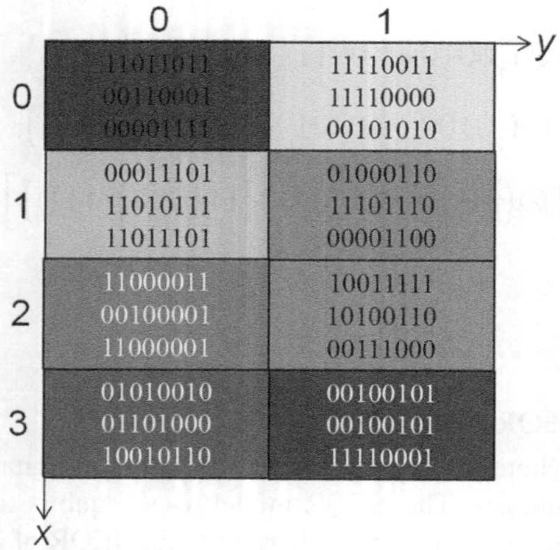

$$\begin{cases} \left|\Psi_B^R\right\rangle = \frac{1}{\sqrt{2^{3+3}}} \sum_{l=0}^{3}\sum_{x=0}^{3}\sum_{y=0}^{1} \left|g_R(x,y)\right\rangle |x\rangle |y\rangle |l\rangle \\ \left|\Psi_B^G\right\rangle = \frac{1}{\sqrt{2^{3+3}}} \sum_{l=0}^{3}\sum_{x=0}^{3}\sum_{y=0}^{1} \left|g_G(x,y)\right\rangle |x\rangle |y\rangle |l\rangle \\ \left|\Psi_B^B\right\rangle = \frac{1}{\sqrt{2^{3+3}}} \sum_{l=0}^{3}\sum_{x=0}^{3}\sum_{y=0}^{1} \left|g_B(x,y)\right\rangle |x\rangle |y\rangle |l\rangle \end{cases}$$

$$\left|\Psi_B^{24}\right\rangle = \frac{1}{\sqrt{3}}\left(\left|\Psi_B^R\right\rangle|01\rangle + \left|\Psi_B^G\right\rangle|10\rangle + \left|\Psi_B^B\right\rangle|11\rangle\right)$$

FIGURE 1.15 A 4 × 2 image and BRQI representation.

where the color information $\left|C_{RGB}^i\right\rangle$ encodes the R, G, and B channel information and is defined as:

$$\begin{aligned} \left|C_{RGB}^i\right\rangle = \cos\theta_R^i|000\rangle + \sin\theta_R^i|100\rangle + \cos\theta_G^i|001\rangle + \sin\theta_G^i|101\rangle + \\ \cos\theta_B^i|010\rangle + \sin\theta_B^i|110\rangle + \cos 0|011\rangle + \sin 0|111\rangle \end{aligned} \tag{1.28}$$

In Equation 1.28, the three angles $\left\{\theta_R^i, \theta_G^i, \theta_B^i\right\} \in [0, \pi/2]$ are used to encode the colors of the R, G, and B channels at the ith pixel position. A color image of size 2 × 2 and the MCQI states are shown in Figure 1.16.

Based on MCQI representation, Song proposed a multi-channel quantum color image representation Color Quantum Image based on Phase Transform (CQIPT)[18] based on phase transformation; Sang et al.[19] proposed the logarithmic polar coordinate representation method Multi-Channel Log-Polar Quantum Image (MCLPQI) for quantum images, which are detailed in Sections 2.3 and 2.5 of Chapter 2.

1.3.1.6 Quantum Image Representation Model NCQI

In 2016, Sang et al.[20] proposed a Novel Quantum representation of Color digital Images (NCQI). For a color image with size $2^n \times 2^n$ the values of red (R), green (G), and blue (B) in the three color channels are $\left[0, 2^q - 1\right]$, where the NCQI is expressed as:

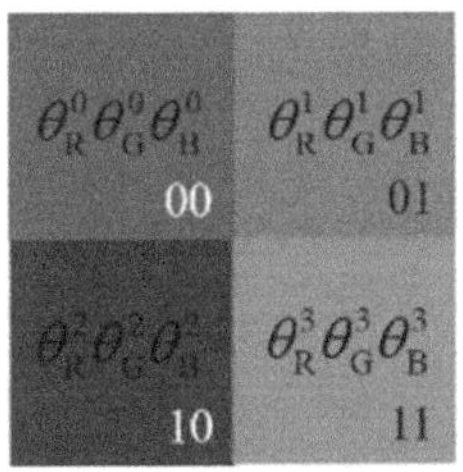

$$
\begin{aligned}
|I(\theta)\rangle = \frac{1}{4}\Big[& \left(\cos\theta_R^0|000\rangle + \cos\theta_G^0|001\rangle + \cos\theta_B^0|010\rangle + \sin\theta_R^0|100\rangle + \sin\theta_G^0|101\rangle + \sin\theta_B^0|110\rangle + \cos 0|011\rangle + \sin 0|111\rangle\right)\otimes|00\rangle + \\
& \left(\cos\theta_R^1|000\rangle + \cos\theta_G^1|001\rangle + \cos\theta_B^1|010\rangle + \sin\theta_R^1|100\rangle + \sin\theta_G^1|101\rangle + \sin\theta_B^1|110\rangle + \cos 0|011\rangle + \sin 0|111\rangle\right)\otimes|01\rangle + \\
& \left(\cos\theta_R^2|000\rangle + \cos\theta_G^2|001\rangle + \cos\theta_B^2|010\rangle + \sin\theta_R^2|100\rangle + \sin\theta_G^2|101\rangle + \sin\theta_B^2|110\rangle + \cos 0|011\rangle + \sin 0|111\rangle\right)\otimes|10\rangle + \\
& \left(\cos\theta_R^3|000\rangle + \cos\theta_G^3|001\rangle + \cos\theta_B^3|010\rangle + \sin\theta_R^3|100\rangle + \sin\theta_G^3|101\rangle + \sin\theta_B^3|110\rangle + \cos 0|011\rangle + \sin 0|111\rangle\right)\otimes|11\rangle\Big]
\end{aligned}
$$

FIGURE 1.16 A 2 × 2 image and MCQI representation.

$$|I\rangle = \frac{1}{2^n}\sum_{y=0}^{2^n-1}\sum_{x=0}^{2^n-1}|C(y,x)\rangle \otimes |yx\rangle \tag{1.29}$$

where $|C(y,x)\rangle$ encodes the corresponding pixel $|yx\rangle$ color value, which is encoded as:

$$|C(y,x)\rangle = \left|\underbrace{R_{q-1}(y,x)\cdots R_0(y,x)}_{Red}\underbrace{G_{q-1}(y,x)\cdots G_0(y,x)}_{Green}\underbrace{B_{q-1}(y,x)\cdots B_0(y,x)}_{Blue}\right\rangle \tag{1.30}$$

The NCQI method needs $3q$ qubits to store the color information of the image, and the specific example diagram and preparation method are described in Section 2.4 of Chapter 2.

Based on NCQI representation, Wang et al.[21] proposed that quantum images represent Quantum Representation model of Color digital Images (QRCI), as shown in Figure 1.17 for a simple image example; Su et al.[22] proposed an improved quantum representation method, Improved NCQI (INCQI), as shown in Figure 1.18.

In 2018, Liu et al.[23] proposed an Optimized Quantum Representation for Color Digital Images (OCQR), which makes full use of quantum superposition to store the RGB value of each pixel. Compared with color digital image quantum representation NCQI, OCQR uses nearly one-third of the qubits to store pixel values. The model uses three-dimensional quantum sequences to store color quantum images:

$$\begin{aligned}|I\rangle &= \frac{1}{2^{n+1}}\sum_{y=0}^{2^n-1}\sum_{x=0}^{2^n-1}|c(x,y)\rangle|ch\_index\rangle|yx\rangle \\ &= \frac{1}{2^{n+1}}\sum_{y=0}^{2^n-1}\sum_{x=0}^{2^n-1}(|r_{yx}\rangle\otimes|00\rangle + |g_{yx}\rangle\otimes|01\rangle + |b_{yx}\rangle\otimes|10\rangle + |s_{yx}\rangle\otimes|11\rangle)|yx\rangle\end{aligned}$$

$$\begin{aligned}|r_{yx}\rangle &= |r_{yx}^{q-1}\cdots r_{yx}^{0}\rangle, |g_{yx}\rangle = |g_{yx}^{q-1}\cdots g_{yx}^{0}\rangle \\ |b_{yx}\rangle &= |b_{yx}^{q-1}\cdots b_{yx}^{0}\rangle, |s_{yx}\rangle = |0_{yx}^{q-1}\cdots 0_{yx}^{0}\rangle\end{aligned}$$

A size 2 × 2 image and OCQR state are shown in Figure 1.19.

1.3.1.7 Quantum Image Other Representation Models

In addition to the quantum image representation model mentioned above, here are some other types of quantum image representation. Venegas-Andraca and Ball[24] proposed Entangled Image, a quantum image representation method stored in the form of entangled states. Li et al.[25] proposed a Normal Arbitrary Quantum Superposition State (NAQSS) (Figure 1.20). Sahin and Yilmaz[26] proposed a Quantum Representation of Multi-Wavelength images（QRMW）method. Khan[27] proposed the Improved Flexible Representation of Quantum Images（IFRQI）method. Xu et al.[28] designed an Order-encoded Quantum Image Model (OQIM). Wang et al.[29] designed a Double Quantum Color Image Representation (DQRCI). Wang et al.[30] proposed a Quantum Indexed Image Representation（QIIR）model (Figure 1.21). Grigoryan and Agaian[31] proposed a quantum image representation model Fourier Transform Qubit Representation (FTQR). Yan et al.[32] proposed a quantum color model Quantum Hue, Saturation, and Lightness (QHSL) based on HSL space (Figure 1.22). Chen et al.[33] proposed a Quantum Image Representation based on HSI Color Space Model (QIRHSI) and its general dimensional representation method Improved QIRHSI (IQIRHSI), and the specific preparation scheme of QIRHSI and IQIRHSI representation is described in Section 2.1 of Chapter 2.

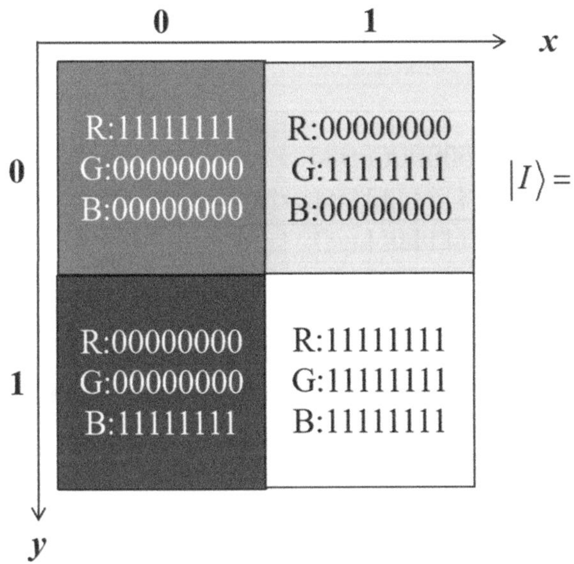

$$
\begin{aligned}
|I\rangle = \frac{1}{\sqrt{2^5}}\{ & |100\rangle(|000\rangle+|001\rangle+|010\rangle+|011\rangle+|100\rangle+|101\rangle+|110\rangle+|111\rangle)|00\rangle+ \\
& |010\rangle(|000\rangle+|001\rangle+|010\rangle+|011\rangle+|100\rangle+|101\rangle+|110\rangle+|111\rangle)|01\rangle+ \\
& |001\rangle(|000\rangle+|001\rangle+|010\rangle+|011\rangle+|100\rangle+|101\rangle+|110\rangle+|111\rangle)|10\rangle+ \\
& |111\rangle(|000\rangle+|001\rangle+|010\rangle+|011\rangle+|100\rangle+|101\rangle+|110\rangle+|111\rangle)|11\rangle\}
\end{aligned}
$$

FIGURE 1.17 A 2×2 image and QRCI representation.

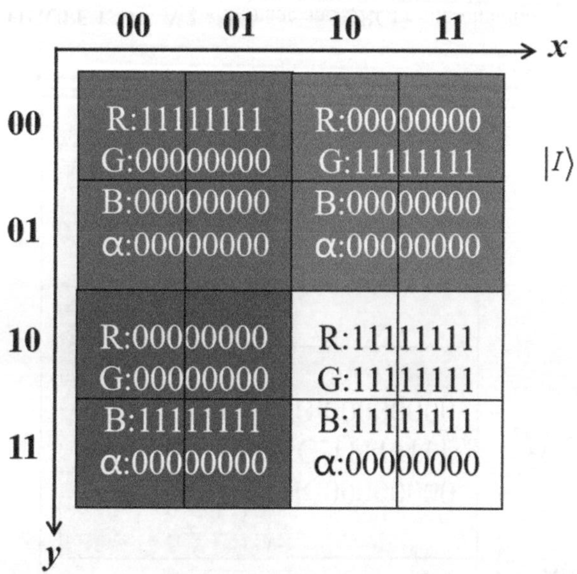

$$
\begin{aligned}
|I\rangle = \frac{1}{\sqrt{2^{2+2}}}\Big[& \big|\underbrace{11111111}_{R}\underbrace{00000000}_{G}\underbrace{00000000}_{B}\underbrace{00000000}_{\alpha}\big\rangle \otimes \left(|0000\rangle + |0001\rangle + |0100\rangle + |0101\rangle\right) + \\
& \big|\underbrace{00000000}_{R}\underbrace{11111111}_{G}\underbrace{00000000}_{B}\underbrace{00000000}_{\alpha}\big\rangle \otimes \left(|0010\rangle + |0011\rangle + |0110\rangle + |0111\rangle\right) + \\
& \big|\underbrace{00000000}_{R}\underbrace{00000000}_{G}\underbrace{11111111}_{B}\underbrace{00000000}_{\alpha}\big\rangle \otimes \left(|1000\rangle + |1001\rangle + |1100\rangle + |1101\rangle\right) + \\
& \big|\underbrace{11111111}_{R}\underbrace{11111111}_{G}\underbrace{11111111}_{B}\underbrace{00000000}_{\alpha}\big\rangle \otimes \left(|1010\rangle + |1011\rangle + |1110\rangle + |1111\rangle\right)\Big]
\end{aligned}
$$

FIGURE 1.18 A 2 × 2 image and INCQI representation.

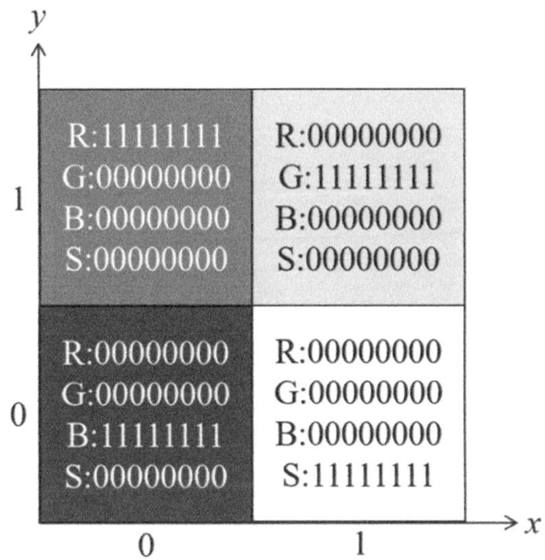

$$
\begin{aligned}
|I\rangle = \frac{1}{2^{1+1}} \Bigg\{ & \Bigg[\Big|\underbrace{11111111}_{\text{G-value}}\Big\rangle \otimes \Big|\underbrace{01}_{\text{G-index}}\Big\rangle + \Big|\underbrace{00000000}_{\text{channel-value}}\Big\rangle \otimes \Big(\Big|\underbrace{00}_{\text{R-index}}\Big\rangle + \Big|\underbrace{10}_{\text{B-index}}\Big\rangle + \Big|\underbrace{11}_{\text{S-index}}\Big\rangle \Big) \Bigg] \otimes |11\rangle + \\
& \Bigg[\Big|\underbrace{11111111}_{\text{B-value}}\Big\rangle \otimes \Big|\underbrace{10}_{\text{B-index}}\Big\rangle + \Big|\underbrace{00000000}_{\text{channel-value}}\Big\rangle \otimes \Big(\Big|\underbrace{00}_{\text{R-index}}\Big\rangle + \Big|\underbrace{01}_{\text{G-index}}\Big\rangle + \Big|\underbrace{11}_{\text{S-index}}\Big\rangle \Big) \Bigg] \otimes |00\rangle + \\
& \Bigg[\Big|\underbrace{11111111}_{\text{B-value}}\Big\rangle \otimes \Big|\underbrace{10}_{\text{B-index}}\Big\rangle + \Big|\underbrace{00000000}_{\text{channel-value}}\Big\rangle \otimes \Big(\Big|\underbrace{00}_{\text{R-index}}\Big\rangle + \Big|\underbrace{01}_{\text{G-index}}\Big\rangle + \Big|\underbrace{11}_{\text{S-index}}\Big\rangle \Big) \Bigg] \otimes |00\rangle + \\
& \Bigg[\Big|\underbrace{00000000}_{\text{S-value}}\Big\rangle \otimes \Big|\underbrace{11}_{\text{S-index}}\Big\rangle + \Big|\underbrace{11111111}_{\text{channel-value}}\Big\rangle \otimes \Big(\Big|\underbrace{00}_{\text{R-index}}\Big\rangle + \Big|\underbrace{01}_{\text{G-index}}\Big\rangle + \Big|\underbrace{10}_{\text{B-index}}\Big\rangle \Big) \Bigg] \otimes |01\rangle \Bigg\}
\end{aligned}
$$

FIGURE 1.19 A 2 × 2 image and OCQR representation.

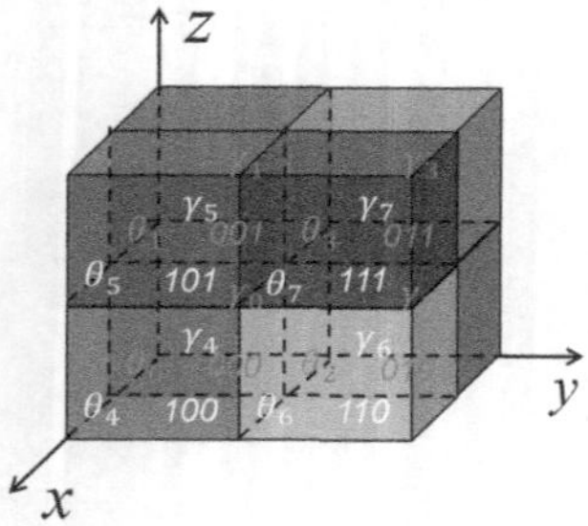

$$\begin{aligned}|I\rangle = \theta_0|000\rangle\otimes(\cos\gamma_0|0\rangle+\sin\gamma_0|1\rangle)+\theta_1|001\rangle\otimes(\cos\gamma_1|0\rangle+\sin\gamma_1|1\rangle)\\ +\theta_2|010\rangle\otimes(\cos\gamma_2|0\rangle+\sin\gamma_2|1\rangle)+\theta_3|011\rangle\otimes(\cos\gamma_3|0\rangle+\sin\gamma_3|1\rangle)\\ +\theta_4|100\rangle\otimes(\cos\gamma_4|0\rangle+\sin\gamma_4|1\rangle)+\theta_5|101\rangle\otimes(\cos\gamma_5|0\rangle+\sin\gamma_5|1\rangle)\\ +\theta_6|110\rangle\otimes(\cos\gamma_6|0\rangle+\sin\gamma_6|1\rangle)+\theta_7|111\rangle\otimes(\cos\gamma_7|0\rangle+\sin\gamma_7|1\rangle)\end{aligned}$$

FIGURE 1.20 NAQSS representation.

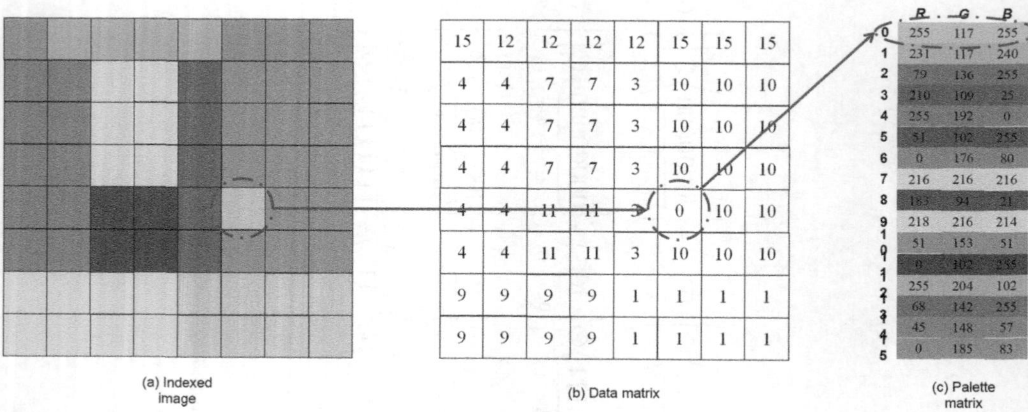

FIGURE 1.21 QIIR representation.

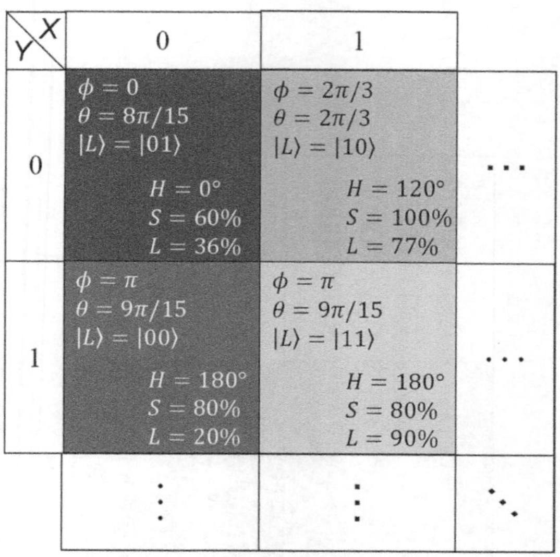

FIGURE 1.22 QHSL representation.

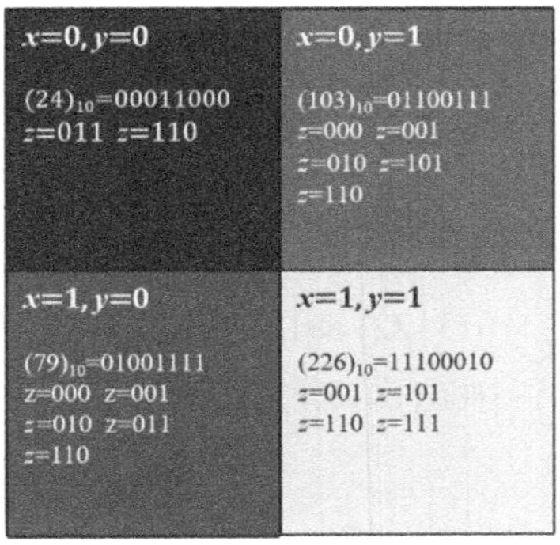

$$
\begin{aligned}
|I_m\rangle = \frac{1}{\sqrt{2^6}}\left(\cos\frac{\pi}{2}|0\rangle + e^{i0}\sin\frac{\pi}{2}|1\rangle\right)
&\{|00011000\rangle \otimes (|011\rangle + |110\rangle)|00\rangle + \\
&|01100111\rangle \otimes (|000\rangle + |001\rangle + |010\rangle + |011\rangle + |110\rangle)|01\rangle + \\
&|01001111\rangle \otimes (|000\rangle + |001\rangle + |010\rangle + |101\rangle + |110\rangle)|10\rangle + \\
&|11100010\rangle \otimes (|001\rangle + |101\rangle + |110\rangle + |111\rangle)|11\rangle\}
\end{aligned}
$$

FIGURE 1.23 A 2 × 2 image and MQIR representation.

In 2021, Zhu et al.[34] proposed a Multimode Quantum Image Representation (MQIR) method. In this representation, the image size can be arbitrary, and the color can be one of several color models, such as Gray, RGB, Lab, HSL. For a grayscale image with a size of $M \times N$, its MQIR is expressed as:

$$
|I_m\rangle = \frac{1}{\sqrt{2}^{n+m+k+1}} \sum_{X=0}^{M-1} \sum_{Y=0}^{N-1} \sum_{Z=0}^{K-1} |m\rangle |C_{ZYX}\rangle |ZYX\rangle
$$

$$
n = \begin{cases} \lceil \log_2^N \rceil, N > 1 \\ 1, N = 1 \end{cases}, m = \begin{cases} \lceil \log_2^M \rceil, M > 1 \\ 1, M = 1 \end{cases}, k = \begin{cases} \lceil \log_2^K \rceil, K > 1 \\ 1, K = 1 \end{cases}
$$

where $|m\rangle = \cos\frac{\theta}{2}|0\rangle + e^{i\phi}\sin\frac{\theta}{2}|1\rangle$, $\theta = \pi, \phi \in \{0, \pi/2, \pi, 3\pi/2\}$, A 2×2 gray-scale image with MQIR state is shown in Figure 1.23.

In 2022, Amankwah et al.[35] introduced a general framework for quantum pixel representations, which captures pixel color and position to a single quantum state, called quantum pixel representations (QPIXL), and showed how FRQI, NEQR, MCRQI, and NCQI image representations can be incorporated into the QPIXL framework.

In 2022, Dong et al.[36] proposed a novel qutrit representation of quantum image (QTRQ). The model uses two entangled quantum sequences to store the grayscale and position information, and stores the entire image in a quantum sequence overlay. Assuming that the gray range of the image is 3^q, the ternary sequence $C_{YX}^0 C_{YX}^1 \cdots C_{YX}^{q-1}$ encodes the gray value $f(Y,X)$ corresponding to pixel (Y,X):

$$
f(Y,X) = C_{YX}^0 C_{YX}^1 \cdots C_{YX}^{q-1}, C_{YX}^i \in \{0,1,2\}, f(Y,X) \in [0, 3^q - 1]
$$

The representative expression for a $3^n \times 3^n$ quantum image of an image can be written as:

$$
|I\rangle = \frac{1}{3^n} \sum_{Y=0}^{3^n-1} \sum_{X=0}^{3^n-1} |f(Y,X)\rangle |YX\rangle = \frac{1}{3^n} \sum_{Y=0}^{3^n-1} \sum_{X=0}^{3^n-1} \bigotimes_{i=0}^{q-1} |C_{YX}^i\rangle |YX\rangle
$$

A 3×3 grayscale image with QTRQ state is shown in Figure 1.24:

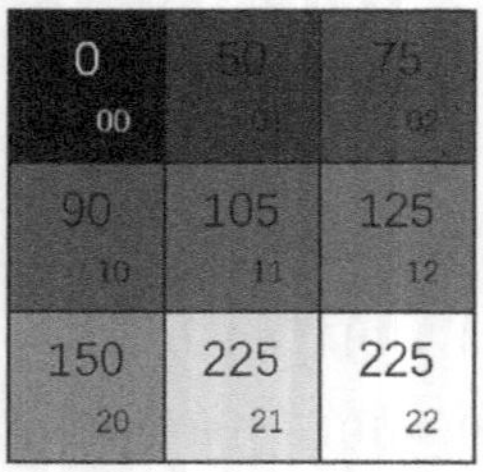

$$
\begin{aligned}
|I\rangle = \frac{1}{3}\big(&|0\rangle\otimes|00\rangle+|50\rangle\otimes|01\rangle+|75\rangle\otimes|02\rangle+|90\rangle\otimes|10\rangle+|105\rangle\otimes|11\rangle+\\
&|125\rangle\otimes|12\rangle+|150\rangle\otimes|20\rangle+|225\rangle\otimes|21\rangle+|255\rangle\otimes|22\rangle\big)\\
= \frac{1}{3}\big(&|000000\rangle\otimes|00\rangle+|001212\rangle\otimes|01\rangle+|002210\rangle\otimes|02\rangle+\\
&|010100\rangle\otimes|10\rangle+|010220\rangle\otimes|11\rangle+|011122\rangle\otimes|12\rangle+\\
&|012120\rangle\otimes|20\rangle+|022100\rangle\otimes|21\rangle+|100110\rangle\otimes|22\rangle\big)
\end{aligned}
$$

FIGURE 1.24 A 3 × 3 image and QTRQ representation.

From the above introduction, if we follow the color model used in quantum images, we can make the following classifications of quantum image representations:

1. *Binary Information-Based Quantum Images:* Binary geometry representation of entangled states[24].
2. *Quantum Image Based on Grayscale Information:* Real Ket[9], FRQI[10], NEQR[13], CQIR[37], FLPI (2-D QSNA)[38], QSMC&QSNC[7], QUALPI (polar coordinates)[14], INEQR[39], GQIR[15], IFRQI[27], QRMMI (multiple)[40], GNEQR[41], OQIM[28], MQIR[34], TMQIR (3D MQIR)[42].
3. *Quantum Image Based on Color Space Model:*
   a. *RGB Color Space:* MCRQI[43], MCQI[17], QSMC&QSNC[7], NAQSS[25], NASSRP[44], CQIPT[18], FQRCI[11], MCLPQI (polar coordinates)[19], NCQI[20], OCQR[23], QRMW[26], FRQCI[12], BRQI[16], QRCI[21], DQRCI (two images)[29], INCQI[22], MQIR[34].
   b. *HSL Color Space:* QHSL[32], MQIR[34].
   c. *HSI Color Space:* QIRHSI[33], IQIRHSI[45].
   d. *Index Image:* QIIR[30].
4. *Quantum Representation of Infrared Images:* SQR[8].

References [46] and [47], respectively, introduce and analyze the quantum image representation methods and development trends. Table 1.4 compares the common quantum image representation methods and characteristics.

1.3.1.8 Quantum Video Representation Model

Quantum video is a continuous set of quantum images. In theory, multiple quantum image representation methods can be used as quantum video representation methods, such as FLPI, MFLPI, QRMMI, and TMQIR. The main quantum video representation method is the quantum video framework proposed by Iliyasu et al. In 2011[48], quantum video framework is proposed based on strip circuit and FRQI quantum gray image representation. In 2015, Yan et al.[49] proposed an extended multi-channel quantum video, namely color video method. In 2016, Wang[50] proposed Quantum Video Based on NEQR（QVNEQR）based on NEQR quantum image representation and Strip circuit.

1.3.2 Quantum Image and Video Processing

A large number of mature and perfect-processing algorithms have been established in the classical image processing field, such as image elementary transformation, image segmentation, and so on. Although the research on quantum image and video processing is in its infancy, many research results have emerged.

TABLE 1.4
Different Quantum Image Representation Models

Quantum Image	Year	Qubit	Color Compilation Mode	Complexity	Access
QLM[6]	2003	2^{2n}	1 angle vectors (grayscale/RGB)	–	–
RKM[9]	2005	–	quantum superposition	–	–
FRQI[10]	2011	$2n+1$	1 angle vectors (grayscale)	$O(2^{4n})$	Probability
MCRQI[43]	2011	$2n+3$	4 angle vectors (RGB)	$O(2^{4n+6}-3\times 2^{2n+3})$	Probability
NEQR[13]	2013	$2n+q$	q qubits sequence (grayscale)	$O(qn2^{2n})$	Certainty
NAQSS[25]	2014	$2n+1$	1 angle vectors (grayscale/RGB)	$O(\log 2^n\times 2^{2n})$	Probability
CQIPT[18]	2014	$2n+3$	4 angle vectors (RGB)	$O(2^{4n+5}-3\times 2^{2n+2})$	Probability
FQRCI[11]	2014	$2n+3$	3 angle vectors(RGB)	$O(3\times 2^{4n})$	Probability
SQR[8]	2014	$2n+q+2$	1 angle vectors(infrared)	$O(2^{2n})$	Probability
GQIR[15]	2015	$h+w+q$	q qubits sequence (grayscale)	$O(2^{4n+2}+qn2^{2n})$	Certainty
NCQI[20]	2016	$2n+3q$	$3q$ qubits sequence (RGB)	$O(6qn\times 2^{2n})$	Certainty
BRQI[16]	2018	$2n+4\,/\,2n+6$	$q/3q$ qubits sequence (grayscale/RGB)	$O(mn\times 2^{2n})$	Certainty
OQIM[28]	2019	$2n+2$	1 angle vectors (grayscale)	$O(2^{4n})$	Probability
QRCI[21]	2019	$2n+6$	q qubits sequence (RGB)	$O(L\times 2^{2n})$	Certainty
FTQR[31]	2020	$r+s$	Fourier transform representation (grayscale/ RGB)	$O(r+s)$	Probability
QHSL[32]	2021	$2n+q+1$	1 angle vectors and q qubits sequence (HSL)	$O(24n+q-10)\cdot 2^{2n}$	Probability/certainty
QIRHSI[33]	2022	$2n+q+2$	2 angle vectors and q qubits sequence (HSI)	$O(2^{4n+2}+qn2^{2n})$	Probability/certainty

1.3.2.1 Quantum Image Color Transformation

Color transformation is a common method in image processing. It takes an image as an element to perform various operations, so as to obtain a new image which is different from the original image. In 2011, Le et al.[51] proposed a color transformation operation for the FRQI of a quantum image; Then, the channel switching operation and single-channel operation of the MCRQI image are given[43]. In 2013, using NOT gate, Zhang et al.[13] proposed the grayscale processing operations of the NEQR model, including global grayscale operations, local grayscale operations, and grayscale statistical operations. In 2016, Sang et al.[20] presented the channel switching operation and single-channel operation of NCQI images.

1.3.2.2 Geometric Transformations of Quantum Images

Within the scope of image processing, the purpose of transforming the shape of an image is achieved by manipulating the position of pixels, which is called geometric transformation. In 2010, with the help of NOT gate, CNOT gate, and Toffoli gate, Le et al.[52] realized the geometric transformation of the FRQI model, including two-point exchange, folded transformation, and right-angle rotation operation. Then, a controlled quantum image geometric transformation operation is proposed to realize the local geometric transformation[53]. In 2014, Wang et al.[54] presented the overall translation and cyclic translation operations of the NEQR model. In 2016, Fan et al.[55] implemented geometric transformation operations for quantum image NASS, such as two-point swap operations and symmetric flip operations, etc. In 2017, Yan et al.[56] proposed a rotation operation based on NEQR quantum image shear transform.

In addition, scaling transformation is an important transformation in geometric transformations. In 2014, Jiang and Wang[57] proposed an improved quantum scaling algorithm with a zoom factor of 2^rin Improved NEQR (INEQR) images based on the nearest neighbor interpolation method. In 2015, an algorithm for integer magnification of quantum image GQIR was proposed[15]. In 2015, Sang et al.[58] proposed a quantum implementation of the nearest neighbor interpolation method for FRQI quantum images and NEQR quantum images. In 2017, Zhou et al.[59] proposed a quantum algorithm based on integer scaling ratio for bilinear interpolation to amplify and reduce image GQIR. In 2018, Li and Liu[60] proposed a bilinear interpolation algorithm for NEQR images based on quantum Fourier transform. In 2020, using bilinear interpolation as a tool, Zhou et al. [61] proposed an image INEQR asymmetric scaling algorithm.

1.3.2.3 Quantum Image Compression

After the image signal is digitized, the amount of data becomes extremely large, and to transmit more image information in a limited channel, it is necessary to compress the image data. In 2013, Li et al.[7] designed an image QSMC compression algorithm. In 2017, Jiang et al.[62] constructed a compression algorithm for arbitrary-size GQIR quantum images based on the Joint Photographic Experts Group (JPEG) of static image compression algorithms.

1.3.2.4 Quantum Image Segmentation

The process of dividing images into areas that have no common areas between them is called image segmentation. In 2010, Venegas-Andraca et al.[24] proposed a segmentation algorithm for the Entangled Image model. In 2015, based on the threshold segmentation method, Caraiman et al.[63] proposed the development of a quantum version suitable for image segmentation. In 2019, with the help of the OTSU algorithm of classical optimal global thresholding, Li et al.[64] proposed a quantum circuit design method based on NEQR image threshold segmentation.

1.3.2.5 Quantum Image Enhancement

The method of using a technique to convert an image into a form suitable for machine analysis to obtain the "needed" information is called image enhancement. In 2015, Jiang et al.[65] proposed an image GQIR pseudo-color coding scheme based on density layering method. In 2017, based on the color image filtering operation of quantum Fourier transform, Li and Xiao[66] realized the quantum circuit of the NCQI image filtering task.

1.3.2.6 Quantum Image Edge Detection and Feature Extraction

One of the most important technologies in the field of image processing is edge detection, which has a great influence on the subsequent research on feature extraction, description, and object recognition. In 2014, Zhang et al.[67] proposed a local feature extraction framework for NEQR images. In 2019, Fan et al.[68] designed edge extraction of NEQR images based on the classical Sobel operator. In 2021, based on the idea of Luminance Contrast (LC), Jiang et al.[69] proposed a scheme based on quantum image points of interest. In 2022, Bao Hualiang and others[70] respectively designed quantum implementation algorithms of the classic Canny edge detection operator and the Marr-Hildreth[71] edge detection operator on NEQR quantum images.

1.3.2.7 Quantum Image Morphological Processing

In 2014, Yuan et al.[70] proposed the operation of two quantum forms, quantum binary and grayscale morphology, based on NEQR quantum images, and designed quantum binary and planar grayscale expansion and corrosion operations and corresponding quantum circuits. In 2018, based on the principle of classical morphological manipulation, Fan et al.[71] proposed the planar gray-scale expansion and corrosion operation of NEQR images, and then realized the morphological gradient operation.

1.3.2.8 Quantum Image Matching

Image matching, also known as correspondence problems, can establish a correspondence between two or more digital images that depict at least two parts of the same scene. In 2015, Yang et al.[72]

proposed a method to match grayscale images composed of quantum information NEQR. In 2016, Jiang et al.[73] proposed a quantum image matching algorithm. In this method, only one pixel (target pixel) is focused on instead of the whole image, and the probability of the pixel is modified based on the Grover algorithm so that the target pixel is measured with a higher probability, and the measurement step is performed only once. In 2018, Luo et al.[74] proposed a fuzzy quantum image matching scheme based on grayscale differences, which was applied to the reference image to locate the target region that is very similar to the template image.

1.3.2.9 Quantum Video Processing

For the processing of quantum video, the main research result is the quantum video moving target detection method proposed by Yan et al.[75, 76], which can detect and locate the moving target in the quantum video frame through the reasonable design of measurement strategies. In 2017, Wang et al.[77] proposed a QVNEQR-based quantum video motion vector extraction algorithm.

1.3.3 Quantum Image and Video Security

1.3.3.1 Quantum Image Watermarking

In 2012, based on non-random geometric transformations, Iliyasu et al.[78] proposed a Watermarking Strategy for Quantum Images (WaQI) for image security, keyless blind watermarking, and authentication on quantum computers. In 2013, Song et al.[79] proposed a watermarking scheme based on quantum wavelet transform QWT; the following year, Song et al.[80] again proposed a watermarking strategy based on the Hadamard transform. In 2014, Yan et al.[81] proposed the Multi-Channel Watermarking strategy for Quantum Images (MC-WaQI), or multi-channel quantum images, which integrates the idea of double key and dual domain to enhance the security of quantum images. In 2016, Heidari and Naseri proposed a new quantum watermarking protocol based on NEQR images, including LSB-based quantum scrambling[82]. In 2017, to solve the problem of watermark embedding in quantum color images, Li et al.[83] proposed an improved scheme using small-scale quantum circuits and color image scrambling. In 2019, Hu et al.[84] proposed an FRQI image watermarking algorithm based on the Haar wavelet transform[84].

1.3.3.2 Quantum Image Scrambling and Diffusion

Image scrambling and diffusion refers to the image that is originally discernible to the human eye into a noise-like image, which is essentially a pixel position scrambling and color diffusion operation, through which the image after this operation is basically unable to distinguish the content of the original image from the image. In 2014, Jiang et al.[85] and Jian and Wang[86] proposed the quantum image scrambling algorithms of Arnold scrambling and Fibonacci scrambling, and designed the quantum circuits of Arnold scrambling and Fibonacci scrambling by ordinary adder and modulo N adder; In the same year, Hilbert's chaotic quantum image scrambling algorithm was proposed, and Hilbert's quantum circuit was designed by recursive and asymptotic layering[87]. In 2015, Zhou et al.[88] proposed a full-color space scrambling strategy based on NEQR images, which utilized Gray code and Bit-plane (GB). Based on the image NEQR, several quantum scrambling methods using different GB knowledge have been designed. In 2016, using CNOT gate and the modulo N adder, Sang et al.[89] designed the Arnold scrambled quantum circuit of quantum image IFRQI.

1.3.3.3 Quantum Image and Video Steganography

Quantum steganography is a technique for hiding secret information in quantum carriers (e.g., quantum images, quantum videos). In 2014, Jiang et al.[90] proposed a blind Least Significant Bit (LSB) steganography algorithm based on NEQR images in two forms of quantum circuits. One is the non-special LSB algorithm, which directly replaces the LSB of the pixel with the help of information bits. Block LSB is another type of algorithm that embeds bits of information into certain pixels belonging to a block of images. In 2016, Sang et al.[91] proposed a Least Significant Qubit (LSQb) information steganography algorithm for the quantum image model NCQI. The proposed method constructs an image of LSQb embedding, which mainly consists of two steps: qubit comparison and qubit embedding.

In quantum video steganography, Qu et al.[92] proposed a large-capacity quantum video steganography scheme based on the unique features of MCQI video frames, Chen and Qu[93] also proposed a large-capacity quantum video steganography scheme based on quantum video QVNEQR, and Wang and Song[77] proposed a motion vector LSQb steganography scheme based on quantum video QVNEQR.

1.3.3.4 Quantum Image Secret Sharing

Quantum secret sharing is the process of dividing information into several parts so that no subset of the parts is sufficient to read that information, but the information is available through the entire set. In 2014, based on the idea of 1 qubit and a strip, Song et al.[94] proposed a Quantum Image Secret Sharing (QISS) scheme. In 2019, an Improved Quantum Image Secret Sharing (Improved QISS, IQISS) scheme based on encryption and measurement of quantum grayscale image FRQI and quantum color image MCQI was proposed again[95].

1.3.3.5 Quantum Image and Video Encryption

In terms of quantum image encryption, Yang et al.[96] proposed a grayscale image encryption scheme in 2013, which uses two types of operations: quantum Fourier transform in the quantum image space domain and double random phase encoding technique in the quantum Fourier transform domain. In 2014, Song et al.[97] designed a quantum image encryption scheme based on controlled geometric transformation and color transformation. The core of this encryption scheme is to first scramble the pixel positions through controlled geometric transformation, and then perform color diffusion operation on the scrambled quantum image through controlled color transformation. In 2015, Zhou et al.[98] proposed an encryption algorithm for quantum image FRQI. Using the generalized Arnold operation to scramble pixel positions and encoding grayscale information through a double random phase operation. In 2016, Gong et al.[99] designed a quantum image encryption algorithm based on quantum image XOR operation. The quantum image XOR operation is designed by controlling the controlled non-operation using the hyper chaotic sequence generated by the Chen hyper chaotic system, which is used to encode grayscale information. In 2017, Wang et al.[100] proposed a quantum image FRQI encryption algorithm based on iterative frameworks of frequency domain transformation and spatial domain transformation. In the same year, Rang et al.[101] proposed a three-pulse injection quantum color image NCQI encryption scheme based on a coupled hyper chaotic Lorenz system. In 2019, Khan and Rasheed[102] designed a special linear unitary transformation related to the arrangement and combination of n different objects, and based on this, proposed a quantum image NCQI encryption algorithm for Chen chaotic dynamical systems based on the special linear unitary transformation. In 2020, Abd El Latif et al.[103] proposed a method for establishing a quantum walk pseudo-random number generator to encrypt color images stored in quantum mechanics systems under the NCQI image model. In 2021, based on Arnold scrambling and logistic mapping, a novel three-level quantum image NEQR encryption algorithm was proposed in reference [104]. To achieve satisfactory encryption results, the original image was subjected to a three-level encryption process of block level diffusion, bit plane diffusion, and pixel level diffusion. In 2022, Chen et al.[105] proposed a quantum color image encryption algorithm based on geometric transformation and intensity channel diffusion using QIRHSI quantum image representation.

In terms of quantum video encryption, in 2015, Yan et al.[106] proposed a method for video encryption and decryption on a quantum computer based on color information transformation of each frame of video content encoding. The proposed method provides a flexible operation for encrypting quantum videos through quantum measurements to enhance video security. In 2020, Song et al.[107]

proposed an efficient and secure quantum video encryption algorithm based on quantum bit plane control XOR operation and improved logistic mapping in multi-layer encryption steps. Three simple cryptographic systems were proposed in the encryption algorithm to complete the entire encryption process: inter-frame diffusion, intra-frame pixel position geometric transformation, and high-four intra-frame quantum bit plane scrambling.

REFERENCES

[1] Nielsen M A, Chuang I L. *Quantum Computation and Quantum Information*. Cambridge: Cambridge University Press, 2005, 57–89, 157–223.

[2] Shende V V, Prasad A K, Markov I L, et al. Synthesis of Reversible Logic Circuits. *IEEE Transactions on Computer-Aided Design of Integrated Circuits & Systems*, 2003, 22(6): 710–722.

[3] Saeedi M, Markov I L. Synthesis and Optimization of Reversible Circuits-A Survey. *ACM Computing Surveys (CSUR)*, 2013, 45(2): 1–34.

[4] Barenco A, Bennett C H, Cleve R, et al. Elementary Gates for Quantum Computation. *Physical Review A*, 1995, 52(5): 1–31.

[5] Vedral V, Barenco A, Ekert A, et al. Quantum Networks for Elementary Arithmetic Operations. *Quantum Physics*, 1995, 54(1): 1–12.

[6] Venegas-Andraca S E, Bose S. Storing, Processing, and Retrieving an Image Using Quantum Mechanics. *The International Society for Optical Engineering,* Orlando, FL, USA, 2003, 5105: 137–147.

[7] Li H S, Zhu Q X, Lan S, et al. Image Storage, Retrieval, Compression and Segmentation in a Quantum System. *Quantum Information Processing*, 2013, 12(6): 2269–2290.

[8] Yuan S Z, Mao X, Xue Y L, et al. SQR: A Simple Quantum Representation of Infrared Images. *Quantum Information Processing*, 2014, 13(6): 1353–1379.

[9] Latorre J I. Image Compression and Entanglement. *Computer Science*, 2005, 1–4.

[10] Le P Q, Dong F Y, Hirota K. A Flexible Representation of Quantum Images for Polynomial Preparation, Image Compression, and Processing Operations. *Quantum Information Processing*, 2011, 10(1): 63–84.

[11] Yang Y G, Jia X, Sun S J, et al. Quantum Cryptographic Algorithm for Color Images Using Quantum Fourier Transform and Double Random-Phase Encoding. *Information Sciences*, 2014, 277: 445–457.

[12] Li P C, Liu X D. Color Image Representation Model and Its Application Based on an Improved FRQI. *International Journal of Quantum Information*, 2018, 16(15): 1850005.

[13] Zhang Y, Lu K, Gao Y H, et al. NEQR: A Novel Enhanced Quantum Representation of Digital Images. *Quantum Information Processing*, 2013, 12(8): 2833–2860.

[14] Zhang Y, Lu K, Gao Y H, et al. A Novel Quantum Representation for Log-Polar Images. *Quantum Information Processing*, 2013, 12(9): 3103–3126.

[15] Jiang N, Wang J, Mu Y. Quantum Image Scaling Up Based on Nearest-Neighbor Interpolation with Integer Scaling Ratio. *Quantum Information Processing*, 2015, 14(11): 4001–4026.

[16] Li H S, Chen X, Xia H, et al. A Quantum Image Representation Based on Bitplanes. *IEEE Access*, 2018, 6: 62396–62404.

[17] Sun B, Iliyasu A M, Yan F, et al. An RGB Multi-Channel Representation for Images on Quantum Computers. *Journal of Advanced Computational Intelligence & Intelligent Informatics*, 2013, 17(3): 404–417.

[18] Song X H, Wang S, Niu X M. Multi-Channel Quantum Image Representation Based on Phase Transform and Elementary Transformations. *Journal of Information Hiding & Multimedia Signal Processing*, 2014, 5(4): 574–585.

[19] Sang J Z, Wang S, Song X H, et al. A Novel Representation for Multi-Channel Log-Polar Quantum Images. *Journal of Information Hiding and Multimedia Signal Processing*, 2015, 6(2): 340–350.

[20] Sang J Z, Wang S, Li Q. A Novel Quantum Representation of Color Digital Images. *Quantum Information Processing*, 2017, 16(2): 42.

[21] Wang L, Ran Q W, Ma J, et al. QRCI: A New Quantum Representation Model of Color Digital Images. *Optics Communications*, 2019, 438: 147–158.

[22] Su J, Guo X C, Liu C Q, et al. An Improved Novel Quantum Image Representation and Its Experimental Test on IBM Quantum Experience. *Scientific Reports*, 2021, 11(1): 1–13.

[23] Liu K, Zhang Y, Lu K, et al. An Optimized Quantum Representation for Color Digital Images. *International Journal of Theoretical Physics*, 2018, 57: 2938–2948.

[24] Venegas-Andraca S E, Ball J L. Processing Images in Entangled Quantum Systems. *Quantum Information Processing*, 2010, 9(1): 1–11.

[25] Li H S, Zhu Q X, Zhou R G, et al. Multi-Dimensional Color Image Storage and Retrieval for a Normal Arbitrary Quantum Superposition State. *Quantum Information Processing*, 2014, 13(4): 991–1011.

[26] Şahin E, Yilmaz I. QRMW: Quantum Representation of Multi Wavelength Images. *Turkish Journal of Electrical Engineering & Computer Sciences*, 2018, 26(2): 768–779.

[27] Khan R A. An Improved Flexible Representation of Quantum Images. *Quantum Information Processing*, 2019, 18(7): 201.

[28] Xu G L, Xu X G, Wang X, et al. Order-Encoded Quantum Image Model and Parallel Histogram Specification. *Quantum Information Processing*, 2019, 18(11): 346.

[29] Wang L, Ran Q W, Ma J. Double Quantum Color Images Encryption Scheme Based on DQRCI. *Multimedia Tools and Applications*, 2020, 79(6/7): 1–27.

[30] Wang B, Hao M Q, Li P C, et al. Quantum Representation of Indexed Images and Its Applications. *International Journal of Theoretical Physics*, 2020, 59(3): 374–402.

[31] Grigoryan A M, Agaian S S. New Look on Quantum Representation of Images: Fourier Transform Representation. *Quantum Information Processing*, 2020, 19(5): 148.

[32] Yan F, Li N Q, Hirota K. QHSL: A Quantum Hue, Saturation, and Lightness Color Model. *Information Sciences*, 2021, 577(4): 196–213.

[33] Chen G L, Song X H, Venegas-Andraca S E, et al. QIRHSI: Novel Quantum Image Representation Based on HSI Color Space Model. *Quantum Information Processing*, 2022, 21(1): 5.

[34] Zhu H H, Chen X B, Yang Y X. A Multimode Quantum Image Representation and Its Encryption Scheme. *Quantum Information Processing*, 2021, 20(9): 315.

[35] Amankwah M G, Camps D, Bethel E W, et al. Quantum Pixel Representations and Compression for *N*-Dimensional Images. *Scientific Reports*, 2022, 12(1): 7712.

[36] Dong H, Lu D, Li C. A Novel Qutrit Representation of Quantum Image. *Quantum Information Processing*, 2022, 21(3): 108.

[37] Caraiman S, Manta V. Image Processing Using Quantum Computing. 2012 *16th International Conference on System Theory, Control and Computing (ICSTCC)*. IEEE, 2012: 1–6.

[38] Wang M, Lu K, Zhang Y, et al. FLPI: Representation of Quantum Images for Log-polar Coordinate. *Fifth International Conference on Digital Image Processing (ICDIP 2013)*. SPIE, 2013, 8878: 77–81.

[39] Jiang N, Wang L. Quantum Image Scaling Using Nearest Neighbor Interpolation. *Quantum Information Processing*, 2015, 14: 1559–1571.

[40] Zhou N, Yan X, Liang H, et al. Multi-Image Encryption Scheme Based on Quantum 3D Arnold Transform and Scaled Zhongtang Chaotic System. *Quantum Information Processing*, 2018, 17: 1–36.

[41] Li H S, Fan P, Xia H Y, et al. Quantum Implementation Circuits of Quantum Signal Representation and Type Conversion. *IEEE Transactions on Circuits and Systems I: Regular Papers*, 2018, 66(1): 341–354.

[42] Zhu H H, Chen X B, Yang Y X. Image Preparations of Multi-Mode Quantum Image Representation and Their Application on Quantum Image Reproduction. *Optik*, 2022, 251: 168321.

[43] Sun B, Le P Q, Iliyasu A M, et al. A Multi-Channel Representation for Images on Quantum Computers Using the RGBα Color Space. *IEEE International Symposium on Intelligent Signal Processing*, Floriana, Malta, 2011: 1–6.

[44] Li H S, Zhu Q, Li M C, et al. Multidimensional Color Image Storage, Retrieval, and Compression Based on Quantum Amplitudes and Phases. *Information Sciences*, 2014, 273: 212–232.

[45] Chen G, Song X. Quantum Color Image Scaling on QIRHSI Model. *Data Science: 7th International Conference of Pioneering Computer Scientists, Engineers and Educators, ICPCSEE 2021, Taiyuan, China, September 17–20, 2021, Proceedings, Part I 7*. Springer Singapore, 2021: 453–467.

[46] Yan F, Iliyasu A M, Venegas-Andraca S E. A Survey of Quantum Image Representations. *Quantum Information Processing*, 2016, 15: 1–35

[47] Su J, Guo X, Liu C, et al. A New Trend of Quantum Image Representations. *IEEE Access*, 2020, 8: 214520–214537.
[48] Iliyasu A M, Le P Q, Dong F, et al. A Framework for Representing and Producing Movies on Quantum Computers. *International Journal of Quantum Information*, 2011, 9(6): 1459–1497.
[49] Yan F, Iliyasu A M, Venegas-Andraca S E, et al. Video Encryption and Decryption on Quantum Computers. *International Journal of Theoretical Physics*, 2015, 54: 2893–2904.
[50] Wang S. Frames Motion Detection of Quantum Video. *Advances in Intelligent Information Hiding and Multimedia Signal Processing: Proceeding of the Twelfth International Conference on Intelligent Information Hiding and Multimedia Signal Processing, November 21–23, 2016, Kaohsiung, Taiwan, Volume 2*. Springer International Publishing, 2017: 145–151.
[51] Le P Q, Iliyasu A M, Dong F Y, et al. Efficient Color Transformations on Quantum Images. *Journal of Advanced Computational Intelligence and Intelligent Informatics*, 2011, 15(6): 698–706.
[52] Le P Q, Iliyasu A M, Dong F Y, et al. Fast Geometric Transformations on Quantum Images. *International Journal of Applied Mathematics*, 2010, 40(3): 113–123.
[53] Le P Q, Iliyasu A M, Dong F Y, et al. Strategies for Designing Geometric Transformations on Quantum Images. *Theoretical Computer Science*, 2011, 412(15): 1406–1418.
[54] Wang J, Jiang N, Wang L. Quantum Image Translation. *Quantum Information Processing*, 2015, 14(5): 1589–1604.
[55] Fan P, Zhou R G, Jing N H, et al. Geometric Transformations of Multidimensional Color Images Based on NASS. *Information Sciences*, 2016, 340: 191–208.
[56] Yan F, Chen K H, Venegas-Andraca S E, et al. Quantum Image Rotation by an Arbitrary Angle. *Quantum Information Processing*, 2017, 16(11): 282.
[57] Jiang N, Wang L. Quantum Image Scaling using Nearest Neighbor Interpolation. *Quantum Information Processing*, 2014, 14(5): 1559–1571.
[58] Sang J Z, Wang S, Niu X M. Quantum Realization of the Nearest-Neighbor Interpolation Method for FRQI and NEQR. *Quantum Information Processing*, 2016, 15(1): 37–64.
[59] Zhou R G, Liu X G, Luo J. Quantum Circuit Realization of the Bilinear Interpolation Method for GQIR. *International Journal of Theoretical Physics*, 2017, 56(9): 2966–2980.
[60] Li P C, Liu X D. Bilinear Interpolation Method for Quantum Images Based on Quantum Fourier Transform. *International Journal of Quantum Information*, 2018, 16(4): 1850031.
[61] Zhou R G, Cheng Y, Qi X F, et al. Asymmetric Scaling Scheme Over the Two Dimensions of a Quantum Image. *Quantum Information Processing*, 2020, 19(9): 343.
[62] Jiang N, Lu X W, Hu H, et al. A Novel Quantum Image Compression Method Based on JPEG. *International Journal of Theoretical Physics*, 2018, 57(3): 611–636.
[63] Caraiman S, Manta V I. Image Segmentation on a Quantum Computer. *Quantum Information Processing*, 2015, 14(5): 1693–1715.
[64] Li P C, Shi T, Zhao Y, et al. Design of Threshold Segmentation Method for Quantum Image. *International Journal of Theoretical Physics*, 2020, 59(2): 514–538.
[65] Jiang N, Wu W Y, Wang L, et al. Quantum Image Pseudocolor Coding Based on the Density-Stratified Method. *Quantum Information Processing*, 2015, 14(5): 1735–1755.
[66] Li P C, Xiao H. An Improved Filtering Method for Quantum Color Image in Frequency Domain. *International Journal of Theoretical Physics*, 2018, 57(5): 1–21.
[67] Zhang Y, Lu K, Xu K, et al. Local Feature Point Extraction for Quantum Images. *Quantum Information Processing*, 2015, 14(5): 1573–1588.
[68] Fan P, Zhou R G, Hu W W, et al. Quantum Image Edge Extraction Based on Classical Sobel Operator for NEQR. *Quantum Information Processing*, 2019, 18(1): 1–24.
[69] Jiang N, Ji Z X, Li H, et al. Quantum Image Interest Point Extraction. *Modern Physics Letters A*, 2021, 36(9): 2150063.
[70] Yuan S Z, Mao X, Li T, et al. Quantum Morphology Operations Based on Quantum Representation Model. *Quantum Information Processing*, 2015, 14(5): 1625–1645.
[71] Fan P, Zhou R G, Hu W W, et al. Quantum Circuit Realization of Morphological Gradient for Quantum Grayscale Image. *International Journal of Theoretical Physics*, 2019, 58(2): 415–435.
[72] Yang Y G, Zhao Q Q, Sun S J. Novel Quantum Gray-Scale Image Matching. *Optik – International Journal for Light and Electron Optics*, 2015, 126(22): 3340–3343.

[73] Jiang N, Dang Y J, Wang J. Quantum Image Matching. *Quantum Information Processing*, 2016, 15(9): 3543–3572.

[74] Luo G F, Zhou R G, Liu X A, et al. Fuzzy Matching Based on Gray-Scale Difference for Quantum Images. *International Journal of Theoretical Physics*, 2018, 57(8): 2447–2460.

[75] Yan F, Iliyasu A M, Khan A R, et al. Moving Target Detection in Multi-Channel Quantum Video. *2015 IEEE 9th International Symposium on Intelligent Signal Processing (WISP) Proceedings*. IEEE, 2015: 1–5.

[76] Yan F, Iliyasu A M, Khan A R, et al. Measurements-Based Moving Target Detection in Quantum Video. *International Journal of Theoretical Physics*, 2016, 55: 2162–2173.

[77] Wang S, Song X. Quantum Video Information Hiding Based on Improved LSQb and Motion Vector. *Journal of Internet Technology*, 2017, 18(6): 1361–1368.

[78] Iliyasu A M, Le P Q, Dong F Y, et al. Watermarking and Authentication of Quantum Images Based on Restricted Geometric Transformations. *Information Sciences*, 2012, 186(1): 126–149.

[79] Song X H, Wang S, Liu S, et al. A Dynamic Watermarking Scheme for Quantum Images Using Quantum Wavelet Transform. *Quantum Information Processing*, 2013, 12(12): 3689–3706.

[80] Song X H, Wang S, Abd El-Latif A A, et al. Dynamic Watermarking Scheme for Quantum Images Based on Hadamard Transform. *Multimedia Systems*, 2014, 20(4): 379–388.

[81] Yan F, Iliyasu A M, Sun B, et al. A Duple Watermarking Strategy for Multi-Channel Quantum Images. *Quantum Information Processing*, 2015, 14(5): 1675–1692.

[82] Heidari S, Naseri M. A Novel LSB Based Quantum Watermarking. *International Journal of Theoretical Physics*, 2016, 55(10): 1–14.

[83] Li P C, Zhao Y, Xiao H, et al. An Improved Quantum Watermarking Scheme Using Small-Scale Quantum Circuits and Color Scrambling. *Quantum Information Processing*, 2017, 16(5): 127.

[84] Hu W W, Zhou R G, El- Rafei A, et al. Quantum Image Watermarking Algorithm Based on Haar Wavelet Transform. *IEEE Access*, 2019, 7: 121303–121320.

[85] Jiang N, Wu W Y, Wang L. The Quantum Realization of Arnold and Fibonacci Image Scrambling. *Quantum Information Processing*, 2014, 13(5): 1223–1236.

[86] Jiang N, Wang L. Analysis and Improvement of the Quantum Arnold Image Scrambling. *Quantum Information Processing*, 2014, 13(7): 1545–1551.

[87] Jiang N, Wang L, Wu W Y. Quantum Hilbert Image Scrambling. *International Journal of Theoretical Physics*, 2014, 53(7): 2463–2484.

[88] Zhou R G, Sun Y J, Fan P. Quantum Image Gray-Code and Bit-Plane Scrambling. *Quantum Information Processing*, 2015, 14(5): 1717–1734.

[89] Sang J Z, Wang S, Shi X, et al. Quantum Realization of Arnold Scrambling for IFRQI. *International Journal of Theoretical Physics*, 2016, 55(8): 3706–3721.

[90] Jiang N, Zhao N, Wang L. LSB Based Quantum Image Steganography Algorithm. *International Journal of Theoretical Physics*, 2016, 55(1): 107–123.

[91] Sang J Z, Wang S, Li Q. Least Significant Qubit Algorithm for Quantum Images. *Quantum Information Processing*, 2016, 15(11): 1–20.

[92] Qu Z, Chen S, Ji S. A Novel Quantum Video Steganography Protocol with Large Payload Based on MCQI Quantum Video. *International Journal of Theoretical Physics*, 2017, 56(11): 3543–3561.

[93] Chen S, Qu Z. Novel Quantum Video Steganography and Authentication Protocol with Large Payload. *International Journal of Theoretical Physics*, 2018, 57(12): 3689–3701.

[94] Song X H, Wang S, Sang J Z, et al. Flexible Quantum Image Secret Sharing Based on Measurement and Strip. *The Tenth International Conference on Intelligent Information Hiding and Multimedia Signal Processing. IEEE,* Kitakyushu, Japan, 2014: 215–218.

[95] Wang H Q, Song X H, Chen L L, et al. A Secret Sharing Scheme for Quantum Gray and Color Images Based on Encryption. *International Journal of Theoretical Physics*, 2019, 58(5): 1626–1650.

[96] Yang Y G, Xia J, Jia X, et al. Novel Image Encryption/Decryption Based on Quantum Fourier Transform and Double Phase Encoding. *Quantum Information Processing*, 2013, 12(11): 3477–3493.

[97] Song X H, Wang S, Abd El-Latif A A, et al. Quantum Image Encryption Based on Restricted Geometric and Color Transformations. *Quantum Information Processing*, 2014, 13(8): 1765–1787.

[98] Zhou N R, Hua T X, Gong L H, et al. Quantum Image Encryption Based on Generalized Arnold Transform and Double Random-Phase Encoding. *Quantum Information Processing*, 2015, 14(4): 1193–1213.

[99] Gong L H, He X T, Cheng S, et al. Quantum Image Encryption Algorithm Based on Quantum Image XOR Operations. *International Journal of Theoretical Physics*, 2016, 55(7): 3234–3250.

[100] Wang H, Wang J, Geng Y C, et al. Quantum Image Encryption Based on Iterative Framework of Frequency-Spatial Domain Transforms. *International Journal of Theoretical Physics*, 2017, 56(8): 1–21.

[101] Ran Q W, Wang L, Ma J, et al. A Quantum Color Image Encryption Scheme Based on Coupled Hyper-Chaotic Lorenz System with Three Impulse Injections. *Quantum Information Processing*, 2018, 17(8): 1–30.

[102] Khan M, Rasheed A. Permutation-Based Special Linear Transforms with Application in Quantum Image Encryption Algorithm. *Quantum Information Processing*, 2019, 18(10): 1–21.

[103] Abd El- Latif A A, Abd- El- Atty B, Venegas-Andraca S E. Controlled Alternate Quantum Walk-Based Pseudo-Random Number Generator and Its Application to Quantum Color Image Encryption. *Physica A: Statistical Mechanics and Its Applications*, 2019, 547: 123869.

[104] Liu X B, Xiao D, Liu C. Three-Level Quantum Image Encryption Based on Arnold Transform and Logistic Map. *Quantum Information Processing*, 2021, 20(1): 23.

[105] Song X, Chen G, Abd El-Latif A A. Quantum Color Image Encryption Scheme Based on Geometric Transformation and Intensity Channel Diffusion. *Mathematics*, 2022, 10(17): 3038.

[106] Yan F, Iliyasu A M, Venegas-Andraca S E, et al. Video Encryption and Decryption on Quantum Computers. *International Journal of Theoretical Physics*, 2015, 54(8): 2893–2904.

[107] Song X H, Wang H Q, Venegas-Andraca S E, et al. Quantum Video Encryption Based on Qubit-Planes Controlled-XOR Operations and Improved Logistic Map. *Physica A: Statistical Mechanics and Its Applications*, 2019, 537: 122660.

2 Quantum Image and Video Representation

Shaocong Lan, Xianhua Song, Ahmed A. Abd El-Latif, and Basma Abd El-Rahiem

2.1 A FRAMEWORK FOR QUANTUM IMAGE REPRESENTATION

According to the theory of quantum computer, the quantum image should be stored in the form of a certain quantum state in the quantum computer[1–27]. In addition, the preparation of quantum images is to use the unitary transformation to transform the qubit initial state into a quantum image representation of the corresponding quantum state, the quantum initial state is generally a number of $|0\rangle$ qubits.

The prevailing quantum image representation methods primarily rely on the quantization of classical image representation, whereby the two components of pixel position and color of an image are encoded into corresponding quantum states. A notable benefit of this approach is its versatility, allowing seamless conversion between classical and quantum computers. This flexibility is particularly advantageous for image-processing tasks, which can be efficiently executed in a quantum computer, and then the classical computer can be utilized for tasks such as data presentation and subsequent processing. The general quantum image representation framework models $|I\rangle$ is

$$|I\rangle = \sum_{i=0}^{N-1} \beta_i |c_i\rangle |p_i\rangle \tag{2.1}$$

In Equation 2.1, $|p_i\rangle$ denotes the pixel position coding, while $|c_i\rangle$ is the quantum coding of the color information, and N is the image size. As the classical image is obtained by sampling and quantization of the analog image, the pixel position and color are discretely encoded. To make the quantum image and the classical image inter-convertible, the proposed quantum image representation framework still uses the discrete finite-value encoding method.

For an image of size $2^m \times 2^n$, the quantum image representation model is

$$|I\rangle = \frac{1}{2^{\frac{m+n}{2}}} \sum_{y=0}^{2^n-1} \sum_{x=0}^{2^m-1} |c_{yx}\rangle \otimes |yx\rangle \tag{2.2}$$

where $|c_{yx}\rangle$ is the color information and $|yx\rangle$ is the pixel coordinate information. The encoding method of $|c_{yx}\rangle$ determines whether the quantum image is a binary, grayscale, or color image. $|yx\rangle$ corresponds to the coordinate system of the quantum image, and the common coordinate systems are the right-angle coordinate system and the polar coordinate system.

DOI: 10.1201/9781003621423-3

The preparation of quantum images has a certain degree of flexibility. For the preparation method of the quantum image, it can generally be accomplished in two steps: first prepare the coordinate system code $|yx\rangle$, which is generally realized by a Hadamard gate with the same number of quantum bits as position $|yx\rangle$. Then for each pixel position of color $|c_{yx}\rangle$, the corresponding controlled gates are designed to implement the quantum encoding of the color.

2.2 QUANTUM COLOR IMAGE REPRESENTATION QIRHSI[1]

2.2.1 QIRHSI Representation

Any color image (e.g., the color image "Peppers" shown in Figure 2.1) can be decomposed into hue (*H*), saturation (*S*), and intensity (*I*) channels. Let the color image size be $2^n \times 2^n$, the intensity channel takes values in the range of $[0, 2^q - 1]$, and using the binary sequence $C_k^0 C_k^1 \cdots C_k^{q-1}$ to encode the intensity I_k at the pixel location, then quantum color image representation based on the HSI color space (QIRHSI) is expressed as shown in Equation 2.3:

$$\begin{aligned} |I(\theta)\rangle &= \frac{1}{2^n} \sum_{k=0}^{2^{2n}-1} |C_k\rangle \otimes |k\rangle \\ &= \frac{1}{2^n} \sum_{k=0}^{2^{2n}-1} |H_k\rangle |S_k\rangle |I_k\rangle \otimes |k\rangle \end{aligned} \tag{2.3}$$

where

$$\begin{aligned} |H_k\rangle &= \cos\theta_{hk}|0\rangle + \sin\theta_{hk}|1\rangle \\ |S_k\rangle &= \cos\theta_{sk}|0\rangle + \sin\theta_{sk}|1\rangle \\ |I_k\rangle &= |C_k^0 C_k^1 \cdots C_k^{q-2} C_k^{q-1}\rangle \end{aligned}$$

$$\begin{gathered} \theta_{hk}, \theta_{sk} \in [0, \pi/2],\ C_k^l \in \{0,1\} \\ l = 0,1,\ldots,q-1 \\ k = 0,1,\ldots,2^{2n}-1 \end{gathered}$$

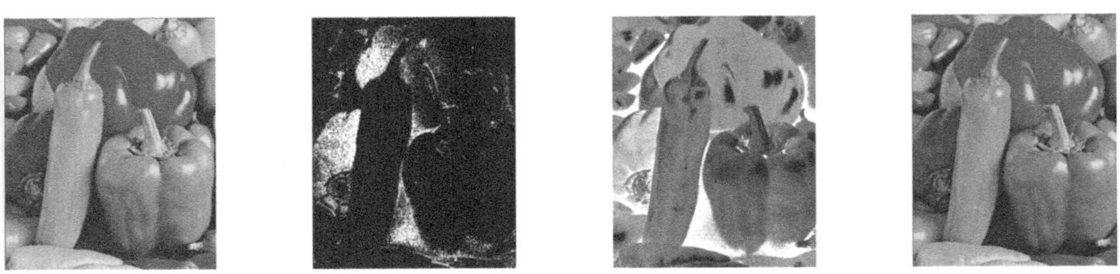

a) Figure Peppers b) Peppers H c) Peppers-S d) Peppers-I

FIGURE 2.1 *H*, *S*, and *I* channels of color image Peppers: (a) Figure Peppers; (b) Peppers-H; (c) Peppers-S; and (d) Peppers-I.

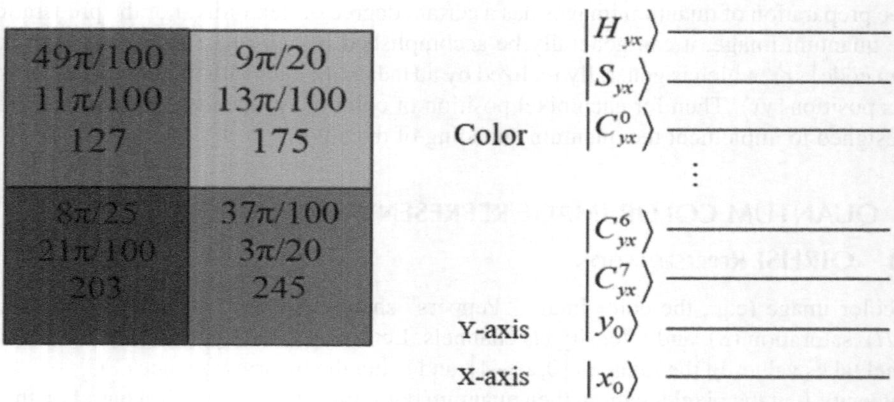

$$
\begin{aligned}
|I(\theta)\rangle = \frac{1}{2}\Big\{ & \left(\cos\frac{49\pi}{100}|0\rangle + \sin\frac{49\pi}{100}|1\rangle\right)\left(\cos\frac{11\pi}{100}|0\rangle + \sin\frac{11\pi}{100}|1\rangle\right)|01111111\rangle \otimes |00\rangle \\
& + \left(\cos\frac{9\pi}{20}|0\rangle + \sin\frac{9\pi}{20}|1\rangle\right)\left(\cos\frac{13\pi}{100}|0\rangle + \sin\frac{13\pi}{100}|1\rangle\right)|10101111\rangle \otimes |01\rangle \\
& + \left(\cos\frac{8\pi}{25}|0\rangle + \sin\frac{8\pi}{25}|1\rangle\right)\left(\cos\frac{21\pi}{100}|0\rangle + \sin\frac{21\pi}{100}|1\rangle\right)|11001011\rangle \otimes |10\rangle \\
& + \left(\cos\frac{37\pi}{100}|0\rangle + \sin\frac{37\pi}{100}|1\rangle\right)\left(\cos\frac{3\pi}{20}|0\rangle + \sin\frac{3\pi}{20}|1\rangle\right)|11110101\rangle \otimes |11\rangle \Big\}
\end{aligned}
$$

FIGURE 2.2 A 2 × 2 QIRHSI image, quantum circuit and quantum state.

$|C_k\rangle$ is used to encode color information and pixel position information is encoded using $|k\rangle$. The information about the hue channel H and the saturation vector S of the HSI color space are encoded by θ_{hk}, θ_{sk}. For an image in a $2n$-qubit system

$$
\begin{gathered}
|k\rangle = |y\rangle|x\rangle = |y_{n-1}\cdots y_1 y_0\rangle|x_{n-1}\cdots x_1 x_0\rangle \\
y, x \in \{0, 1, \ldots, 2^n - 1\} \\
|y_j\rangle, |x_j\rangle \in \{|0\rangle, |1\rangle\}, j = 0,1,\ldots,n-1
\end{gathered}
\tag{2.4}
$$

Here, $|y\rangle = |y_{n-1}\cdots y_1 y_0\rangle$ encodes the previous qubit along the vertical axis and the last n qubits are encoded using $|x\rangle = |x_{n-1}\cdots x_1 x_0\rangle$ along the horizontal axis.

A simple 2 × 2 QIRHSI image, quantum lines, and quantum states are shown in Figure 2.2.

2.2.2 Preparation of QIRHSI

Within quantum computing, a quantum computer is first initialized to an initial state. Therefore, it is necessary to give the complete preparation process for the transition from the initialized state to the quantum image. The polynomial preparation theorem (PPT) given in Theorem 2.1 shows an efficient realization of the preparation process.

Lemma 2.1: Given 2^{2n} binary sequences $C_k^0 C_k^1, \ldots, C_k^{q-1}$, $(k = 0,1,\ldots,2^{2n}-1)$ and 2 vectors $\theta_x = \left(\theta_{x_0}, \theta_{x_1} \theta\cdots, \theta_{x_{2^{2n}-1}}\right)$, $x \in \{h, s\}$ satisfying Equation 2.3. There exists a unitary transformation P

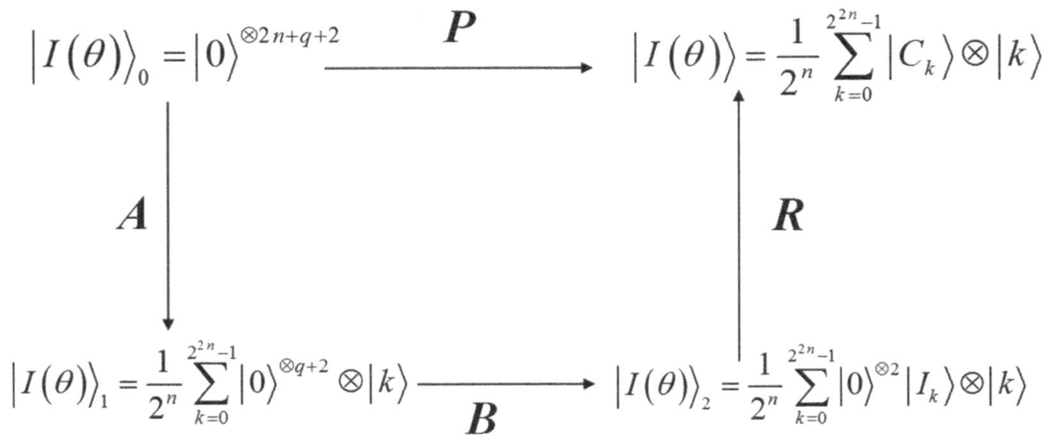

FIGURE 2.3 Workflow for preparing QIRHSI quantum image models.

that converts the quantum computer from an initial state $|0\rangle^{\otimes 2n+q+2}$ to a QIRHSI state $|I(\theta)\rangle$, where P consists of Hadamard transforms, CNOT transforms and controlled rotational transforms.

Proof: In the QIRHSI model, the unitary transform P can be constructed in three steps, which in turn initializes a set of qubits (see Figure 2.3). Figure 2.3 depicts the workflow for preparing a new quantum model from the initial state. The whole preparation process is divided into three steps: A, B, and R. First, step A encodes the pixels using the Hadamard transform; second, step B generates the intensity values with the help of a controlled non-transform; and finally, step R generates the hue and saturation using a controlled rotational transform.

Step A: Pixel Coding: Given two 1-qubit gates, the second order constant matrix I and the Hadamard matrix H.

$$I = \begin{pmatrix} 1 & 0 \\ 0 & 1 \end{pmatrix},\quad H = \frac{1}{\sqrt{2}}\begin{pmatrix} 1 & 1 \\ 1 & -1 \end{pmatrix} \tag{2.5}$$

We denote the tensor product of the $2n$ Hadamard matrices by $H^{\otimes 2n}$ and apply the transformation

$$A = I^{\otimes q+2} \otimes H^{\otimes 2n} \tag{2.6}$$

to the initial state $|0\rangle^{\otimes 2n+q+2}$, then we can obtain an intermediate state $|I(\theta)\rangle_1$, that is,

$$\begin{aligned} A\left(|0\rangle^{\otimes 2n+q+2}\right) &= \left(I|0\rangle\right)^{\otimes q+2} \otimes \left(H|0\rangle\right)^{\otimes 2n} \\ &= \frac{1}{2^n}\sum_{k=0}^{2^{2n}-1}|0\rangle^{\otimes q+2} \otimes |k\rangle \\ &\triangleq |I(\theta)\rangle_1 \end{aligned} \tag{2.7}$$

Step B: Intensity Value Encoding: To prepare a quantum image, color values are also prepared for each pixel. On the intermediate state $|I(\theta)\rangle_1$, all pixels have been stored inside the superposition state of the qubit sequence in the model. For each pixel m, the quantum operation B_m is

$$B_m = I^{\otimes q+2} \otimes \sum_{k=0,k\neq m}^{2^{2n}-1} |k\rangle\langle k| + I^{\otimes 2} \otimes \Lambda_m \otimes |m\rangle\langle m| \tag{2.8}$$

Λ_m is the intensity value generation operation for the pixel m, and

$$\Lambda_m = \bigotimes_{i=0}^{q-1} \Lambda_m^i \tag{2.9}$$

As the QIRHSI representation represents the intensity value in q qubits, the Λ_m consists of q quantum black boxes as shown in Equation 2.10:

$$\Lambda_m^i : |0\rangle \rightarrow |0 \oplus C_m^i\rangle \tag{2.10}$$

As can be seen from Equation 2.10, Λ_m^i is a quantum heterodyne XOR operation. Therefore, the quantum transformation Λ_m which sets the intensity value for the pixel is given by Equation 2.11:

$$\Lambda_m |0\rangle^{\otimes q} = \bigotimes_{i=0}^{q-1} \left(\Lambda_m^i |0\rangle\right) = \bigotimes_{i=0}^{q-1} |0 \oplus C_m^i\rangle = \bigotimes_{i=0}^{q-1} |C_m^i\rangle = |I_m\rangle \tag{2.11}$$

Then, the intermediate state $|I(\theta)\rangle_1$ is converted to $|I(\theta)\rangle_2$, see Equation 2.12:

$$\begin{aligned} B_m\left(|I(\theta)\rangle_1\right) &= \frac{1}{2^n} B_m \left\{ \sum_{k=0,k\neq m}^{2^{2n}-1} |0\rangle^{\otimes q+2} \otimes |k\rangle + |0\rangle^{\otimes q+2} \otimes |m\rangle \right\} \\ &= \frac{1}{2^n} \left\{ \sum_{k=0,k\neq m}^{2^{2n}-1} \left(I|0\rangle\right)^{\otimes q+2} \otimes |k\rangle + \left(I|0\rangle\right)^{\otimes 2} \left(\Lambda_m |0\rangle^{\otimes q}\right) \otimes |m\rangle \right\} \\ &= \frac{1}{2^n} \left\{ \sum_{k=0,k\neq m}^{2^{2n}-1} |0\rangle^{\otimes q+2} \otimes |k\rangle + |0\rangle^{\otimes 2} |I_m\rangle \otimes |m\rangle \right\} \end{aligned} \tag{2.12}$$

Acting Equation 2.8 twice gives Equation 2.13:

$$\begin{aligned} B_l B_m\left(|I(\theta)\rangle_1\right) &= \frac{1}{2^n} \left\{ \sum_{k=0,k\neq m,l}^{2^{2n}-1} \left(I|0\rangle\right)^{\otimes q+2} \otimes |k\rangle + \left(I|0\rangle\right)^{\otimes 2} \left(\Lambda_m |0\rangle^{\otimes q}\right) \otimes |m\rangle \right. \\ &\quad \left. + \left(I|0\rangle\right)^{\otimes 2} \left(\Lambda_l |0\rangle^{\otimes q}\right) \otimes |l\rangle \right\} \\ &= \frac{1}{2^n} \left\{ \sum_{k=0,k\neq m,l}^{2^{2n}-1} |0\rangle^{\otimes q+2} \otimes |k\rangle + |0\rangle^{\otimes 2} |I_m\rangle \otimes |m\rangle + |0\rangle^{\otimes 2} |I_l\rangle \otimes |l\rangle \right\} \end{aligned} \tag{2.13}$$

From Equation 2.13, we can obtain

$$B\left(|I(\theta)\rangle_1\right) = \left(\prod_{k=0}^{2^{2n}-1} B_k\right)\left(|I(\theta)\rangle_1\right) \triangleq |I(\theta)\rangle_2 \tag{2.14}$$

Step R: Hue and Saturation Coding: Next, consider the following rotational transformations (rotations with angles $2\theta_{sl}$ and $2\theta_{hl}$ around the y axis),

$$R_y\left(2\theta_{sl}\right)=\begin{pmatrix}\cos\theta_{sl} & -\sin\theta_{sl}\\ \sin\theta_{sl} & \cos\theta_{sl}\end{pmatrix} \tag{2.15}$$

$$R_y\left(2\theta_{hl}\right)=\begin{pmatrix}\cos\theta_{hl} & -\sin\theta_{hl}\\ \sin\theta_{hl} & \cos\theta_{hl}\end{pmatrix} \tag{2.16}$$

Based on the above transformations, two controlled rotation matrices are constructed R_{sl}, R_{hl} ,$l = 0,1,\cdots,2^{2n}-1$.

$$R_{sl} = I^{\otimes q+2} \otimes \sum_{k=0,k\neq l}^{2^{2n}-1} |k\rangle\langle k| + I \otimes R_y\left(2\theta_{sl}\right) \otimes I^{\otimes q} \otimes |l\rangle\langle l| \tag{2.17}$$

$$R_{hl} = I^{\otimes q+2} \otimes \sum_{k=0,k\neq l}^{2^{2n}-1} |k\rangle\langle k| + R_y\left(2\theta_{hl}\right) \otimes I \otimes I^{\otimes q} \otimes |l\rangle\langle l| \tag{2.18}$$

When $l = i, j$, acting R_{si} on $|I(\theta)\rangle_2$ gets:

$$\begin{aligned}
R_{si}\left(|I(\theta)\rangle_2\right) &= \frac{1}{2^n}\Bigg\{(I|0\rangle)^{\otimes 2} \otimes \left(\sum_{k=0,k\neq i}^{2^{2n}-1} |I_k\rangle \otimes |k\rangle\langle k|\right)\left(\sum_{k=0}^{2^{2n}-1}|k\rangle\right) \\
&\quad + |0\rangle\left(R_y\left(2\theta_{si}\right)|0\rangle\right)|I_i\rangle \otimes |i\rangle\langle i|\left(\sum_{k=0}^{2^{2n}-1}|k\rangle\right)\Bigg\} \\
&= \frac{1}{2^n}\Bigg\{|0\rangle^{\otimes 2} \otimes \left(\sum_{k=0,k\neq i}^{2^{2n}-1} |I_k\rangle \otimes |k\rangle\right) \\
&\quad + |0\rangle\left(\cos\theta_{si}|0\rangle + \sin\theta_{si}|1\rangle\right)|I_i\rangle \otimes |i\rangle\Bigg\} \\
&= \frac{1}{2^n}\Bigg\{|0\rangle^{\otimes 2} \otimes \left(\sum_{k=0,k\neq i}^{2^{2n}-1} |I_k\rangle \otimes |k\rangle\right) + |0\rangle|S_i\rangle|I_i\rangle \otimes |i\rangle\Bigg\}
\end{aligned} \tag{2.19}$$

Applying Equation 2.19 twice, we can obtain:

$$\begin{aligned}
&R_{sj}R_{si}\left(|I(\theta)\rangle_2\right) \\
&= \frac{1}{2^n}\Bigg\{|0\rangle^{\otimes 2} \otimes \left(\sum_{k=0,k\neq i,j}^{2^{2n}-1} |I_k\rangle \otimes |k\rangle\right) + |0\rangle\left(\cos\theta_{si}|0\rangle + \sin\theta_{si}|1\rangle\right)|I_i\rangle \otimes |i\rangle \\
&\quad + |0\rangle\left(\cos\theta_{sj}|0\rangle + \sin\theta_{sj}|1\rangle\right)|I_j\rangle \otimes |j\rangle\Bigg\} \\
&= \frac{1}{2^n}\Bigg\{|0\rangle^{\otimes 2} \otimes \left(\sum_{k=0,k\neq i,j}^{2^{2n}-1} |I_k\rangle \otimes |k\rangle\right) + |0\rangle|S_i\rangle|I_i\rangle \otimes |i\rangle + |0\rangle|S_j\rangle|I_j\rangle \otimes |j\rangle\Bigg\}
\end{aligned} \tag{2.20}$$

From Equation 2.20, it follows that

$$R_s\left(\left|I(\theta)\right\rangle_2\right)=\left(\prod_{k=0}^{2^{2n}-1}R_{sk}\right)\left(\left|I(\theta)\right\rangle_2\right)=\frac{1}{2^n}\sum_{k=0}^{2^{2n}-1}|0\rangle\left|S_k\right\rangle\left|I_k\right\rangle\otimes|k\rangle \tag{2.21}$$

Likewise,

$$\begin{aligned}
&R_{hi}\left(\left|I(\theta)\right\rangle_2\right)\\
&=\frac{1}{2^n}\left\{I|0\rangle\otimes\left(\sum_{k=0,k\neq i}^{2^{2n}-1}\left|S_k\right\rangle\left|I_k\right\rangle\otimes|k\rangle\langle k|\right)\left(\sum_{k=0}^{2^{2n}-1}|k\rangle\right)\right.\\
&\quad\left.+\left(R_y\left(2\theta_{hi}\right)|0\rangle\right)\left|S_i\right\rangle\left|I_i\right\rangle\otimes|i\rangle\langle i|\left(\sum_{k=0}^{2^{2n}-1}|k\rangle\right)\right\}\\
&=\frac{1}{2^n}\left\{|0\rangle\otimes\left(\sum_{k=0,k\neq i}^{2^{2n}-1}\left|S_k\right\rangle\left|I_k\right\rangle\otimes|k\rangle\right)+\left(\cos\theta_{hi}|0\rangle+\sin\theta_{hi}|1\rangle\right)\left|S_i\right\rangle\left|I_i\right\rangle\otimes|i\rangle\right\}\\
&=\frac{1}{2^n}\left\{|0\rangle\otimes\left(\sum_{k=0,k\neq i}^{2^{2n}-1}\left|S_k\right\rangle\left|I_k\right\rangle\otimes|k\rangle\right)+\left|H_i\right\rangle\left|S_i\right\rangle\left|I_i\right\rangle\otimes|i\rangle\right\}
\end{aligned} \tag{2.22}$$

Applying Equation 2.22 twice, we get:

$$\begin{aligned}
&R_{hj}R_{hi}\left(\frac{1}{2^n}\sum_{k=0}^{2^{2n}-1}|0\rangle\left|S_k\right\rangle\left|I_k\right\rangle\otimes|k\rangle\right)\\
&=\frac{1}{2^n}\left\{|0\rangle\otimes\left(\sum_{k=0,k\neq i,j}^{2^{2n}-1}\left|S_k\right\rangle\left|I_k\right\rangle\otimes|k\rangle\right)+\left(\cos\theta_{hi}|0\rangle+\sin\theta_{hi}|1\rangle\right)\left|S_i\right\rangle\left|I_i\right\rangle\otimes|i\rangle\right.\\
&\quad\left.+\left(\cos\theta_{hj}|0\rangle+\sin\theta_{hj}|1\rangle\right)\left|S_j\right\rangle\left|I_j\right\rangle\otimes|j\rangle\right\}\\
&=\frac{1}{2^n}\left\{|0\rangle\otimes\left(\sum_{k=0,k\neq t,l}^{2^{2n}-1}\left|S_k\right\rangle\left|I_k\right\rangle\otimes|k\rangle\right)+\left|H_i\right\rangle\left|S_i\right\rangle\left|I_i\right\rangle\otimes|i\rangle+\left|H_j\right\rangle\left|S_j\right\rangle\left|I_j\right\rangle\otimes|j\rangle\right\}\\
&=\frac{1}{2^n}\left\{|0\rangle\otimes\left(\sum_{k=0,k\neq i,j}^{2^{2n}-1}\left|S_k\right\rangle\left|I_k\right\rangle\otimes|k\rangle\right)+\left|C_i\right\rangle\otimes|i\rangle+\left|C_j\right\rangle\otimes|j\rangle\right\}
\end{aligned} \tag{2.23}$$

From Equation 2.23, we can obtain

$$R\left(\left|I(\theta)\right\rangle_2\right)=R_h\left(R_s\left(\left|I(\theta)\right\rangle_2\right)\right)=\left(\prod_{k=0}^{2^{2n}-1}R_{hk}\right)\left(R_s\left(\left|I(\theta)\right\rangle_2\right)\right)=\left|I(\theta)\right\rangle \tag{2.24}$$

Thus, $P=RBA$ is the unitary transformation from the initial state $|0\rangle^{\otimes 2n+q+2}$ to the quantum image $\left|I(\theta)\right\rangle$.

□

In the quantum line model, complex transformations can be decomposed into basic quantum gates, that is, 1-qubit gates (NOT gates, Hadamard gates as shown in Table 1.1) and controlled 2-qubit gates (CNOT gates as shown in Table 1.2).

Corollary 2.1: Given 2^{2n} binary sequences $C_k^0 C_k^1 \cdots C_k^{q-1}$, $(k = 0,1,\cdots,2^{2n}-1)$and two angle vectors $\theta_x = \left(\theta_{x_0}, \theta_{x_1}, \ldots, \theta_{x_{2^{2n}-1}}\right)$, $x \in \{h,s\}$, the unitary transformation P described in Lemma 2.1 is able to be performed by Hadamard gates, CNOT gate, Controlled revolving doors $C^{2n}\left(R_y\left(\frac{2\theta_{sk}}{2^{2n}-1}\right)\right)$ and $C^{2n}\left(R_y\left(\frac{2\theta_{hk}}{2^{2n}-1}\right)\right)$, where $R_y\left(\frac{2\theta_{sk}}{2^{2n}-1}\right)$ and $R_y\left(\frac{2\theta_{hk}}{2^{2n}-1}\right)$ rotate at an angle of $\frac{2\theta_{sk}}{2^{2n}-1}$, $\frac{2\theta_{hk}}{2^{2n}-1}$ about axis y, $k = 0,1,\ldots,2^{2n}-1$.

Proof: It follows from the proof of Lemma 2.1 that the transformation P consists of RBA. Transformations A, B, and R can be each made by $2n$ Hadamard gates. Not exceeding $qn2^{2n}$ CNOT gates, 2^{2n} controlled revolving doors R_{sk} and 2^{2n} controlled revolving doors or R_{hk} and $C^{2n}(2\theta_{sk})$ operations. In addition, the controlled rotation operations R_{sk} and R_{hk} can be realized by the $C^{2n}(2\theta_{sk})$, $C^{2n}(2\theta_{hk})$, and NOT operations.

Furthermore, the $C^{2n}\left(R_y(2\theta_k)\right)$ operations can be decomposed into $2^{2n}-1$ basic operations $R_y\left(\frac{2\theta_k}{2^{2n}-1}\right)$, $R_y\left(-\frac{2\theta_k}{2^{2n}-1}\right)$, and $2^{2n}-2$ CNOT operations.

It should be noted that quantum operation A is the operation in step A. From Equation 2.6, the basic operand in step A is $2n$. However, steps B and R are more complex than step A. In step B, the entire quantum operation B is divided into $2n$ sub-operations B_m to prepare intensity information for each pixel. From Theorem 2.1 in reference [2], it follows that the number of quantum elementary operations consumed by B_m is qn. As the intensity value of each pixel is set independently in step B, the basic number of operations required for the whole operation B is $qn2^{2n}$. In step R, the entire quantum operation R is divided into operations R_s and R_h. From Corollary 2.1, it is clear that the $R_y(2\theta_{sl})$ operations in Equation 2.17 can be divided into $2^{2n}-1$ basic operations and $2^{2n}-2$ CNOT gate operations. Because the saturation information of each pixel is independent of each other, the basic number of operations required for the R_s operation in Equation 2.21 is $2^{2n} \times (2^{2n}-1+2^{2n}-2)$; therefore, the basic operations required in step R are $2 \times 2^{2n} \times (2^{2n}-1+2^{2n}-2)$.

In summary, the total number of basic operations used to prepare QIRHSI quantum states is

$$2n + qn2^{2n} + 2 \times 2^{2n} \times (2^{2n}-1+2^{2n}-2) = 2^{4n+2} + qn2^{2n} + 2n - 3 \times 2^{2n+1}$$

which is the squared relation of 2×2^{2n} angular numbers, thus the polynomial-level complexity of the preparation process is verified.

□

Theorem 2.1 (QIRHSI Preparation Theorem):

Given 2^{2n} binary sequences $C_k^0 C_k^1 \cdots C_k^{q-1}$, $(k = 0,1,\ldots,2^{2n}-1)$ and two angle vectors $\theta_x = \left(\theta_{x_0}, \theta_{x_1} \theta \ldots, \theta_{x_{2^{2n}-1}}\right)$, $x \in \{h,s\}$, there exists a unitary transformation P that allows a quantum computer to transition from an initial state $|0\rangle^{\otimes 2n+q+2}$ to a QIRHSI state and P can be decomposed into elementary quantum gates of polynomial order.

Proof: This conclusion follows directly from Lemma 2.1 and Corollary 2.1.

□

Quantum Image Representation based on HSI Color Space (QIRHSI) combines the advantages of FRQI and NEQR representations as an improvement of quantum image representation. QIRHSI is able to enhance FRQI and NEQR by:

1. In preparing a quantum image, QIRHSI represents H and S with an angle vector, respectively, and I with a binary sequence of q bits, which not only reduces the number of qubits required to represent H and S, but also facilitates various color-dependent operations on I.
2. Combining FRQI and NEQR models with HSI color space provides superior visual compatibility with the human eye compared to the common RGB space.

2.2.3 Improved QIRHSI[3]

The QIRHSI model can only process color images of size $2^n \times 2^n$. In some image processing, such as image scale scaling, the image size is no longer of size $2^n \times 2^n$. To solve this problem, in this section, the Improved QIRHSI (IQIRHSI) model is proposed to be able to represent quantum images of size $2^m \times 2^n$.

$$\begin{gathered} |I(\theta)\rangle = \frac{1}{\sqrt{2^{m+n}}} \sum_{y=0}^{2^m-1} \sum_{x=0}^{2^n-1} |C_{yx}\rangle \otimes |yx\rangle = \frac{1}{\sqrt{2^{m+n}}} \sum_{y=0}^{2^m-1} \sum_{x=0}^{2^n-1} |H_{yx}\rangle |S_{yx}\rangle |I_{yx}\rangle \otimes |yx\rangle \\ |yx\rangle = |y\rangle|x\rangle = |y_0 y_1 \cdots y_{m-2} y_{m-1}\rangle |x_0 x_1 \cdots x_{n-2} x_{n-1}\rangle, y_j, x_j \in \{0,1\} \end{gathered} \tag{2.25}$$

IQIRHSI uses $q+2+m+n$ qubits to represent a $2^m \times 2^n$ color image with a range of intensity values of 2^q. Figure 2.4 shows a simple 4 × 2 IQIRHSI image and its representation.

2.3 QUANTUM COLOR IMAGE REPRESENTATION CQIPT[4]

In the field of classical computers, there exists a variety of ways to encode images to suit different applications. Similarly, in quantum computers, image representations should be flexible and diverse. In particular, many algorithms for quantum image processing and security are closely related to quantum image representation. For the representation of color images, the Multi-Channel Representation for Quantum Image (MCRQI) method and the Normal Arbitrary Quantum Superposition State (NAQSS) method, both of which are based on a revolving door construction. However, many quantum techniques are based on other quantum gate operations. Therefore, a Color

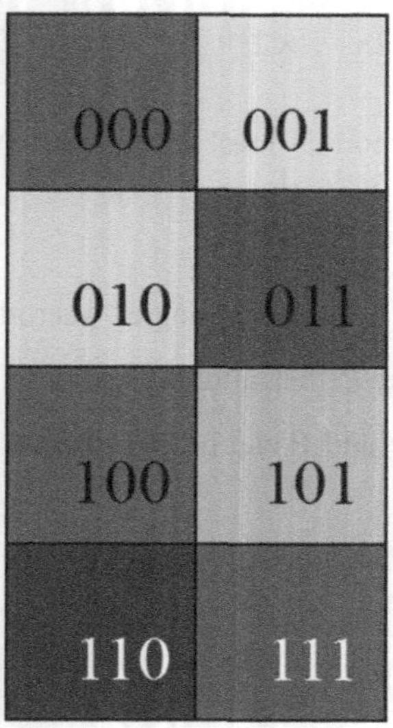

$$\begin{aligned} |I(\theta)\rangle = \frac{1}{\sqrt{2^3}} \Big\{ & \left(\cos\frac{49\pi}{100}|0\rangle + \sin\frac{49\pi}{100}|1\rangle\right)\left(\cos\frac{\pi}{2}|0\rangle + \sin\frac{\pi}{2}|1\rangle\right)|01111111\rangle \otimes |000\rangle \\ & + \left(\cos\frac{9\pi}{20}|0\rangle + \sin\frac{9\pi}{20}|1\rangle\right)\left(\cos\frac{\pi}{2}|0\rangle + \sin\frac{\pi}{2}|1\rangle\right)|01111111\rangle \otimes |001\rangle \\ & + \left(\cos\frac{9\pi}{50}|0\rangle + \sin\frac{9\pi}{50}|1\rangle\right)\left(\cos\frac{19\pi}{100}|0\rangle + \sin\frac{19\pi}{100}|1\rangle\right)|10100011\rangle \otimes |010\rangle \\ & + \left(\cos\frac{37\pi}{100}|0\rangle + \sin\frac{37\pi}{100}|1\rangle\right)\left(\cos\frac{\pi}{2}|0\rangle + \sin\frac{\pi}{2}|1\rangle\right)|01011000\rangle \otimes |011\rangle \\ & + \left(\cos\frac{8\pi}{25}|0\rangle + \sin\frac{8\pi}{25}|1\rangle\right)\left(\cos\frac{\pi}{2}|0\rangle + \sin\frac{\pi}{2}|1\rangle\right)|01111000\rangle \otimes |100\rangle \\ & + \left(\cos\frac{23\pi}{50}|0\rangle + \sin\frac{23\pi}{50}|1\rangle\right)\left(\cos\frac{\pi}{2}|0\rangle + \sin\frac{\pi}{2}|1\rangle\right)|01111111\rangle \otimes |101\rangle \\ & + \left(\cos\frac{31\pi}{100}|0\rangle + \sin\frac{31\pi}{100}|1\rangle\right)\left(\cos\frac{\pi}{2}|0\rangle + \sin\frac{\pi}{2}|1\rangle\right)|01100000\rangle \otimes |110\rangle \\ & + \left(\cos\frac{23\pi}{100}|0\rangle + \sin\frac{23\pi}{100}|1\rangle\right)\left(\cos\frac{8\pi}{25}|0\rangle + \sin\frac{8\pi}{25}|1\rangle\right)|01101000\rangle \otimes |111\rangle \Big\} \end{aligned}$$

FIGURE 2.4 A 4 × 2 IQIRHSI quantum color image and its representation.

Quantum Image based on Phase Transform (CQIPT) based on Controlled Phase Gate representation is proposed in this section. It has the following features:

1. Each basis state has the same probability amplitude.
2. Polynomial time preparation complexity and application flexibility: greater applicability to image processing and security algorithms based on phase transformations. For example, the combination of double random-phase encoding (DRPE) and certain transform methods has been applied to many areas of image security, such as image encryption and information hiding. Then the representation based on phase transformation will be more suitable for such processing.

2.3.1 CQIPT Representation

Let the color image size be $2^n \times 2^n$ and the multichannel quantum image representation CQIPT is of the form:

$$|I(\theta)\rangle = \frac{1}{2^{n+3/2}} \sum_{k=0}^{2^{2n}-1} \left|C_{RGB\alpha}^k\right\rangle \otimes |k\rangle \tag{2.26}$$

In Equation 2.26, $\left|C_{RGB\alpha}^k\right\rangle$ is the quantum state characterizing the color channels R, G, and B, and the transparency α, which is defined as

$$\begin{aligned}\left|C_{RGB\alpha}^k\right\rangle &= |000\rangle + e^{i\theta_{Rk}}|001\rangle + |010\rangle + e^{i\theta_{Gk}}|011\rangle \\ &+ |100\rangle + e^{i\theta_{Bk}}|101\rangle + |110\rangle + e^{i\theta_{\alpha k}}|111\rangle\end{aligned}$$

where $i = \sqrt{-1}, \theta_{Xk} \in \left[0, \frac{\pi}{2}\right], X \in \{R, G, B, \alpha\}, k = 0, 1, \cdots, 2^{2n}-1$ encode color information, $|k\rangle$ encodes the corresponding location information. The position information includes both vertical and horizontal components. For a color image stored in a $2n + 3$ qubit system, there are:

$$|k\rangle = |y\rangle|x\rangle = |y_{n-1}y_{n-2}\cdots y_0\rangle|x_{n-1}x_{n-2}\cdots x_0\rangle, x, y \in \{0, 1, \ldots, 2^n - 1\},$$

$$|y_j\rangle, |x_j\rangle \in \{|0\rangle, |1\rangle\}, j = 0, 1, \cdots, n-1$$

where $|y\rangle$ encodes n qubit positions in the vertical direction and $|x\rangle$ encodes n qubit positions in the horizontal direction. Further, the CQIPT satisfies the canonicalization requirement of the quantum state, that is:

$$\begin{aligned}\left\||I(\theta)\rangle\right\| &= \frac{1}{2^{n+3/2}} \sqrt{\sum_{j=0}^{2^{2n}-1} \left(1 + \left|e^{i\theta_{Rj}}\right|^2 + 1 + \left|e^{i\theta_{Gj}}\right|^2 + 1 + \left|e^{i\theta_{Bj}}\right|^2 + 1 + \left|e^{i\theta_{\alpha j}}\right|^2\right)} \\ &= \frac{1}{2^{n+3/2}} \sqrt{2^{2n} \times 2^3} = 1\end{aligned}$$

A quantum image demonstration diagram of CQIPT for a color image of size 2 × 2 is shown in Figure 2.5. By setting the number of qubits in the registers representing the x, y axes to m, n, respectively, CQIPT is easily extended to represent any color image of size $2^m \times 2^n$.

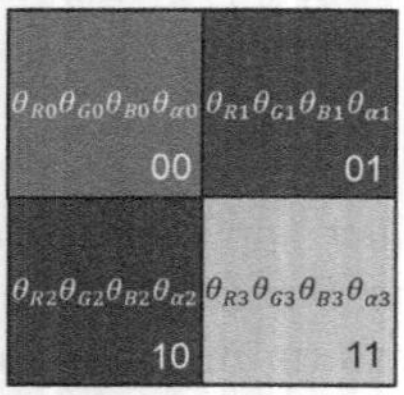

$$
\begin{aligned}
|I(\theta)\rangle = \frac{1}{\sqrt{2^5}}\Big[&\big(|000\rangle + e^{i\theta_{R0}}|001\rangle + |010\rangle + e^{i\theta_{G0}}|011\rangle + |100\rangle + e^{i\theta_{B0}}|101\rangle + |110\rangle + e^{i\theta_{\alpha 0}}|111\rangle\big)\otimes|00\rangle \\
&+\big(|000\rangle + e^{i\theta_{R1}}|001\rangle + |010\rangle + e^{i\theta_{G1}}|011\rangle + |100\rangle + e^{i\theta_{B1}}|101\rangle + |110\rangle + e^{i\theta_{\alpha 1}}|111\rangle\big)\otimes|01\rangle \\
&+\big(|000\rangle + e^{i\theta_{R2}}|001\rangle + |010\rangle + e^{i\theta_{G2}}|011\rangle + |100\rangle + e^{i\theta_{B2}}|101\rangle + |110\rangle + e^{i\theta_{\alpha 2}}|111\rangle\big)\otimes|10\rangle \\
&+\big(|000\rangle + e^{i\theta_{R3}}|001\rangle + |010\rangle + e^{i\theta_{G3}}|011\rangle + |100\rangle + e^{i\theta_{B3}}|101\rangle + |110\rangle + e^{i\theta_{\alpha 3}}|111\rangle\big)\otimes|11\rangle\Big]
\end{aligned}
$$

FIGURE 2.5 A 2×2 CQIPT image and quantum states.

2.3.2 Preparation of CQIPT

As in Section 2.2.2, the initial state of the register is the tensor product state of the base state $|0\rangle$. Therefore, the initial state of the quantum computer is $|0\rangle^{\otimes 2n+3}$, and the purpose of the preparation is to study the method of transforming the initial state into a multichannel quantum image representation state $|I(\theta)\rangle$ using the unitary transformation.

Lemma 2.2: Given a set of angular vectors $\theta_X = (\theta_{X0}, \theta_{X1}, \cdots, \theta_{X(2^{2n}-1)})$, $X \in \{R, G, B, \alpha\}$, there exists a unitary transformation P such that the quantum computer is able to transform from the initial state $|0\rangle^{\otimes 2n+3}$ to the target state $|I(\theta)\rangle$, and the unitary transformation can be decomposed into Hadamard gates and controlled phase gates.

Proof: The whole preparation process can be divided into two steps A and B. Two 1-qubit gates are assumed, the identity operation and the Hadamard gate, as in Equation 2.5.

First, applying the H to every qubit of $|0\rangle^{\otimes 2n+3}$, that is, $A = H^{\otimes 2} \otimes H \otimes H^{\otimes 2n}$ acting on $|0\rangle^{\otimes 2n+3} = |0\rangle^{\otimes 2} \otimes |0\rangle \otimes |0\rangle^{\otimes 2n}$, we can obtain

$$
\begin{aligned}
A(|0\rangle^{\otimes 2n+3}) &= \frac{1}{2^{n+3/2}} \sum_{l=0}^{3} |l\rangle \otimes (|0\rangle + |1\rangle) \otimes \sum_{k=0}^{2^{2n}-1} |k\rangle \\
&\triangleq |W\rangle
\end{aligned}
$$

Then, four 8×8 controlled phase matrices P_{Ri}, P_{Gi}, P_{Bi}, $P_{\alpha i}$ are constructed as shown in Equation 2.27:

$$
P_{Xi} = \left(\sum_{j=0, j \neq f(X)}^{3} |j\rangle\langle j| \otimes I \right) + |f(X)\rangle\langle f(X)| \otimes P(\theta_{Xi}) \tag{2.27}
$$

where $P(\theta_{Xj})$ is a phase transformation and $P(\theta_{Xj}) = \begin{pmatrix} 1 & 0 \\ 0 & e^{i\theta_{Xj}} \end{pmatrix}$, $X \in \{R, G, B, \alpha\}$,

$$
f(X) = \begin{cases} 0, & X = R \\ 1, & X = G \\ 2, & X = B \\ 3, & X = \alpha \end{cases}
$$

Therefore, we can obtain $P_i' = P_{Ri} P_{Gi} P_{Bi} P_{\alpha i}$, $i = 0, 1, \ldots, 2^{2n} - 1$. Further, constructing transformations

$$
P_i = I^{\otimes 3} \otimes \sum_{k=0, k \neq i}^{2^{2n}-1} |k\rangle\langle k| + P_i' \otimes |i\rangle\langle i| \tag{2.28}
$$

It is easy to prove that P_i is a unitary matrix. Thus, acting P_i on $|W\rangle$ yields:

$$\begin{aligned} P_i(|W\rangle) &= \frac{1}{2^{n+3/2}}\left\{\left[\left(I^{\otimes 2}\left(\sum_{l=0}^{3}|l\rangle\right)\right)\otimes I(|0\rangle+|1\rangle)\right]\otimes\left[\left(\sum_{j=0,j\neq i}^{2^{2n}-1}|j\rangle\langle j|\right)\left(\sum_{k=0}^{2^{2n}-1}|k\rangle\right)\right]+\right. \\ &\left.+\left[P_i'\left(\sum_{l=0}^{3}|l\rangle\otimes(|0\rangle+|1\rangle)\right)\right]\otimes\left[(|i\rangle\langle i|)\left(\sum_{k=0}^{2^{2n}-1}|k\rangle\right)\right]\right\} \\ &= \frac{1}{2^{n+3/2}}\left[\sum_{l=0}^{3}|l\rangle\otimes(|0\rangle+|1\rangle)\otimes\sum_{j=0,j\neq i}^{2^{2n}-1}|j\rangle+\left|C_{RGB\alpha}^{i}\right\rangle\otimes|i\rangle\right] \end{aligned} \tag{2.29}$$

$$\begin{aligned} P_mP_i(|W\rangle) &= P_m\left(P_i|W\rangle\right) \\ &= \frac{1}{2^{n+3/2}}\left[\sum_{l=0}^{3}|l\rangle\otimes(|0\rangle+|1\rangle)\otimes\sum_{j=0,j\neq k,i}^{2^{2n}-1}|j\rangle+\left|C_{RGB\alpha}^{i}\right\rangle\otimes|i\rangle+\left|C_{RGB\alpha}^{m}\right\rangle\otimes|m\rangle\right] \end{aligned} \tag{2.30}$$

Continuing to apply the transformation on the resultant state after the action of Equation 2.30, there is:

$$B(|W\rangle) = \left(\prod_{i=0}^{2^{2n}-1}P_i\right)|W\rangle = |I(\theta)\rangle \tag{2.31}$$

From Equation 2.31, it can be obtained that the unitary transformation $P = BA$ is able to transform the quantum computer from the initial state to the normalized quantum image state CQIPT.

□

Next, consider the feasibility of decomposing the transformation P into simple quantum gates such as NOT gates, Hadamard gates, and CNOT gates.

Corollary 2.2: The unitary transformation P in Lemma 2.1 can be constructed from Hadamard gates, CNOT gates, and controlled phase gate $C^{2n+2}\left(P\left(\frac{\theta_{Xi}}{2^{2n+1}}\right)\right)$, $i = 0,1,\cdots,2^{2n}-1$.

Proof: From the proof of Lemma 2.2, it follows that the transformation P consists of BA, the transformation A can be realized directly by $2n + 3$ Hadamard gates, and the transformation B can be realized by $\prod_{i=0}^{2^{2n}-1}P_i$, where

$$P_i = I^{\otimes 3}\otimes\sum_{j=0,j\neq i}^{2^{2n}-1}|j\rangle\langle j| + P_i'\otimes|i\rangle\langle i| \tag{2.32}$$

$$P_i' = P_{Ri}P_{Gi}P_{Bi}P_{\alpha i}$$

$$P_{Xi} = C^2\left(P\left(\theta_{Xi}\right)\right)$$

Thus, P_i in Equation 2.32 can be decomposed into $C^{2n+2}\left(P\left(\theta_{Xi}\right)\right)$ and NOT gates. Further, $C^{2n+2}\left(P\left(\theta_{Xi}\right)\right)$ can be decomposed into $2^{2n+2}-1$ simple gates, $P\left(\frac{\theta_{Xi}}{2^{2n+1}}\right)$ and $2^{2n+2}-2$ CNOT gates.

Thus, the total number of simple operations required to prepare a color image representation ~CQIPT~ is

$$\begin{aligned}&2n+3+4\times 2^{2n}\times(2^{2(n+1)}-1+2^{2(n+1)}-2)\\&=2\times(4\times 2^{2n})^2-3\times(4\times 2^{2n})+2n+3\end{aligned}$$

Here, the total number of angles is of the order of the square of the image size 4×2^{2n}, that is, $O\left(N^2\right)$.

□

Theorem 2.2 (CQIPT Preparation Theorem):

Given four vectors $\theta_X=(\theta_{X0},\theta_{X1},\ldots,\theta_{X(2^{2n}-1)}),X\in\{R,G,B,\alpha\}$, there exists a unitary transformation P that transforms a quantum computer from an initial state $|0\rangle^{\otimes 2n+3}$ to a multichannel quantum image representation of a CQIPT state $|I(\theta)\rangle=\frac{1}{2^{n+3/2}}\sum_{j=0}^{2^{2n}-1}\left|c_{RGB\alpha}^{j}\right\rangle\otimes|j\rangle$ and requires a number of elementary quantum gates of polynomial order of magnitude in the size of the problem.

Proof: This conclusion follows directly from Lemma 2.2 and Corollary 2.2.

□

2.4 QUANTUM COLOR IMAGE REPRESENTATION NCQI[5]

2.4.1 NCQI Representation

This section introduces the quantum color image representation model, novel color quantum image (NCQI); the specific form is given below. Assuming a classical color image, its R, G, B three-channel color value range of $[0,2^q-1]$, where q is the color depth of each color channel (e.g., R, G, B three color channel value range are [0, 255], then the color depth of $q=8$), then its NCQI quantum image representation model can be expressed as equation (2.33):

$$\begin{aligned}|I(x,y)\rangle&=\frac{1}{2^n}\sum_{k=0}^{2^{2n}-1}\left|C_k\right\rangle\otimes|k\rangle\\&=\frac{1}{2^n}\sum_{y=0}^{2^n-1}\sum_{x=0}^{2^n-1}|C(y,x)\rangle\otimes|yx\rangle\end{aligned}\tag{2.33}$$

Here $\left|C(x,y)\right\rangle$ denotes the color value of the corresponding pixel, the form of which is given by Equation 2.34:

$$|C(y,x)\rangle=\left|\underbrace{R_{q-1}\cdots R_0}_{R}\underbrace{G_{q-1}\cdots G_0}_{G}\underbrace{B_{q-1}\cdots B_0}_{B}\right\rangle\tag{2.34}$$

Equation 2.33 shows that the whole color image is stored as a normalized quantum superposition state, and this quantum state contains two parts, namely, color information $\left|C(y,x)\right\rangle$ and position information $|yx\rangle$, and the tensor product of these 2-qubit sequences constitutes the NCQI quantum superposition state. Equation 2.34) shows that the NCQI model uses a sequence of ground state qubits to store the R, G, and B color values of each pixel. A color image of size $2^n \times 2^n$ with each color channel taking a value of $[0, 2^q - 1]$ uses $2n + 3q$ qubits to store the information for the entire image.

Examples are given below to explain the proposed NCQI quantum image representation model, a color image of 4×4 size with three color channels taking values in the interval $[0, 2^q - 1]$, that is, $n = 2$, $q = 8$, whose NCQI quantum image expression is shown in the right-hand side equation of Figure 2.6, from which it can also be seen that the entire image is stored as a normalized quantum superposition state.

2.4.2 Preparation of NCQI

Under the quantum mechanism condition, the image information is stored as an initialized quantum superposition state. First, an initialized quantum register is prepared, which stores a number of qubits of $2n + 3q$ and stores an initialized quantum superposition state of $|I\rangle_0 = 0^{\otimes 2n+3q}$. The preparation process of the NCQI quantum color image representation model consists of two steps.

1. Construct a blank quantum image state of size $2^n \times 2^n$, denoted as the intermediate state $|I\rangle_1$, and unitary operator U_1 can evolve the initial state $|I\rangle_0$ into the intermediate state $|I\rangle_1$, U_1 as shown in Equation 2.35, which is constructed from the fundamental quantum gates I and H.

$$U_1 = I^{\otimes 3q} \otimes H^{2n} \tag{2.3}$$

Equation 2.36 gives the specific process by which the unitary operator U_1 evolves the initial state $|I\rangle_0$ into the intermediate state $|I\rangle_1$.

$$\begin{aligned} U_1\left(|I\rangle_0\right) &= I^{\otimes 3q} \otimes H^{\otimes 2n}\left(|0\rangle^{2n+3q}\right) \\ &= |0\rangle^{3q} \otimes \frac{1}{\sqrt{2^n}} \sum_{y=0}^{2^n-1} |y\rangle \otimes \frac{1}{\sqrt{2^n}} \sum_{x=0}^{2^n-1} |x\rangle \\ &= \frac{1}{2^n} \sum_{y=0}^{2^n-1} \sum_{x=0}^{2^n-1} |0\rangle^{3q} \otimes |y\rangle |x\rangle \\ &= |I\rangle_1 \end{aligned} \tag{2.36}$$

As the color-coded qubit of intermediate state $|I\rangle_1$ takes the value of $|0\rangle^{3q}$, and the position of each pixel is stored in an entangled superposition, intermediate state $|I\rangle_1$ is a blank quantum image of size $2^n \times 2^n$.

2. Evolution of the intermediate state $|I\rangle_1$ to the target state $|I(x, y)\rangle$ using the unitary transformation U_2.

The difference between the intermediate state $|I\rangle_1$ and the target state $|I(x, y)\rangle$ is reflected in the fact that each pixel of the target state $|I(x, y)\rangle$ has a specific color value, while the intermediate state $|I\rangle_1$ is a blank quantum image, so the key of this step is to assign a value to each pixel of the intermediate state image $|I\rangle_1$. As the image size is $2^n \times 2^n$, 2^{2n}sub-operations are required to set the color for each pixel. For pixel (x, y), the unitary operation to set the color value for it is Ω_{yx}. Its specific form is shown in Equation 2.37 and contains a total of $3q$ quantum black-box operations Ω_{yx}^i, as shown in Equation 2.38. The unitary operator Ω_{yx} uses the operator Ω_{yx}^i to process each qubit $|C_i\rangle$ in the color qubit sequence $|C(y, x)\rangle$ of the pixel (x, y), $i = 0, 1, \cdots, 3q - 1$. Specifically, when $C_i = 1$, Ω_{yx}^i is a $2n$-CNOT gate and the ith qubit $|C_i\rangle$ of $|C(y, x)\rangle$ is acted $2n$-CNOT gate; when $C_i = 0$, Ω_{yx}^i is a

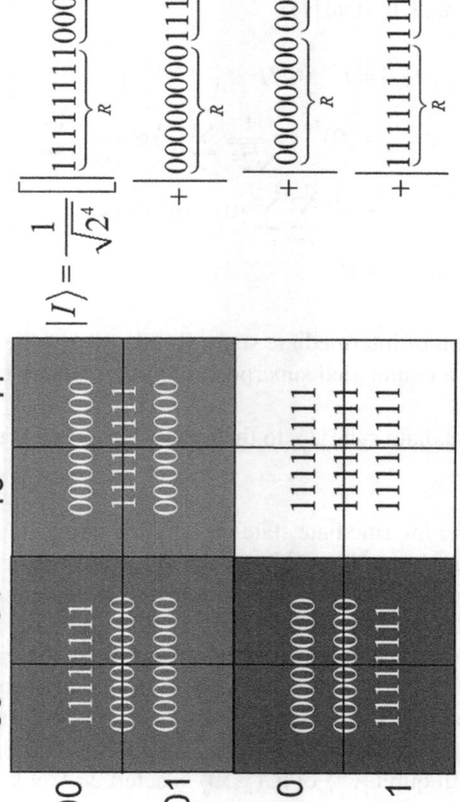

$$
\begin{aligned}
|I\rangle = \frac{1}{\sqrt{2^4}}\Big[& |\underbrace{11111111}_{R}\underbrace{00000000}_{G}\underbrace{00000000}_{B}\rangle \otimes (|0000\rangle + |0001\rangle + |0100\rangle + |0101\rangle) \\
+ & |\underbrace{00000000}_{R}\underbrace{11111111}_{G}\underbrace{00000000}_{B}\rangle \otimes (|0010\rangle + |0011\rangle + |0110\rangle + |0111\rangle) \\
+ & |\underbrace{00000000}_{R}\underbrace{00000000}_{G}\underbrace{11111111}_{B}\rangle \otimes (|1000\rangle + |1001\rangle + |1100\rangle + |1101\rangle) \\
+ & |\underbrace{11111111}_{R}\underbrace{11111111}_{G}\underbrace{11111111}_{B}\rangle \otimes (|1010\rangle + |1011\rangle + |1110\rangle + |1111\rangle)\Big]
\end{aligned}
$$

FIGURE 2.6 A 4 × 4 color image and its NCQI quantum state.

unit gate and acts as a unit gate for qubit $|C_i\rangle$. Ω_{yx} acting on the color-coded qubit $|0\rangle^{3q}$ is shown in Equation 2.39:

$$\Omega_{yx} = \bigotimes_{i=0}^{3q-1} \Omega_{yx}^{i} \tag{2.37}$$

$$\Omega_{yx}^{i} : |0\rangle \rightarrow |0 + C_i\rangle \tag{2.38}$$

$$\Omega_{yx} : \bigotimes_{0}^{3q-1} |0\rangle \rightarrow \bigotimes_{0}^{3q-1} |0 + C_i\rangle = \bigotimes_{0}^{3q-1} |C_i\rangle = |c(y,x)\rangle \tag{2.39}$$

Here, $|C_i\rangle = |B_i\rangle$, $i = 0,1,\ldots,q-1$ denotes the blue channel, $|C_i\rangle = |G_i\rangle$, $i = q, q+1,\ldots,2q-1$ denotes the green channel, $|C_i\rangle = |R_i\rangle$, $i = 2q, 2q+1,\ldots,3q-1$ denotes the red channel.

A controlled transformation U_{yx} may be constructed based on the unitary transformation Ω_{yx} for setting the color value of a fixed pixel in the image while not changing the color values of other pixels:

$$U_{yx} = \left(I^{\otimes 3q} \otimes \sum_{j=0}^{2^n-1} \sum_{i=0, ji \neq yx}^{2^n-1} |ji\rangle\langle ji| \right) + \Omega_{yx} \otimes |yx\rangle\langle yx| \tag{2.40}$$

When the controlled transformation U_{yx} is applied to the intermediate state $|I\rangle_1$, the evolution process can be expressed as Equation 2.41. From Equation 2.41 it can be seen that the controlled transform U_{yx} assigns a color value only to the pixel at position (x, y).

$$\begin{aligned} & U_{yx}(|I\rangle_1) \\ &= \left[\left(I^{\otimes 3q} \otimes \sum_{j=0}^{2^n-1} \sum_{i=0, ji \neq yx}^{2^n-1} |ji\rangle\langle ji| \right) + \Omega_{yx} \otimes |yx\rangle\langle yx| \right] \left(\frac{1}{2^n} \sum_{y=0}^{2^n-1} \sum_{x=0}^{2^n-1} |0\rangle^{\otimes 3q} |yx\rangle \right) \\ &= \frac{1}{2^n} \sum_{y=0}^{2^n-1} \sum_{x=0}^{2^n-1} |0\rangle^{\otimes 3q} |ji\rangle + \Omega_{yx} |0\rangle^{\otimes 3q} |yx\rangle \\ &= \frac{1}{2^n} \sum_{y=0}^{2^n-1} \sum_{x=0}^{2^n-1} |0\rangle^{\otimes 3q} |ji\rangle + |c(y,x)\rangle |yx\rangle \end{aligned} \tag{2.41}$$

To set the color values for all the pixels in the image, unitary transformation U_2 is designed.

$$U_2 = \prod_{y=0}^{2^n-1} \prod_{x=0}^{2^n-1} U_{yx} \tag{2.42}$$

The role of the unitary transformation U_2 can be described by Equation 2.43:

$$\begin{aligned} & U_2(|I\rangle_1) \\ &= \prod_{y=0}^{2^n-1} \prod_{x=0}^{2^n-1} U_{yx}(|I\rangle_1) \\ &= \prod_{s=0}^{2^n-1} \prod_{t=0, st \doteq yx}^{2^n-1} U_{st} \left(\frac{1}{2^n} \sum_{j=0}^{2^n-1} \sum_{i=0\, ji \neq yx}^{2^n-1} |0\rangle^{\otimes 3q} |ji\rangle + |c(y,x)\rangle |yx\rangle \right) \\ &= \frac{1}{2^n} \sum_{y=0}^{2^n-1} \sum_{x=0}^{2^n-1} |c(y,x)\rangle |yx\rangle \\ &= |I(x,y)\rangle \end{aligned} \tag{2.43}$$

From Equation 2.43, it can be seen that the unitary transformation U_2 achieves the purpose of evolving the intermediate state $|I\rangle_1$ into the target state $|I(x,y)\rangle$.

After the above two steps, the preparation process of quantum image NCQI has been completed, that is, the corresponding NCQI quantum image state is prepared. In order to prove the validity of the quantum image representation model NCQI, the preparation complexity of the NCQI model needs to be calculated, and the relevant theorem and proof are given below.

Theorem 2.3 (NCQI Preparation Theorem):

The complexity of the entire preparation and storage process when storing NCQI quantum image states for a color image of size $2^n \times 2^n$ with color channels *R*, *G*, and *B* all taking values in the interval $[0, 2^q - 1]$ does not exceed $O(3q + 2n + 6qn \cdot 2^{2n})$.

Proof: The preparation of the quantum image representation model NCQI consists of two steps, and its preparation complexity is analyzed as follows.

In step 1, the unitary operator U_1 is used, and it can be seen from Equation 2.35 that the complexity of U_1 is $O(3q + 2n)$ due to the inclusion of $3q + 2n$ elementary quantum gates.

In step 2, the unitary operator U_2 is used with the purpose of setting color values for all pixels in the quantum image. Equation 2.42 indicates that U_2 contains 2^{2n} controlled transforms U_{yx}. Each controlled transform U_{yx} utilizes a quantum operation Ω_{yx} to set the color value for the corresponding pixel. In quantum operation Ω_{yx}, there are a total of $3q$ quantum black-box operations Ω^i_{yx} that apply a $2n$-CNOT gate or a unitary gate to the *i*th qubit $|C_i\rangle$ of the color qubit sequence $\bigotimes_{i=0}^{3q-1} |C_i\rangle$. The $2n$-CNOT gate can be decomposed into no more than $O(2n)$ simple quantum gates, so the complexity of the controlled transformation U_{yx} does not exceed $O(3q \cdot 2n)$. The unitary operator U_2 has 2^{2n} controlled transformations U_{yx}, so the complexity of step 2 does not exceed $O(3q \cdot 2n \cdot 2^{2n})$.

Based on the above analysis, it can be concluded that the complexity of NCQI quantum image state preparation for a color image of size $2^n \times 2^n$, with each color channel *R*, *G*, and *B* taking values in the range of $[0, 2^q - 1]$ intervals, is no more than $O(3q + 2n + 6qn \cdot 2^{2n})$, which is obviously polynomial in the order of the image size of 2^{2n}, and therefore the preparation process is effective.

□

2.5 QUANTUM COLOR IMAGE REPRESENTATION MCLPQI[6]

2.5.1 MCLPQI Representation

Most of the existing quantum color image representation models can represent color images in Cartesian coordinate system, but there are some image processing algorithms, such as image alignment, which are more easily designed in polar coordinate system. In this section, we introduce a quantum color image representation model MCLPQI in polar coordinate system based on revolving gates, which contains the *R*, *G*, and *B* channel color information of the image, as well as the transparency information α with the following features:

1. Color information for each channel using angle coding;
2. Preparation complexity is of polynomial order, and the model is more flexible in its application.

The specific form of the MCLPQI model is given first.

For a color image of size $2^m \times 2^n$, that is, the resolutions of polar diameter ρ and polar angle θ are 2^m and 2^n, respectively, the quantum color image representation model MCLPQI in polar coordinates is shown in Equation 2.44.

$$|I(\rho,\theta)\rangle = \frac{1}{\sqrt{2^{2+m+n}}} \sum_{\rho=0}^{2^m-1} \sum_{\theta=0}^{2^n-1} |C^{\rho\theta}_{RGB\alpha}\rangle \otimes |\rho\theta\rangle \tag{2.44}$$

where $|C_{RGB\alpha}^{\rho\theta}\rangle$ denotes color information, α denotes the transparency of the pixel, $|\rho\theta\rangle$ denotes position information, $\rho = 0,1,\ldots,2^m - 1, \theta = 0,1,\ldots,2^n - 1$.

$$\begin{aligned}|R_{RGB\alpha}^{\rho\theta}\rangle = \cos\theta_R^{\rho\theta}|000\rangle + \cos\theta_G^{\rho\theta}|001\rangle + \cos\theta_B^{\rho\theta}|010\rangle + \cos\theta_\alpha^{\rho\theta}|011\rangle \\ + \sin\theta_R^{\rho\theta}|100\rangle + \sin\theta_G^{\rho\theta}|101\rangle + \sin\theta_B^{\rho\theta}|110\rangle + \sin\theta_\alpha^{\rho\theta}|111\rangle\end{aligned} \quad (2.45)$$

where $\theta_X^{\rho\theta} \in [0, \frac{\pi}{2}]$, $X \in \{R, G, B, \alpha\}$.

Examples are given below to explain the proposed MCLPQI quantum image representation model for an $\rho \times \theta = 2 \times 2$-size color image whose MCLPQI quantum image expression is shown in the right-hand side equation of Figure 2.7.

2.5.2 Preparation of MCLPQI

In quantum computing, the register initial state is generally assumed to be a number of $|0\rangle$ tensor product states. In the preparation of the MCLPQI model, the initial state is assumed to be $|I\rangle_0 = |0\rangle^{\otimes 3+m+n}$, and the preparation process, that is, the preparation process, explores which kind of unitary transformations can be utilized to evolve the initial state into the final state MCLPQI. Similar to the preparation of the NCQI model, the preparation process of the MCLPQI model is divided into two steps.

1. Using the unitary transformation U_1, (2.46) the initial state $|I\rangle_0 = |0\rangle^{\otimes 3+m+n}$ is evolved into the intermediate state $|I\rangle_1$, where $|I\rangle_1$ is a blank quantum image of size $2^m \times 2^n$.

$$U_1 = I \otimes H^{\otimes 2+m+n} \quad (2.46)$$

The specific process by which U_1 evolves the initial state $|I\rangle_0$ into the intermediate state $|I\rangle_1$ is shown in Equation 2.47:

$$\begin{aligned}U_1(|I\rangle_0) &= |0\rangle \otimes \left(\frac{1}{\sqrt{2^2}}\sum_{i=0}^{3}|i\rangle\right) \otimes \left(\frac{1}{\sqrt{2^m}}\sum_{\rho=0}^{2^m-1}|\rho\rangle\right) \otimes \left(\frac{1}{\sqrt{2^n}}\sum_{\theta=0}^{2^n-1}|\theta\rangle\right) \\ &= \frac{1}{\sqrt{2^{2+m+n}}}\left(|0\rangle\sum_{i=0}^{3}|i\rangle\right) \otimes \sum_{\rho=0}^{2^m-1}\sum_{\theta=0}^{2^n-1}|\rho\theta\rangle \\ &\triangleq |I\rangle_1\end{aligned} \quad (2.47)$$

From Equation 2.47, it can be seen that the intermediate state $|I\rangle_1$ is a blank quantum image of size $2^m \times 2^n$.

2. Set the RGB color value for each pixel of the intermediate state $|I\rangle_1$.
As the size of intermediate state $|I\rangle_1$ is $2^m \times 2^n$, 2^{m+n} suboperations are needed to set the RGB color value of each pixel in the image, and suboperation $U_{\rho\theta}$ is constructed first.

$$U_{\rho\theta} = I^{\otimes 3} \otimes \sum_{j=0}^{2^m-1}\sum_{i=0, ji\neq\rho\theta}^{2^n-1}|ji\rangle\langle ji| + \Omega_{\rho\theta} \otimes |\rho\theta\rangle\langle\rho\theta| \quad (2.48)$$

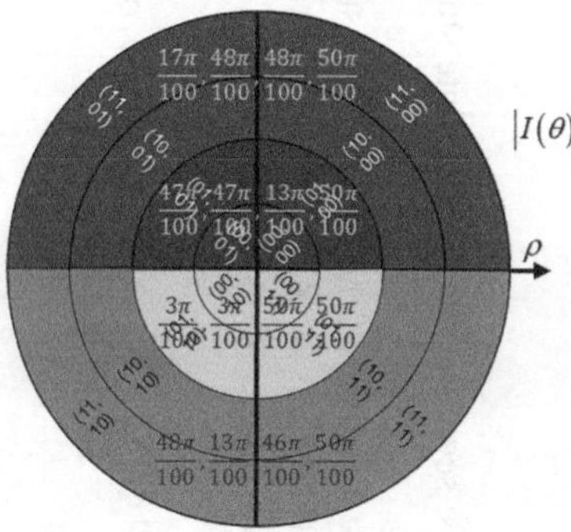

$$
\begin{aligned}
|I(\theta)\rangle = \frac{1}{8}\Bigg[& \left(\cos\frac{47\pi}{100}|000\rangle + \cos\frac{47\pi}{100}|001\rangle + \cos\frac{13\pi}{100}|010\rangle + \sin\frac{50\pi}{100}|100\rangle + \sin\frac{47\pi}{100}|101\rangle + \sin\frac{47\pi}{100}|110\rangle + \cos\frac{13\pi}{100}|011\rangle + \sin\frac{50\pi}{100}|111\rangle\right) \otimes (|0000\rangle + |0001\rangle + |0100\rangle + |0101\rangle) \\
& + \left(\cos\frac{3\pi}{100}|000\rangle + \cos\frac{3\pi}{100}|001\rangle + \cos\frac{50\pi}{100}|010\rangle + \sin\frac{50\pi}{100}|100\rangle + \sin\frac{3\pi}{100}|101\rangle + \sin\frac{3\pi}{100}|110\rangle + \cos\frac{50\pi}{100}|011\rangle + \sin\frac{50\pi}{100}|111\rangle\right) \otimes (|0010\rangle + |0011\rangle + |0110\rangle + |0111\rangle) \\
& + \left(\cos\frac{17\pi}{100}|000\rangle + \cos\frac{17\pi}{100}|001\rangle + \cos\frac{48\pi}{100}|010\rangle + \sin\frac{48\pi}{100}|100\rangle + \sin\frac{17\pi}{100}|101\rangle + \sin\frac{48\pi}{100}|110\rangle + \cos\frac{48\pi}{100}|011\rangle + \sin\frac{50\pi}{100}|111\rangle\right) \otimes (|1000\rangle + |1001\rangle + |1100\rangle + |1101\rangle) \\
& + \left(\cos\frac{48\pi}{100}|000\rangle + \cos\frac{13\pi}{100}|001\rangle + \cos\frac{46\pi}{100}|010\rangle + \sin\frac{50\pi}{100}|100\rangle + \sin\frac{48\pi}{100}|101\rangle + \sin\frac{13\pi}{100}|110\rangle + \cos\frac{46\pi}{100}|011\rangle + \sin\frac{50\pi}{100}|111\rangle\right) \otimes (|1010\rangle + |1011\rangle + |1110\rangle + |1111\rangle)\Bigg]
\end{aligned}
$$

FIGURE 2.7 A 4 × 8 color image and its NCQI quantum state.

where

$$\Omega_{\rho\theta} = R_B^{\rho\theta} R_G^{\rho\theta} R_R^{\rho\theta} R_\alpha^{\rho\theta}$$

$$R_R^{\rho\theta} = I \otimes \sum_{i=1}^{3} |i\rangle\langle i| + R_y(2\theta_R^{\rho\theta}) \otimes |0\rangle\langle 0|$$

$$R_G^{\rho\theta} = I \otimes \sum_{i=0, i\neq 1}^{3} |i\rangle\langle i| + R_y(2\theta_G^{\rho\theta}) \otimes |1\rangle\langle 1|$$

$$R_B^{\rho\theta} = I \otimes \sum_{i=0, i\neq 2}^{3} |i\rangle\langle i| + R_y(2\theta_B^{\rho\theta}) \otimes |2\rangle\langle 2|$$

$$R_\alpha^{\rho\theta} = I \otimes \sum_{i=0}^{2} |i\rangle\langle i| + R_y(2\theta_\alpha^{\rho\theta}) \otimes |3\rangle\langle 3|$$

The process of suboperation $U_{\rho\theta}$ acting on intermediate state $|I\rangle_1$ can be expressed as Equation 2.49.

$$\begin{aligned}
U_{\rho\theta}(|I\rangle_1) &= \left(I^{\otimes 3} \otimes \sum_{j=0}^{2^m-1} \sum_{i=0, ji\neq\rho\theta}^{2^n-1} |ji\rangle\langle ji| + \Omega_{\rho\theta} \otimes |\rho\theta\rangle\langle\rho\theta| \right) \left(\frac{1}{\sqrt{2^{2+m+n}}} \left(|0\rangle \sum_{i=0}^{3} |i\rangle \right) \otimes \left(\sum_{\rho=0}^{2^m-1} \sum_{\theta=0}^{2^n-1} |\rho\theta\rangle \right) \right) \\
&= \frac{1}{\sqrt{2^{2+m+n}}} \left\{ \left(I^{\otimes 3} \left(|0\rangle \sum_{i=0}^{3} |i\rangle \right) \right) \otimes \left(\sum_{j=0}^{2^m-1} \sum_{i=0, ji\neq\rho\theta}^{2^n-1} |ji\rangle\langle ji| \left(\sum_{\rho=0}^{2^m-1} \sum_{\theta=0}^{2^n-1} |\rho\theta\rangle \right) \right) \right. \\
&\quad \left. + \left(\Omega_{\rho\theta} \left(|0\rangle \sum_{i=0}^{3} |i\rangle \right) \right) \otimes \left(|\rho\theta\rangle\langle\rho\theta| \left(\sum_{\rho=0}^{2^m-1} \sum_{\theta=0}^{2^n-1} |\rho\theta\rangle \right) \right) \right\} \\
&= \frac{1}{\sqrt{2^{2+m+n}}} \left(|0\rangle \sum_{i=0}^{3} |i\rangle \right) \otimes \sum_{j=0}^{2^m-1} \sum_{i=0, ji\neq\rho\theta}^{2^n-1} |ji\rangle + R_{RGB\alpha}^{\rho\theta} \otimes |\rho\theta\rangle
\end{aligned} \tag{2.49}$$

It is clear from Equation 2.49 that suboperation $U_{\rho\theta}$ sets the pixel at position (ρ, θ) to its corresponding RGB color value without changing the pixel values at the other positions. Notice that the derivation of Equation 2.49, where the second equality sign to the third equality sign holds, is as follows:

$$\begin{aligned}
&\Omega_{\rho\theta} \left(|0\rangle \sum_{i=0}^{3} |i\rangle \right) \\
&= R_B^{\rho\theta} R_G^{\rho\theta} R_R^{\rho\theta} R_\alpha^{\rho\theta} \left(|0\rangle \sum_{i=0}^{3} |i\rangle \right) \\
&= R_B^{\rho\theta} R_G^{\rho\theta} R_R^{\rho\theta} \left(I \otimes \sum_{i=0}^{2} |i\rangle\langle i| + R_y(2\theta_\alpha^{\rho\theta}) \otimes |3\rangle\langle 3| \right) \left(|0\rangle \sum_{i=0}^{3} |i\rangle \right) \\
&= R_B^{\rho\theta} R_G^{\rho\theta} R_R^{\rho\theta} \left(|0\rangle \sum_{i=0}^{2} |i\rangle + \cos\theta_\alpha^{\rho\theta} |011\rangle + \sin\theta_\alpha^{\rho\theta} |111\rangle \right) \\
&= R_B^{\rho\theta} R_G^{\rho\theta} \left(I \otimes \sum_{i=1}^{3} |i\rangle\langle i| + R_y(2\theta_R^{\rho\theta}) \otimes |0\rangle\langle 0| \right) \left(|0\rangle \sum_{i=0}^{2} |i\rangle + \cos\theta_\alpha^{\rho\theta} |011\rangle + \sin\theta_\alpha^{\rho\theta} |111\rangle \right) \\
&= R_B^{\rho\theta} R_G^{\rho\theta} \left(|0\rangle \sum_{i=1}^{2} |i\rangle + \cos\theta_\alpha^{\rho\theta} |011\rangle + \sin\theta_\alpha^{\rho\theta} |111\rangle + \cos\theta_R^{\rho\theta} |000\rangle + \sin\theta_R^{\rho\theta} |100\rangle \right) \\
&= R_B^{\rho\theta} \left(|0\rangle|2\rangle + \cos\theta_R^{\rho\theta} |000\rangle + \sin\theta_R^{\rho\theta} |100\rangle + \cos\theta_\alpha^{\rho\theta} |011\rangle + \sin\theta_\alpha^{\rho\theta} |111\rangle + \cos\theta_G^{\rho\theta} |001\rangle + \sin\theta_G^{\rho\theta} |101\rangle \right) \\
&= R_{RGB\alpha}^{\rho\theta}
\end{aligned}$$

To set color values for all pixels of the intermediate state $|I\rangle_1$, a specific form of the unitary transformation U_2 is given:

$$U_2 = \prod_{\rho=0}^{2^m-1}\prod_{\theta=0}^{2^n-1} U_{\rho\theta} \tag{2.50}$$

By the action of the unitary transformation U_2, all pixels are set to their corresponding color values, the evolution of which is shown in Equation 2.51, so that the final state $|I(\rho,\theta)\rangle$ can be prepared:

$$\begin{aligned} U_2\left(|I\rangle_1\right) &= \prod_{\rho=0}^{2^m-1}\prod_{\theta=0}^{2^n-1} U_{\rho\theta}\left(|I\rangle_1\right) \\ &= \frac{1}{\sqrt{2^{2+m+n}}}\sum_{\rho=0}^{2^m-1}\sum_{\theta=0}^{2^n-1}\left|C_{RGB\alpha}^{\rho\theta}\right\rangle\otimes|\rho\theta\rangle \\ &= |I(\rho,\theta)\rangle \end{aligned} \tag{2.51}$$

By the above two steps, the whole preparation process of quantum image state MCLPQI has been completed, and the theorem and corollary are given below to calculate the computational complexity of the MCLPQI preparation process.

Theorem 2.4: Given four angle vectors $\theta_X = \left(\theta_{X_0}, \theta_{X_1}, \ldots, \theta_{X_{2^{n+m}-1}}\right)$, $X \in \{R, G, B, \alpha\}$ and satisfying Equation 2.45, with $n+m$ representing the number of quantum bits required to encode the image position, there exists a unitary transformation U consisting of a Hadamard gate and a series of controlled rotating gates that evolves a quantum initial state $|0\rangle^{\otimes 3+m+n}$ into a target state $|I(\rho,\theta)\rangle$.

Proof: To evolve the quantum initial state into the final quantum state $|I(\rho,\theta)\rangle$, the preparation process of the MCLPQI model consists of two steps. Step 1 uses the unitary transformation U_1 and step 2 uses the unitary transformation U_2, that is, $U = U_2U_1$, and thus the theorem is proved.

□

Corollary 2.3: Given four angle vectors $\theta_X = \left(\theta_{X_0}, \theta_{X_1}, \ldots, \theta_{X_{2^{n+m}-1}}\right)$, $X \in \{R, G, B, \alpha\}$, the unitary transformations in Theorem 2.4 can be constructed from Hadamard gates, CNOT gates, $R_y\left(\frac{2\theta_X^{\rho\theta}}{2^{m+n}-1}\right)$ and $R_y\left(-\frac{2\theta_X^{\rho\theta}}{2^{m+n}-1}\right)$.

Proof: From the proof of Theorem 2.4, it follows that the unitary transform U consists of U_2 and U_1. The unitary transform U_1 can be constructed by a unitary gate and $2+m+n$ Hadamard gate, and the unitary transform U_2 is constructed by 2^{m+n} suboperations $U_{\rho\theta}$, where $U_{\rho\theta}$ is given by Equation 2.48. The suboperations $U_{\rho\theta}$ can be constructed with $C^{m+n+2}\left(R_y\left(2\theta_X^{\rho\theta}\right)\right)$ and NOT gates. The unitary transformation $C^{m+n+2}\left(R_y\left(2\theta_X^{\rho\theta}\right)\right)$ can be decomposed into $2^{2+m+n}-1$ simple operations $R_y\left(\frac{2\theta_X^{\rho\theta}}{2^{m+n}-1}\right)$, $R_y\left(-\frac{2\theta_X^{\rho\theta}}{2^{m+n}-1}\right)$ and $2^{2+m+n}-2$ CNOT gates. Therefore, the number of simple elementary gates used in the preparation of MCLPQI quantum states is

$$\begin{aligned} &3+m+n+4\times 2^{m+n}\left(2^{m+n+2}-1+2^{m+n+2}-2+1\right) \\ &= 32\times 2^{2(m+n)} - 8\times 2^{m+n} + (m+n+3) \end{aligned} \tag{2.52}$$

This value is squarely related to the scale 2^{m+n} of the polar angle and polar diameter, which indicates to some extent the effectiveness of the MCLPQI quantum state preparation.

□

Finally, a polynomial preparation theorem for MCLPQI is given.

Theorem 2.5 (MCLPQI Preparation Theorem):

A color image of size $2^m \times 2^n$, with color channels R, G, B and transparency channel α defined by angle vectors $\theta_X = \left(\theta_{X_0}, \theta_{X_1}, \cdots, \theta_{X_{2^{n+m}-1}}\right)$, $X \in \{R, G, B, \alpha\}$, in polar coordinates, has a polynomial level of complexity for the entire preparation storage when stored as an MCLPQI quantum state.

Proof: From Theorem 2.4 and Corollary 2.3, the preparation complexity of the MCLPQI quantum state is $32 \cdot 2^{2(m+n)} - 8 \cdot 2^{m+n} + (m+n+3)$, which is a polynomial level of the image size $2^m \times 2^n$, so the theorem holds.

□

2.6 QUANTUM VIDEO REPRESENTATION

Quantum video representation is characterized by flexibility and diversity, and the design of the representation needs to be beneficial to the research of quantum video processing and security and confidentiality algorithms. The generalized quantum video representation contributes to the diversity of video processing algorithm research. The general framework structure of quantum video representation is given below, and then the quantum video model Quantum Video based on NEQR (QVNEQR) quantum image representation is introduced.

2.6.1 Quantum Video Representation Framework

In 2011[7]. quantum video framework is proposed based on strip circuit and FRQI quantum gray image representation. In 2015, Yan et al.[8] proposed an extended multi-channel quantum video, namely color video method. In 2016, Wang[9] proposed QVNEQR based on NEQR quantum image representation and Strip circuit.

The general form of a single quantum video representation that currently exists is:

$$|V\rangle = \sum_{j=0}^{2^m-1} \alpha_j \left|I_j\right\rangle |j\rangle \tag{2.53}$$

$$\left|I_j\right\rangle = \sum_{i=0}^{N-1} \beta_{ji} \left|c_{ji}\right\rangle \left|p_{ji}\right\rangle$$

where $\left|I_j\right\rangle, j = 0, 1, \ldots, m-1$ is the m-frame quantum image shown in Equation 2.1. Specifically, $|j\rangle$ represents the location where each frame of the image is stored in the video $|V\rangle$, $\left|c_{ji}\right\rangle$ encodes the color information of the quantum image $\left|I_j\right\rangle$, $\left|p_{ji}\right\rangle$ encodes the pixel position information of the quantum image $\left|I_j\right\rangle$, and N is the size metric of the quantum image $\left|I_j\right\rangle$. $\alpha_j, \beta_{ji} \in C$ is the probability amplitude of the corresponding ground state and satisfies the normalization condition, that is, $\sum_{j=0}^{m-1}\sum_{i=0}^{N-1}\left|\alpha_j \beta_{ji}\right|^2 = 1$. Different quantum image representations will encode the ground state $\left|c_{ji}\right\rangle\left|p_{ji}\right\rangle$ and the probability amplitude in different ways.

2.6.2 Quantum Video Representation QVNEQR (QVNCQI)[9]

Drawing on the FRQI-based quantum video representation proposed by A. M. Iliyasu, it is possible to construct a quantum video existing as a sequence of images for almost any quantum image representation. Taking NEQR (NCQI) as an example, the specific representation scheme is as follows:

By adding m quantum lines, 2^m frames of quantum color images can be prepared as a quantum video superposition state. Each frame is a binary-coded quantum color image of size $2^n \times 2^n$, which is represented as follows:

$$|V\rangle = \frac{1}{2^m} \sum_{j=0}^{2^m-1} \left|F_j\right\rangle \otimes |j\rangle$$

$$\left|F_j\right\rangle = \frac{1}{2^n} \sum_{i=0}^{2^{2n}-1} \left|c_{j,i}\right\rangle \otimes |i\rangle = \frac{1}{2^n} \sum_{i=0}^{2^{2n}-1} \left|c_{q-1}^{j,i} c_{q-2}^{j,i} \ldots c_0^{j,i}\right\rangle |i\rangle$$

$$c_s^{j,i} \in \{0,1\}, s = 0,1,\ldots,q-1, \quad q = 8(grayscale) \quad \text{or} \quad q = 24(colorscale)$$

In the expression, each video frame contains either grayscale ($q = 8$) or *R, G, B* color information ($q = 24$).

For this quantum video representation, the design idea is to superimpose all frame images by adding m quantum lines, which can be easily prepared according to the quantum image pixel position preparation method introduced in Sections 2.1–2.5.

2.7 SUMMARY OF THIS CHAPTER

This chapter first introduces the general method of quantum color image representation, and then specifically introduces several quantum color image representations, including the QIRHSI model based on the HSI color space, the CQIPT model based on the RGB color space, the NCQI model, and the MCLPQI model, where the MCLPQI model is the color image representation model in polar coordinates. After introducing each representation, the preparation methods for converting the quantum initial state into the corresponding quantum color image representation and the preparation complexity of each representation are given. Finally, the general framework structure of quantum video representation is given, and the representation and preparation strategy of quantum color video is introduced using QVNEQR as an example.

REFERENCES

[1] Chen G L, Song X H, Venegas-Andraca S E, et al. QIRHSI: Novel Quantum Image Representation Based on HSI Color Space Model. *Quantum Information Processing*, 2022, 21(1): 5.

[2] Zhang Y, Lu K, Gao Y H, et al. NEQR: A Novel Enhanced Quantum Representation of Digital Images. *Quantum Information Processing*, 2013, 12(8): 2833–2860.

[3] Chen G, Song X. Quantum Color Image Scaling on QIRHSI Model. *Data Science: 7th International Conference of Pioneering Computer Scientists, Engineers and Educators, ICPCSEE 2021, Taiyuan, China, September 17–20, 2021, Proceedings, Part I 7*. Springer Singapore, 2021: 453–467.

[4] Song X H, Wang S, Niu X M. Multi-Channel Quantum Image Representation Based on Phase Transform and Elementary Transformations. *Journal of Information Hiding & Multimedia Signal Processing*, 2014, 5(4): 574–585.

[5] Sang J Z, Wang S, Song X H, et al. A Novel Representation for Multi-Channel Log-Polar Quantum Images. *Journal of Information Hiding and Multimedia Signal Processing*, 2015, 6(2): 340–350.

[6] Sang J Z, Wang S, Li Q. A Novel Quantum Representation of Color Digital Images. *Quantum Information Processing*, 2017, 16(2): 42.

[7] Iliyasu A M, Le P Q, Dong F, et al. A Framework for Representing and Producing Movies on Quantum Computers. *International Journal of Quantum Information*, 2011, 9(06): 1459–1497.

[8] Yan F, Iliyasu A M, Venegas-Andraca S E, et al. Video Encryption and Decryption on Quantum Computers. *International Journal of Theoretical Physics*, 2015, 54: 2893–2904.

[9] Wang S. Frames Motion Detection of Quantum Video. *Advances in Intelligent Information Hiding and Multimedia Signal Processing: Proceeding of the Twelfth International Conference on Intelligent Information Hiding and Multimedia Signal Processing, November, 21–23, 2016, Kaohsiung, Taiwan, Volume 2*. Springer International Publishing, 2017: 145–151.

[10] Yang Y G, Jia X, Sun S J, et al. Quantum Cryptographic Algorithm for Color Images Using Quantum Fourier Transform and Double Random-Phase Encoding. *Information Sciences*, 2014, 277: 445–457.

[11] Li P C, Liu X D. Color Image Representation Model and Its Application Based on an Improved FRQI. *International Journal of Quantum Information*, 2018, 16(15): 1850005.

[12] Zhang Y, Lu K, Gao Y H, et al. NEQR: A Novel Enhanced Quantum Representation of Digital Images. *Quantum Information Processing*, 2013, 12(8): 2833–2860.

[13] Zhang Y, Lu K, Gao Y H, et al. A Novel Quantum Representation for Log-Polar Images. *Quantum Information Processing*, 2013, 12(9): 3103–3126.

[14] Jiang N, Wang J, Mu Y. Quantum Image Scaling Up Based on Nearest-Neighbor Interpolation with Integer Scaling Ratio. *Quantum Information Processing*, 2015, 14(11): 4001–4026.

[15] Li H S, Chen X, Xia H, et al. A Quantum Image Representation Based on Bitplanes. *IEEE Access*, 2018, 6: 62396–62404.

[16] Sun B, Iliyasu A M, Yan F, et al. An RGB Multi-Channel Representation for Images on Quantum Computers. *Journal of Advanced Computational Intelligence & Intelligent Informatics*, 2013, 17(3): 404–417.

[17] Wang L, Ran Q W, Ma J, et al. QRCI: A New Quantum Representation Model of Color Digital Images. *Optics Communications*, 2019, 438: 147–158.

[18] Su J, Guo X C, Liu C Q. et al. An Improved Novel Quantum Image Representation and Its Experimental Test on IBM Quantum Experience. *Scientific Reports*, 2021, 11(1): 1–13.

[19] Liu K, Zhang Y, Lu K, et al. An Optimized Quantum Representation for Color Digital Images. *International Journal of Theoretical Physics*, 2018, 57: 2938–2948.

[20] Li H S, Zhu Q X, Zhou R G, et al. Multi-Dimensional Color Image Storage and Retrieval for a Normal Arbitrary Quantum Superposition State. *Quantum Information Processing*, 2014, 13(4): 991–1011.

[21] Şahin E, Yilmaz I. QRMW: Quantum Representation of Multi Wavelength Images. *Turkish Journal of Electrical Engineering & Computer Sciences*, 2018, 26(2): 768–779.

[22] Khan R A. An Improved Flexible Representation of Quantum Images. *Quantum Information Processing*, 2019, 18(7): 201.

[23] Xu G L, Xu X G, Wang X, et al. Order-Encoded Quantum Image Model and Parallel Histogram Specification. *Quantum Information Processing*, 2019, 18(11): 346.

[24] Wang L, Ran Q W, Ma J. Double Quantum Color Images Encryption Scheme Based on DQRCI. *Multimedia Tools and Applications*, 2020, 79(6–7): 1–27.

[25] Wang B, Hao M Q, Li P C, et al. Quantum Representation of Indexed Images and Its Applications. *International Journal of Theoretical Physics*, 2020, 59(3): 374–402.

[26] Grigoryan A M, Agaian S S. New Look on Quantum Representation of Images: Fourier Transform Representation. *Quantum Information Processing*, 2020, 19(5): 148.

[27] Yan F, Li N Q, Hirota K. QHSL: A Quantum Hue, Saturation, and Lightness Color Model. *Information Sciences*, 2021, 577(4): 196–213.

3 Basic Operations of Quantum Images

Zhixin Zhang, Xianhua Song, Ahmed A. Abd El-Latif, Bassem Abd-El-Atty, and Basma Abd El-Rahiem

3.1 ARITHMETIC OPERATION OF QIRHSI

3.1.1 Quantum Comparator

Represent a and b, respectively, as quantum states with two groups of n-bit binary numbers: $|a_{n-1}\cdots a_1 a_0\rangle$ and $|b_{n-1}\cdots b_1 b_0\rangle$. Figure 3.1 shows an n-bit Quantum Comparator (nQC) [1–4]. The initial states of u and v are both set to $|0\rangle$, which are used to record the comparison result of a and b. If uv is equal to 10, then $a > b$; if uv is equal to 01, then $a < b$; if uv is equal to 00, then $a = b$.

3.1.2 Intensity-based Inversion Operation

The intensity of the quantum color image quantum color image representation based on the HSI color space (QIRHSI) is stored in the form of a binary sequence [5]. Therefore, the inversion operation is to reverse all the qubits representing the intensity information of the image. The corresponding unitary matrix U_R for the intensity-based inversion operation is shown in Equation 3.1 as follows:

$$U_R = I^{\otimes 2} \otimes X^{\otimes q} \otimes I^{\otimes 2n} \tag{3.1}$$

For the quantum superposition state $|I(\theta)\rangle$, the operation of U_R can use q NOT gates to, respectively, implement the inversion operation of each qubit in the intensity channel, as shown in Equation 3.2:

$$\begin{aligned}
U_R(|I(\theta)\rangle) &= \frac{1}{2^n} U_R \left\{ \sum_{k=0}^{2^{2n}-1} |H_k\rangle |S_k\rangle |I_k\rangle \otimes |k\rangle \right\} \\
&= \frac{1}{2^n} \sum_{k=0}^{2^{2n}-1} |H_k\rangle |S_k\rangle \otimes \left(\bigotimes_{j=0}^{q-1} \left(X |C_k^j\rangle \right) \right) \otimes |k\rangle \\
&= \frac{1}{2^n} \sum_{k=0}^{2^{2n}-1} |H_k\rangle |S_k\rangle \otimes \left(\bigotimes_{j=0}^{q-1} |\overline{C}_k^j\rangle \right) \otimes |k\rangle \\
&= \frac{1}{2^n} \sum_{k=0}^{2^{2n}-1} |H_k\rangle |S_k\rangle |2^q - 1 - I_k\rangle \otimes |k\rangle
\end{aligned} \tag{3.2}$$

Among them, $I_K \in [0, 2^q - 1]$. Figure 3.2 presents the quantum circuit of the U_R operation on the QIRHSI model, taking the image "Peppers" as an example for illustration.

DOI: 10.1201/9781003621423-4

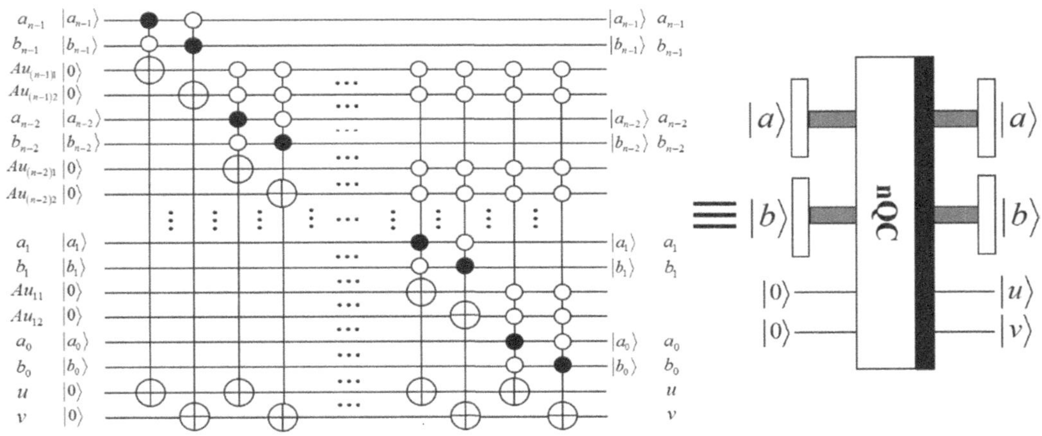

FIGURE 3.1 *n*-bit quantum comparator.

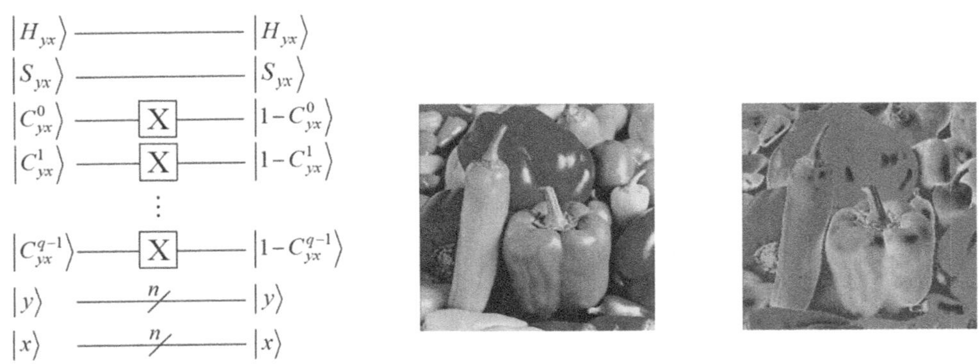

a) The quantum circuit of the intensity inversion operation U_R

b) image Peppers

c) The result of image b) after the operation of U_R

FIGURE 3.2 Intensity inversion operation based on the QIRHSI model.

3.1.3 Intensity-based Complementation Operation

The intensity information of the quantum image is encoded and stored in a binary sequence [6–9], and the highest qubit in the sequence is reversed, and the remaining low qubit remains the original value. The intensity complement operator U_C is shown in Equation 3.3:

$$U_C = I^{\otimes 2} \otimes X \otimes I^{\otimes q-1} \otimes I^{\otimes 2n} \tag{3.3}$$

For the quantum superposition state $|I(\theta)\rangle$, when the operator U_c acts on $|I(\theta)\rangle$, the operation process is shown in Equation 3.4:

$$\begin{aligned} U_C\left(|I(\theta)\rangle\right) &= \frac{1}{2^n} U_C \left\{ \sum_{k=0}^{2^{2n}-1} |H_k\rangle |S_k\rangle |I_k\rangle \otimes |k\rangle \right\} \\ &= \frac{1}{2^n} \sum_{k=0}^{2^n-1} |H_k\rangle |S_k\rangle \otimes \left(X|C_k^0\rangle\right) \left(\bigotimes_{j=1}^{q-1} \left(X|C_k^j\rangle\right) \right) \otimes |k\rangle \end{aligned}$$

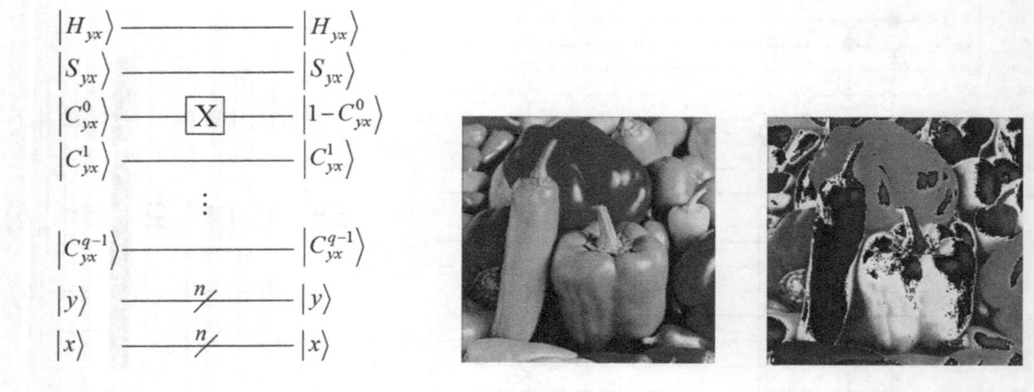

a) The quantum circuit of the intensity complementation operation Uc

b) image Peppers

c) The result of image b) after the operation of Uc

FIGURE 3.3 Intensity complementation operation based on the QIRHSI model.

$$=\frac{1}{2^n}\sum_{k=0}^{2^n-1}|H_k\rangle|S_k\rangle\otimes|\bar{C}_k^0\rangle\left(\bigotimes^{q-1}|C_k^j\rangle\right)\otimes|k\rangle$$

$$=\frac{1}{2^n}\sum_{k=0}^{2^{2n}-1}|H_k\rangle|S_k\rangle|(I_k+2^{q-1})\bmod 2^q\rangle\otimes|k\rangle \tag{3.4}$$

Among them, $I_k\in[0,2^q-1]$. Figure 3.3 shows the quantum circuit of the U_c operation in the QIRHSI model, taking the image Peppers as an example.

3.1.4 Intensity-based Addition Operation

Suppose the sizes of two images are both $2^n\times 2^n$, and their representations [10–13] are shown in Equation 3.5:

$$|I_A(\theta)\rangle=\frac{1}{2^n}\sum_{k=0}^{2^n-1}|C_k^A\otimes|k\rangle=\frac{1}{2^n}\sum_{y=0}^{2^n-1}\sum_{x=0}^{2^n-1}|H_{yx}^A\rangle|S_{yx}^A\rangle|I_{yx}^A\rangle\otimes\langle yx|$$

$$|I_B(\theta)\rangle=\frac{1}{2^n}\sum_{k=0}^{2^{2n}-1}|C_k^B\otimes|k\rangle=\frac{1}{2^n}\sum_{y=0}^{2^n-1}\sum_{x=0}^{2^n-1}|H_{yx}^B|S_{yx}^B|I_{yx}^B\otimes|yx\rangle \tag{3.5}$$

For images $|I_A(\theta)\rangle$ and $|I_B(\theta)\rangle$, at any pixel position $|yx\rangle$ has $|H_{yx}^A\rangle=|H_{yx}^B\rangle$, $|S_{yx}^A\rangle=|S_{yx}^B\rangle$, $I_{yx}^A,I_{yx}^B\in[0,2^q-1]$. Therefore, the intensity addition operation ADD in the QIRHSI model is defined as

$$|I_C(\theta)\rangle=ADD\{|I_A(\theta)\rangle,|I_B(\theta)\rangle\}=\frac{1}{2^n}\sum_{y=0}^{2^n-1}\sum_{x=0}^{2^n-1}|C_{yx}^C\rangle\otimes|yx\rangle$$

$$=\frac{1}{2^n}\sum_{y=0}^{2^n-1}\sum_{x=0}^{2^n-1}\left(|C_{yx}^A\rangle+|C_{yx}^B\rangle\right)\otimes|yx\rangle$$

$$
= \begin{cases} \dfrac{1}{2^n} \sum_{y=0}^{2^n-1} \sum_{x=0}^{2^n-1} \left|H_{yx}^C\right\rangle \left|S_{yx}^C\right\rangle \left(\left|I_{yx}^A\right\rangle + \left|I_{yx}^B\right\rangle\right) \otimes |yx\rangle, & 0 \le I_{yx}^A + I_{yx}^B \le 2^q - 1 \\ \dfrac{1}{2^n} \sum_{y=0}^{2^n-1} \sum_{x=0}^{2^n-1} \left|H_{yx}^C\right\rangle \left|S_{yx}^C\right\rangle \left|2^q - 1\right\rangle \otimes |yx\rangle, & 2^q - 1 < I_{yx}^A + I_{yx}^B \le 2^{q+1} - 2 \end{cases} \tag{3.6}
$$

Figure 3.4 shows the quantum circuit of the intensity addition operation ADD in the QIRHSI model. Figure 3.5 gives examples of the addition operation ADD using Peppers images with different intensities.

The intensity addition operation ADD is achieved through three steps: Step 1: Use the nQC twice (see Figure 3.1) to locate the color information of the images $|I_A(\theta)\rangle$ and $|I_B(\theta)\rangle$ at the same pixel position $|yx\rangle$. Step 2: Use a common adder (see Figure 1.4) to add the intensities $\left|I_{yx}^A\right\rangle$ and $\left|I_{yx}^B\right\rangle$ of the images $|I_A(\theta)\rangle$ and $|I_B(\theta)\rangle$ at the same pixel position $|yx\rangle$. Step 3: Compare the sum of $\left|I_{yx}^A\right\rangle$ and $\left|I_{yx}^B\right\rangle$ with $|2^q - 1\rangle$ and then output the color values of the image $|I_C(\theta)\rangle$ under different conditions. The following will give a detailed description of the three steps of the intensity addition operation ADD.

Step 1: As the images $|I_A(\theta)\rangle$ and $|I_B(\theta)\rangle$ are stored in the form of quantum superposition states, to locate the same pixel positions of the images $|I_A(\theta)\rangle$ and $|I_B(\theta)\rangle$, for any pixel $|yx\rangle = |y\rangle|x\rangle = |y_{n-1}\cdots y_1 y_0\rangle|x_{n-1}\cdots x_1 x_0\rangle$, $|y\rangle$ and $|x\rangle$ must be compared, respectively. Use the n-qubit comparator n-QC twice. When $|de\rangle$ is equal to $|00\rangle$, the pixel coordinates of the images $|I_A(\theta)\rangle$ and $|I_B(\theta)\rangle$ satisfy $|y^A\rangle = |y^B\rangle$, that is, $\left|y_{n-1}^A \cdots y_1^A y_0^A\right\rangle = \left|y_{n-1}^B \cdots y_1^B y_0^B\right\rangle$; when $|fg\rangle$ is equal to $|00\rangle$, the pixel coordinates of the images $|I_A(\theta)\rangle$ and $|I_B(\theta)\rangle$ satisfy $|x^A\rangle = |x^B\rangle$, that is, $\left|x_{n-1}^A \cdots x_1^A x_0^A\right\rangle = \left|x_{n-1}^B \cdots x_1^B x_0^B\right\rangle$. In other words, when $|defg\rangle$ is equal to $|0000\rangle$, we have $|y^A x^A\rangle = |y^B x^B\rangle$, that is, the images $|I_A(\theta)\rangle$ and $|I_B(\theta)\rangle$ have been located at the same pixel position (see Step 1 in Figure 3.4).

Step 2: In Step 1, we used the nQC twice to locate the intensity information of the images $|I_A(\theta)\rangle$ and $|I_B(\theta)\rangle$ at the same pixel position $|yx\rangle$. At this time, $|defg\rangle$ is equal to $|0000\rangle$. Then, we use the four-controlled ADDER operator to add the intensities of the images $|I_A(\theta)\rangle$ and $|I_B(\theta)\rangle$ to obtain $\left|I_{yx}^A\right\rangle + \left|I_{yx}^B\right\rangle$ (see Step 2 in Figure 3.4).

Step 3: In Step 2, we have already obtained $\left|I_{yx}^A\right\rangle + \left|I_{yx}^B\right\rangle$. To output $\left|I_{yx}^C\right\rangle$ under different conditions, it is necessary to determine the magnitude relationship between $\left|I_{yx}^A\right\rangle + \left|I_{yx}^B\right\rangle$ and $|2^q - 1\rangle$. As $\left|I_{yx}^A\right\rangle + \left|I_{yx}^B\right\rangle$ is a $(q+1)$-bit binary number and $|2^q - 1\rangle$ is a q-bit binary number, we only need to check the value of the most significant bit $|w\rangle$ of $\left|I_{yx}^A\right\rangle + \left|I_{yx}^B\right\rangle$. When $|w\rangle$ is equal to $|0\rangle$, this falls under Case 1, where $\left|I_{yx}^A\right\rangle + \left|I_{yx}^B\right\rangle$ is less than or equal to $|2^q - 1\rangle$, and we can directly output the value of $\left|I_{yx}^C\right\rangle$ as $\left|I_{yx}^A\right\rangle + \left|I_{yx}^B\right\rangle$. When $|w\rangle$ is equal to $|1\rangle$, this falls under Case 2, where $\left|I_{yx}^A\right\rangle + \left|I_{yx}^B\right\rangle$ is greater than $|2^q - 1\rangle$, and we can directly output the value of $\left|I_{yx}^C\right\rangle$ as $2^q - 1$. At the same time, the hue and saturation remain unchanged, that is, $\left|H_{yx}^C\right\rangle = \left|H_{yx}^A\right\rangle$ and $\left|S_{yx}^C\right\rangle = \left|S_{yx}^A\right\rangle$ (see Step 3 in Figure 3.4).

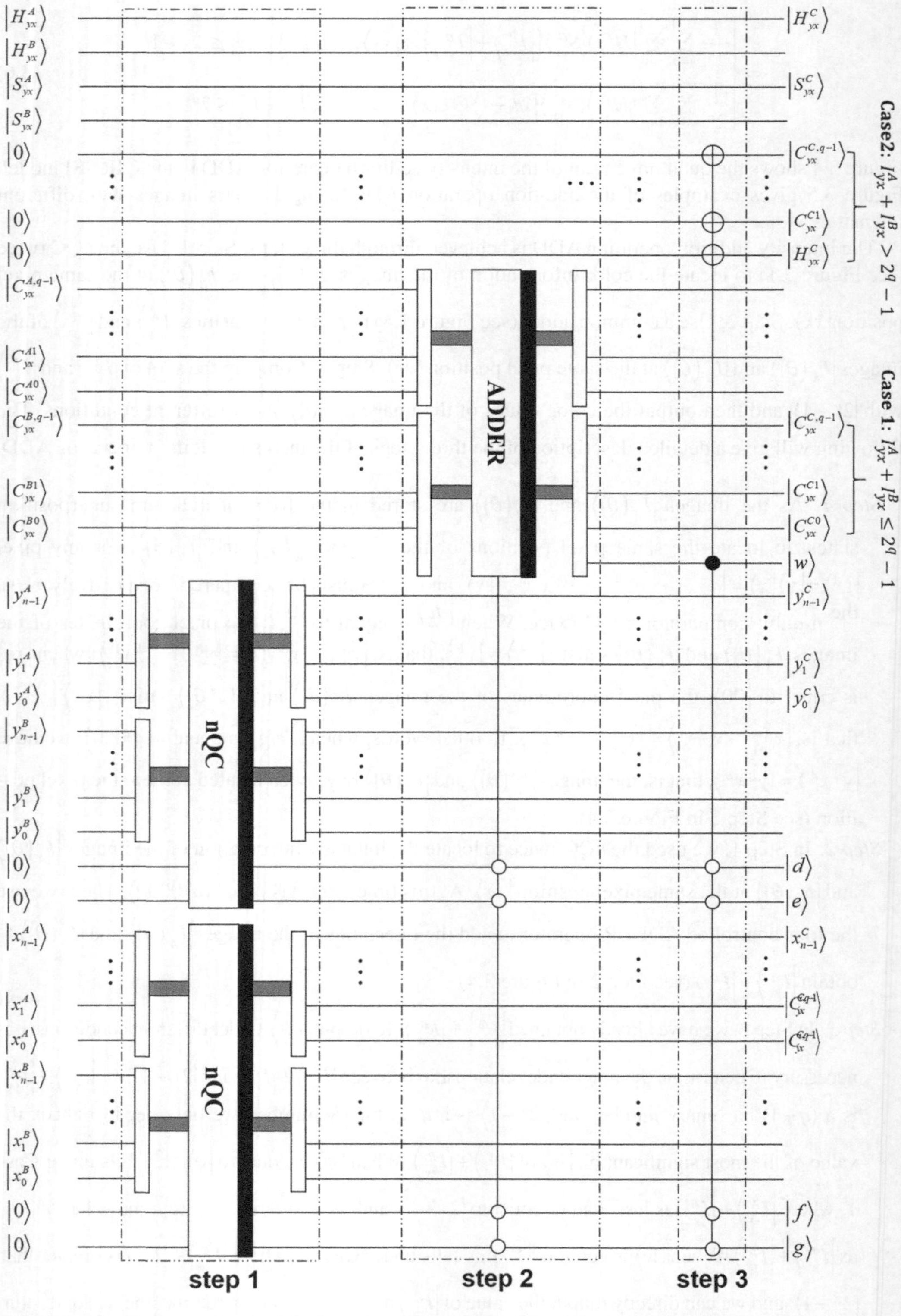

FIGURE 3.4 Quantum circuit of the intensity addition operation ADD in the QIRHSI model.

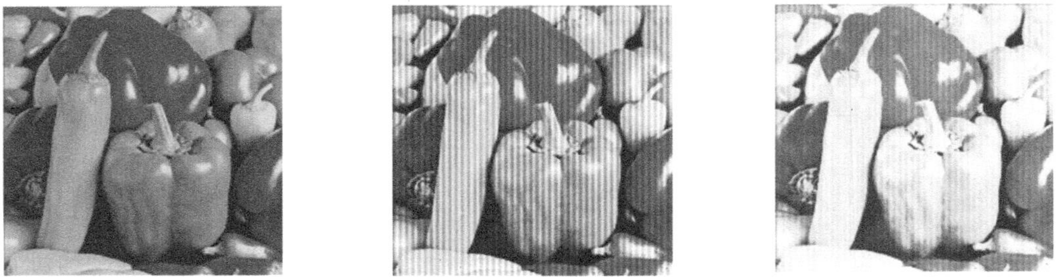

a) image $|I_A(\theta)\rangle$ b) image $|I_B(\theta)\rangle$ c) image $|I_C(\theta)\rangle$

FIGURE 3.5 The operator ADD acts on the images $|I_A(\theta)\rangle$, $|I_B(\theta)\rangle$ to obtain the image $|I_C(\theta)\rangle$.

3.1.5 Intensity-Based Subtraction Operation

Let the size of the two images be $2^n \times 2^n$, and their expression is shown as in Equation 3.7

$$|I_A(\theta)\rangle = \frac{1}{2^n}\sum_{k=0}^{2^n-1}|C_k^A\rangle \otimes |k\rangle = \frac{1}{2^n}\sum_{y=0}^{2^n-1}\sum_{x=0}^{2^n-1}|H_{yx}^A\rangle|S_{xx}^A\rangle|I_{yx}^A\rangle \otimes |yx\rangle$$

$$|I_B(\theta)\rangle = \frac{1}{2^n}\sum_{k=0}^{2^{2n}-1}|C_k^B\rangle \otimes |k\rangle = \frac{1}{2^n}\sum_{y=0}^{2^n-1}\sum_{x=0}^{2^n-1}|H_{yx}^B\rangle|S_{xx}^B\rangle|I_{xx}^s\rangle \otimes |yx\rangle \tag{3.7}$$

For the images $|I_A(\theta)\rangle$ and $|I_B(\theta)\rangle$, at any pixel $|yx\rangle$, we have $|H_{yx}^A\rangle = |H_{yx}^B\rangle$, $|S_{yx}^A\rangle = |S_{yx}^B\rangle$, and $I_{yx}^A, I_{yx}^B \in [0, 2^q - 1]$. Therefore, the intensity subtraction operation SUB based on the QIRHSI model [14–16] is defined as

$$|I_C(\theta)\rangle = SUB\{|I_A(\theta)\rangle, |I_B(\theta)\rangle\} = \frac{1}{2^n}\sum_{y=0}^{2^n-1}\sum_{x=0}^{2^n-1}|C_{yx}^C\rangle \otimes |yx\rangle$$

$$= \frac{1}{2^n}\sum_{y=0}^{2^n-1}\sum_{x=0}^{2^n-1}\left(|C_{yx}^A\rangle - |C_{yx}^B\rangle\right) \otimes |yx\rangle$$

$$= \begin{cases} \frac{1}{2^n}\sum_{y=0}^{2^n-1}\sum_{x=0}^{2^n-1}|H_{yx}^C\rangle|S_{yx}^C\rangle\left(|I_{yx}^A\rangle - |I_{yx}^B\rangle\right) \otimes |yx\rangle, & I_{yx}^A > I_{yx}^B \\ \frac{1}{2^n}\sum_{y=0}^{2^n-1}\sum_{x=0}^{2^n-1}|H_{yx}^C\rangle|S_{yx}^C\rangle|0\rangle \otimes |yx\rangle, & I_{yx}^A \le I_{yx}^B \end{cases} \tag{3.8}$$

Figure 3.6 presents the quantum circuit of the intensity subtraction operation SUB in the QIRHSI model. Figure 3.7 shows an example of the subtraction operation SUB for Peppers images with different intensities.

The intensity subtraction operation SUB can be implemented through three steps:

Step 1: Use the nQC twice (see Figure 3.1) to locate the color information of the images $|I_A(\theta)\rangle$ and $|I_B(\theta)\rangle$ at the same pixel position $|yx\rangle$. Step 2: Use the qQC ($n = q$ in Figure 3.1) to compare the

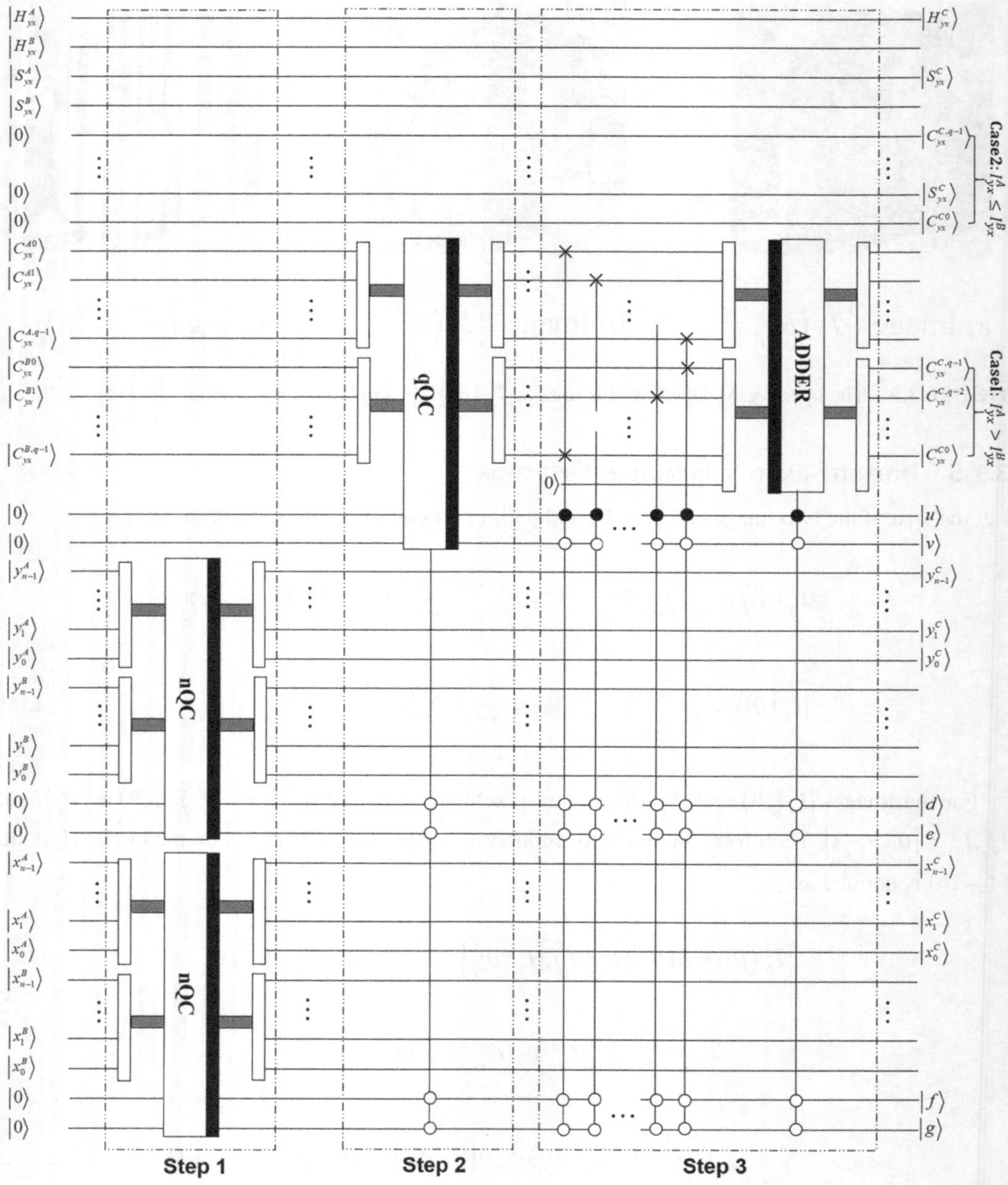

FIGURE 3.6 Quantum circuit of the intensity subtraction operation SUB in the QIRHSI model it will not be repeated here. (See Step 1 in Figure 3.6.)

magnitude relationship between the intensity values $\left|I_{yx}^{A}\right\rangle$) and $\left|I_{yx}^{B}\right\rangle$ of the images $\left|I_{A}(\theta)\right\rangle$ and $\left|I_{B}(\theta)\right\rangle$ at the same pixel position $|yx\rangle$. Step 3: Output the intensity $\left|I_{yx}^{C}\right\rangle$ of the image $\left|I_{C}(\theta)\right\rangle$ according to the magnitude relationship between $\left|I_{yx}^{A}\right\rangle$ and $\left|I_{yx}^{B}\right\rangle$. The following gives a detailed description of the three steps based on the intensity subtraction operation SUB.

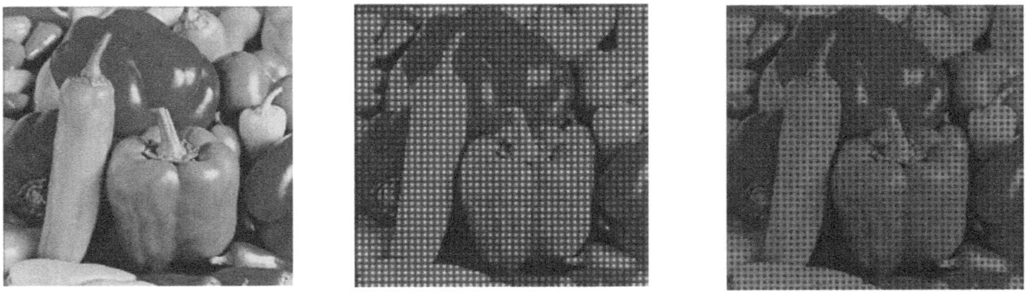

a) image $|I_A(\theta)\rangle$ b) image $|I_B(\theta)\rangle$ c) image $|I_C(\theta)\rangle$

FIGURE 3.7 The operator SUB acts on the images $|I_A(\theta)\rangle$, $|I_B(\theta)\rangle$ to obtain the image $|I_C(\theta)\rangle$.

Step 1: This step is the same as Step 1 of the intensity addition operation ADD.

Step 2: In Step 1, the nQC has been used twice to locate the color information of the images $|I_A(\theta)\rangle$ and $|I_B(\theta)\rangle$ at the same pixel position $|yx\rangle$. At this moment, $|defg\rangle$ is equal to $|0000\rangle$. Immediately afterwards, a four-controlled qQC is used to compare the magnitude relationship between the intensities $|I_{yx}^A\rangle$ and $|I_{yx}^B\rangle$ of the images $|I_A(\theta)\rangle$ and $|I_B(\theta)\rangle$ at the pixel position $|yx\rangle$. When $|uv\rangle$ is equal to $|10\rangle$, $|I_{yx}^A\rangle > |I_{yx}^B\rangle$; when $|uv\rangle$ is equal to $|01\rangle$ or $|00\rangle$, $|I_{yx}^A\rangle \le |I_{yx}^B\rangle$ (see Step 2 in Figure 3.6).

Step 3: In the previous step, the magnitude relationship between the intensities $|I_{yx}^A\rangle$ and $|I_{yx}^B\rangle$ of the images $|I_A(\theta)\rangle$ and $|I_B(\theta)\rangle$ at the same pixel position $|yx\rangle$ has been compared. Next, it is only necessary to output the intensity $|I_{yx}^C\rangle$ of the image $|I_C(\theta)\rangle$ at the pixel position $|yx\rangle$ under different conditions.

It should be noted that when using the *n*QC shown in Figure 3.1 to compare the magnitude relationship between two binary numbers a and b, the higher bits are on the top and the lower bits are at the bottom. The comparison is carried out successively from the higher bits to the lower bits, and finally the magnitude relationship between a and b is obtained. When a common adder network calculates the sum of two $(n + 1)$-bit binary numbers a and b, the lower bits are on the top and the higher bits are at the bottom. In addition, a most significant bit b_{n+1} needs to be added to output the sum of a and b.

Therefore, when $|uv\rangle$ is equal to $|10\rangle$, the intensity of the pixels at the position $|yx\rangle$ in the images $|I_A(\theta)\rangle$ and $|I_B(\theta)\rangle$ satisfies $|I_{yx}^A\rangle > |I_{yx}^B\rangle$. To calculate the difference between $|I_{yx}^A\rangle$ and $|I_{yx}^B\rangle$, first, q four – controlled swap gates are needed to swap the intensity bits of $|I_{yx}^A\rangle$ and $|I_{yx}^B\rangle$ in sequence, and a highest – order bit $|0\rangle$ is inserted to enable the ADDER to perform the inverse addition operation. The calculated value of $|I_{yx}^A\rangle - |I_{yx}^B\rangle$ is $|I_{yx}^C\rangle$, which is Case 1. When $|uv\rangle$ is equal to $|01\rangle$ or $|00\rangle$, the intensity of the pixels at the position $|yx\rangle$ in the images $|I_A(\theta)\rangle$ and $|I_B(\theta)\rangle$ satisfies $|I_{yx}^A\rangle \le |I_{yx}^B\rangle$, and directly output $|0\rangle$, which is $|I_{yx}^C\rangle$, and this is Case 2 (see Step 3 in Figure 3.6).

3.1.6 Complexity Analysis of Quantum Networks

The quantum circuits for intensity inversion and intensity complement operations are relatively simple. Figure 3.2 shows that q NOT gates are needed to implement the intensity inversion operation. Therefore, the complexity of the quantum gates for the intensity inversion operation is $O(q)$. Figure 3.3 shows that only 1 NOT gate is required to implement the intensity complement operation. Thus, the complexity of the quantum gates for the complement operation is $O(1)$.

Next, let us analyze the complexity of the quantum gates for the intensity addition and subtraction operations. In Figures 3.4 and 3.6, the nQC is involved. We first need to analyze the complexity of the nQC.

Theorem 3.1: The complexity of the quantum gates [17] required for nQC is $O(n^2)$, where the number of input bits is $2n$ and the number of auxiliary bits required is $2n$.

Proof: In Figure 3.1, the number of input bits is $2n$, and the number of auxiliary bits required is $2n$. For two n-bit binary numbers, at most n comparisons are needed, that is, each bit is compared once in sequence. Therefore, for a multi-controlled NOT gate, the total number of CNOTs is

$$2\left(2\Lambda_2(X)+2\Lambda_4(X)+\cdots+2\Lambda_{2n}(X)\right)=4\left(\Lambda_2(X)+\Lambda_4(X)+\cdots+\Lambda_{2n}(X)\right) \tag{3.9}$$

According to Lemmas 1.1 and 1.2, a $\Lambda_4(X)$ gate is equivalent to two Toffoli gates and two $\Lambda_3(X)$ gates. Similarly, a $\Lambda_3(X)$ gate is equivalent to four Toffoli gates, and a Toffoli gate can be simulated by six CNOT gates.

To simplify all the $\Lambda_{2k}(X)(k=3,4,5,\ldots,n)$ gates, we can obtain the following by using Lemma 1.3

$$\begin{aligned}&\Lambda_6(X)+\Lambda_8(X)+\cdots+\Lambda_{2n}(X)\\&=8\times(3+5+\cdots 2n-3)\ \text{Toffoli}\\&=8n^2-16n\ \text{Toffoli}\end{aligned}$$

Therefore, Equation 3.9 can be simplified to

$$\begin{aligned}&4\times\left(\Lambda_2(X)+\Lambda_4(X)+\Lambda_6(X)+\cdots+\Lambda_{2n}(X)\right)\\&=4\times\left(1+10+8n^2-16n\right)\Lambda_2(X)\\&=32n^2-64n+44\Lambda_2(X)\\&=192n^2-384n+264\ \text{CNOT}\end{aligned}$$

The number of NOT gates required in Figure 3.1 is $8n-4$.

Therefore, the nQC can be simulated by $8n-4$ NOT gates and $192n^2-384n+264\cdot$ CNOT gates. In other words, the complexity of the quantum gates required for the nQC is $O\left(192n^2-376n+260\right)\approx O\left(n^2\right)$

□

Theorem 3.2: The complexity of the quantum gates required for the intensity addition operation ADD based on the QIRHSI model is $O(n^2+q)$, and the number of auxiliary quantum bits required is $4n+2q+2$.

Proof: In Figure 3.4, to analyze the complexity of the quantum gates required for the addition operation ADD based on the QIRHSI model, it is only necessary to analyze the number of quantum gates required in the three steps.

In Step 1, nQC is used twice. From the proof process of Theorem 3.1, it is known that $16n-8$ NOT gates and $384n^2-768n+528$ CNOT gates are required, and the number of auxiliary bits required is $4n$. Next, analyze the number of basic quantum gates required in Steps 2 and 3.

In Step 2, by means of a common adder with four control bits, the q-bit $\left|I_{yx}^{A}\right\rangle$ and $\left|I_{yx}^{B}\right\rangle$ are added together. It is necessary to use $2q-1$ carry operations with four control bits (see Figure 1.8a), q sum operations with four control bits (see Figure 1.8b), and one $\Lambda_5(X)$ gate. A carry operation with four control bits can be decomposed into two $\Lambda_6(X)$ gates and one $\Lambda_5(X)$ gate, and a sum operation with four control bits can be decomposed into two $\Lambda_5(X)$ gates.

A $\Lambda_6(X)$ gate and a $\Lambda_5(X)$ gate are equivalent to 24 and 16 Toffoli gates, respectively. Therefore, a carry operation with 4 control bits is equivalent to 64 Toffoli gates, and a sum operation with 4 control bits is equivalent to 32 Toffoli gates.

In conclusion, the number of basic quantum gates required in Step 2 is

$$64(2q-1)+32q+16\text{Toffoli}+4\text{NOT}$$

$$=160q-48\,\text{Toffoli}+4\text{NOT}$$

$$=960q-288\text{CNOT}+4\text{NOT}$$

Among them, the number of required auxiliary bits is $q+2$.

In Step 3, q $\Lambda_5(X)$ gates and 4 NOT gates are used. Therefore, the number of basic gates required in Step 3 is

$$q\Lambda_5(X)+4\text{NOT}=16q\ \text{Toffoli}+4\text{NOT}=96q\text{CNOT}+4\text{NOT}$$

Among them, the number of auxiliary bits required is q.

Therefore, the complexity of the quantum gates required for the intensity addition operation ADD based on the QIRHSI model is

$$O\big((384n^2-768n+528\text{CNOT}+16n-8\text{NOT})(\text{step 1})$$

$$+(960q-288\text{CNOT}+4\text{NOT})(\text{step 2})+(96q\text{CNOT}+4\text{NOT})(\text{step 3}))$$

$$=O\left(16n\text{NOT}+384n^2-768n+1056q+240\text{CNOT}\right)$$

$$=O\left(384n^2-752n+1056q+240\right)$$

$$\approx O\left(n^2+q\right)$$

□

Theorem 3.3: The complexity of the quantum gates required for the intensity subtraction operation SUB based on the QIRHSI model is $O(n^2+q^2)$, and the number of auxiliary quantum bits required is $4n+2q+2$.

Proof: Similar to the steps of the intensity addition operation ADD in the QIRHSI model, in Step 1 of the intensity subtraction operation SUB shown in Figure 3.6, $16n-8$ NOT gates and $384n^2-768n+528$ CNOT gates are required, and the number of auxiliary bits required is $4n$.

In Step 2, to compare the magnitude relationship between $\left|I_{yx}^{A}\right\rangle$ and $\left|I_{yx}^{B}\right\rangle$ at the pixel position $|yx\rangle$, a four-controlled qQC is used. Therefore, the number of multi-controlled NOT gates required is

$$
\begin{aligned}
&2\left(2\Lambda_6(X)+2\Lambda_8(X)+\cdots+2\Lambda_{2q+4}(X)\right)\\
&=4\left(\Lambda_6(X)+\Lambda_8(X)+\cdots+\Lambda_{2q+4}(X)\right)\\
&=32\left(3+5+7+\ldots 2q+1\right)\text{ Toffoli}\\
&=32q^2+64q\text{ Toffoli}\\
&=192q^2+384q\text{ CNOT}
\end{aligned}
$$

Among them, the number of auxiliary bits required is $2q$.

In addition, the number of NOT gates required is $8q$. Therefore, in Step 2, $8q$ NOT gates and $192q^2+384q$ CNOT gates are required.

In Step 3, first, q swap gates with six control bits are used. One swap gate with 6 control bits is equivalent to three $\Lambda_7(X)$ gates, and one $\Lambda_7(X)$ gate is equivalent to 32 Toffoli gates. Therefore, a total of $96q$ Toffoli gates are used.

Then, a q-bit ordinary adder with six control bits is used. An ordinary adder with six control bits employs $2q-1$ carry operators with six control bits (see Figure 1.5a), q sum operators with six control bits (see Figure 1.5b), and one $\Lambda_7(X)$ gate.

A carry operator with six control bits can be decomposed into two $\Lambda_8(X)$ gates and one $\Lambda_7(X)$ gate. A sum operator with six control bits can be decomposed into two $\Lambda_7(X)$ gates. One $\Lambda_8(X)$ gate is equivalent to 40 Toffoli gates.

Therefore, the basic quantum gates used by an ordinary adder with six control bits are

$$
\begin{aligned}
&=(2q-1)\times\left(2\Lambda_8(X)+\Lambda_7(X)\right)+\cdots+2q\Lambda_7(X)+\Lambda_7(X)\\
&=(2q-1)\times\left(2\times 40+32\right)+64q+32\text{ Toffoli}\\
&=288q+80\text{ Toffoli}\\
&=172q+480\text{ CNOT}
\end{aligned}
$$

Therefore, the basic quantum gates required in Step 3 are

$$
\begin{aligned}
&5\text{NOT}+96q\text{ Toffoli}+1728q-480\text{CNOT}\\
&=5\text{NOT}+576q\text{ CNOT}+1728q-480\text{CNOT}\\
&=5\text{NOT}+2304q-480\text{CNOT}
\end{aligned}
$$

Among them, the number of required auxiliary bits is 2q + 2.

In conclusion, the complexity of the basic quantum gates required for the intensity subtraction operation SUB based on the QIRHSI model is

$$
\begin{aligned}
&= O((384n^2 \_768n + 528\,\mathrm{CNOT} + 16n - 8\,\mathrm{NOT})(\text{step } 1) \\
&\quad + (192q^2 + 384q\,\mathrm{CNOT} + 8q\,\mathrm{NOT})(\text{step } 2) \\
&\quad + (2304q \_480\,\mathrm{CNOT} + 5\,\mathrm{NOT})(\text{step } 3)) \\
&= O(384n^2 - 768n + 192q^2 + 2688q + 48\,\mathrm{CNOT} + 16n + 8q - 3\,\mathrm{NOT}) \\
&= O(384n^2 - 752n + 192q^2\ 2696q + 45) \\
&\approx O(n^2 + q^2)
\end{aligned}
$$

□

3.1.7 Experimental Examples

Taking the simple 4 × 4 image intensity component shown in Figure 3.8a as an example to illustrate the intensity inversion and complement operation [18]. Here, $n = 2$ and $q = 4$.

The original image is

$$|I(\theta)\rangle = \frac{1}{2^2}\sum_{y=0}^{3}\sum_{x=0}^{3}|H_{yx}\rangle|S_{yx}\rangle|I_{yx}\rangle\otimes|yx\rangle = \frac{1}{4}\sum_{y=0}^{3}\sum_{x=0}^{3}|H_{yx}\rangle||S_{yx}\rangle|C_{yx}^{0}C_{yx}^{1}C_{yx}^{1}C_{yx}^{3}\rangle\otimes|yx\rangle$$

The corresponding intensity values are shown in Figure 3.8a.

The image after inversion is

$$
\begin{aligned}
U_R(|I(\theta)\rangle) &= \frac{1}{2^2}\sum_{y=0}^{3}\sum_{x=0}^{3}|H_{yx}\rangle|S_{yx}\rangle|2^4 - 1 - I_{yx}\rangle\otimes|yx\rangle \\
&= \frac{1}{4}\sum_{y=0}^{3}\sum_{x=0}^{3}|H_{yx}\rangle|S_{yx}\rangle|15 - I_{yx}\rangle\otimes|yx\rangle
\end{aligned}
$$

The corresponding intensity values are shown in Figure 3.8b.

The image after taking the complement is

$$
\begin{aligned}
U_C(I|\theta\rangle) &= \frac{1}{2^2}\sum_{y=0}^{3}\sum_{x=0}^{3}|H_{yx}\rangle|S_{yx}\rangle|(I_{yx} + 2^3)\,\mathrm{mod}\,2^4\rangle\otimes|yx\rangle \\
&= \frac{1}{4}\sum_{y=0}^{3}\sum_{x=0}^{3}|H_{yx}\rangle|S_{yx}\rangle|(I_{yx} + 8)\,\mathrm{mod}\,16\rangle\otimes|yx\rangle
\end{aligned}
$$

	00	01	10	11
00	0	1	2	3
01	4	5	6	7
10	8	9	10	11
11	12	13	14	15

a) The intensity component of the original image

	00	01	10	11
00	15	14	13	12
01	11	10	9	8
10	7	6	5	4
11	3	2	1	0

b) The intensity component of the inverted image

	00	01	10	11
00	8	9	10	11
01	12	13	14	15
10	0	1	2	3
11	4	5	6	7

c) The intensity component of the image after taking the complement

FIGURE 3.8 The results of the inversion and complement operations are on the intensity components of the 4 × 4 quantum image.

The corresponding intensity values are shown in Figure 3.8c.

The original image $|I_A(\theta)\rangle$ and $|I_B(\theta)\rangle$ are

$$|I_A(\theta)\rangle = \frac{1}{2^2}\sum_{y=0}^{3}\sum_{x=0}^{3}|H_{yx}\rangle|S_{yx}\rangle|I_{yx}^{A}\rangle \otimes |yx\rangle = \frac{1}{4}\sum_{y=0}^{3}\sum_{x=0}^{3}|H_{yx}\rangle|S_{yx}\rangle|C_{yx}^{A0}C_{yx}^{A1}C_{yx}^{A2}C_{yx}^{A3}\rangle \otimes |yx\rangle$$

$$|I_B(\theta)\rangle = \frac{1}{2^2}\sum_{y=0}^{3}\sum_{x=0}^{3}|H_{yx}\rangle|S_{yx}\rangle|I_{yx}^{B}\rangle \otimes |yx\rangle = \frac{1}{4}\sum_{y=0}^{3}\sum_{x=0}^{3}|H_{yx}\rangle|S_{yx}\rangle|C_{yx}^{B0}C_{yx}^{B1}C_{yx}^{B2}C_{yx}^{B3}\rangle \otimes |yx\rangle$$

Correspondingly, the intensity values are shown in Figure 3.9a and b.

The image obtained by the addition operation ADD of the images $|I_A(\theta)\rangle$ and $|I_B(\theta)\rangle$ is $|I_C(\theta)\rangle$, as shown below:

$$|I_C(\theta)\rangle = ADD\{|I_A(\theta)\rangle, |I_B(\theta)\rangle\} = \frac{1}{2^2}\sum_{y=0}^{3}\sum_{x=0}^{3}|C_{yx}^{C}\rangle \otimes |yx\rangle$$

$$= \begin{cases} \frac{1}{4}\sum_{y=0}^{3}\sum_{x=0}^{3}|H_{yx}\rangle|S_{yx}\rangle\left(|I_{yx}^{A}\rangle + |I_{yx}^{B}\rangle\right) \otimes |yx\rangle, & 0 \le I_{yx}^{A} + I_{yx}^{B} \le 15 \\ \frac{1}{4}\sum_{y=0}^{3}\sum_{x=0}^{3}|H_{yx}\rangle|S_{yx}\rangle|15\rangle \otimes |yx\rangle & 15 < I_{yx}^{A} + I_{yx}^{B} \le 30 \end{cases}$$

In addition, the intensity values of the image $|I_C(\theta)\rangle$ are shown in Figure 3.9c.

The image obtained by the subtraction operation SUB of the images $|I_A(\theta)\rangle$ and $|I_B(\theta)\rangle$ is $|I_C(\theta)\rangle$, as shown below:

	00	01	10	11
00	6	3	11	7
01	14	8	5	15
10	1	2	4	13
11	9	10	12	0

a) The intensity component of the image $|I_A(\theta)\rangle$

	00	01	10	11
00	7	1	15	13
01	0	2	14	6
10	10	12	11	4
11	8	3	9	5

b) The intensity component of the image $|I_B(\theta)\rangle$

	00	01	10	11
00	13	4	15	15
01	14	10	15	15
10	11	14	15	15
11	15	13	15	5

c) The intensity component after adding the images $|I_A(\theta)\rangle$ and $|I_B(\theta)\rangle$

	00	01	10	11
00	0	2	0	0
01	14	6	0	9
10	6	0	0	9
11	1	7	3	0

d) The intensity component after SUB of the images $|I_A(\theta)\rangle$ and $|I_B(\theta)\rangle$

FIGURE 3.9 The results of the addition and subtraction operations are on the intensity components of the 4×4 quantum image.

$$|I_C(\theta)\rangle = SUB\{|I_A(\theta)\rangle, |I_B(\theta)\rangle\} = \frac{1}{2^2}\sum_{y=0}^{3}\sum_{x=0}^{3}|C_{yx}^{C}\rangle \otimes |yx\rangle$$

$$= \begin{cases} \frac{1}{4}\sum_{y=0}^{3}\sum_{x=0}^{3}|H_{yx}\rangle|S_{yx}\rangle\left(|I_{yx}^{A}\rangle - |I_{yx}^{B}\rangle\right)\otimes|yx\rangle, & I_{yx}^{A} > I_{yx}^{B} \\ \frac{1}{4}\sum_{y=0}^{3}\sum_{x=0}^{3}|H_{yx}\rangle|S_{yx}\rangle|0\rangle\otimes|yx\rangle, & I_{yx}^{A} \le I_{yx}^{B} \end{cases}$$

Also, the intensity values of the image $|I_C(\theta)\rangle$ are shown in Figure 3.9d).

3.2 QUANTUM NOISE IMAGE

Quantum noise images are often applied in the evaluation of the quality of quantum images and the testing of quantum image processing algorithms [19, 20]. This section introduces a common method for preparing a type of noise image, that is, the preparation method of salt-and-pepper noise images.

The noise in quantum images mainly originates from the image transmission and processing processes, and it is mainly additive noise. A quantum noise image can be expressed as

$$|G\rangle = |F\rangle - |\eta\rangle$$

Among them, $|F\rangle$ represents the quantum image, $|\eta\rangle$ represents the additive noise image, and $|G\rangle$ represents the degraded image after superimposing the noise. Gaussian noise and salt-and-pepper noise are two common and major types of additive noise. Through the degradation function, it can be obtained that the noise process is equivalent to the preparation of a quantum noise image.

Let the quantum salt-and-pepper noise image have the following NEQR format:

$$|I\rangle_z = \left(\sqrt{p_a}|c_a\rangle|a\rangle + \sqrt{p_b}|c_b\rangle|b\rangle + \frac{\sqrt{1-p_a-p_b}}{\sqrt{2^{2n}-2}}\sum_{i\neq a,b}^{2^{2n}-1}|c_i\rangle|i\rangle\right)$$

Here, p_a and p_b are, respectively, the values of the probability density functions at the two positions of $|C_a\rangle$ and $|C_b\rangle$. Furthermore, this representation satisfies the normalized quantum state, that is,

$$\left\||I\rangle_z\right\| = \sqrt{p_a - p_b + \frac{(1-p_a-p_b)(2^{2n}-2)}{2^{2n}-2}} = 1$$

The preparation process of the quantum salt-and-pepper noise image is as follows:

Step 1: First, assume that the initial quantum state is a superposition state of $q+4n-2$ $|0\rangle$ states:

$$|\Phi\rangle_0 = |0\rangle^{\otimes q+4n-2} = |0\rangle^{\otimes q}|0\rangle^{\otimes n}|0\rangle^{\otimes n}|0\rangle^{\otimes 2n-2}$$

Step 2: Construct a unitary transformation. The unitary transformation U_1 is constructed as:

$$U_1 = I^{\otimes q} \otimes \left(\sqrt{p_a}I\right)^n \otimes \left(\sqrt{p_b}I\right)^n \otimes H^{\otimes 2n-2}$$

Step 3: After applying U_1, the initial state $|\Phi_0\rangle$ is transformed into

$$\begin{aligned}|\Psi\rangle_1 &= U_1\left(|\Psi\rangle_0\right)\\ &= \left(I^{\otimes q}\otimes\left(\sqrt{p_a}I\right)^{\otimes n}\otimes\left(\sqrt{p_b}I\right)^{\otimes n}\otimes H^{\otimes 2n-2}\right)\left(|0\rangle^{\otimes q+4n-2}\right)\\ &= \left(I^{\otimes q}\otimes\left(\sqrt{p_a}I\right)^{\otimes n}\otimes\left(\sqrt{p_b}I\right)^{\otimes n}\otimes H^{\otimes 2n-2}\right)\left(|0\rangle^{\otimes q}|0\rangle^{\otimes n}|0\rangle^{\otimes n}|0\rangle^{\otimes 2n-2}\right)\\ &= |0\rangle^{\otimes q}\sqrt{p_a}|a\rangle\sqrt{p_b}|b\rangle\otimes\frac{1}{2^{2n}-2}\sum_{i=0,i\neq a,b}^{2^{2n}-2}|i\rangle\end{aligned}$$

Step 4: After applying U_2

$$U_2 = I^{\otimes q}\otimes I^{\otimes n}\otimes I^{\otimes n}\otimes\left(\sqrt{1-p_a-p_b}I\right)^{\otimes n}$$

to the quantum state $|\Psi_1\rangle$, get $|\Psi_2\rangle$

$$\begin{aligned}|\Psi\rangle_2 &= U_2\left(|\Psi\rangle_1\right)\\ &= \left(I^{\otimes q}\otimes I^{\otimes n}\otimes I^{\otimes n}\otimes\left(\sqrt{1-p_a-p_b}I\right)^{\otimes n}\right)\left(|0\rangle^{\otimes q}\sqrt{p_a}|a\rangle\sqrt{p_b}|b\rangle\otimes\frac{1}{2^{2n}-2}\sum_{i=0,i\neq a,b}^{2^{2n}-2}|i\rangle\right)\\ &= |0\rangle^{\otimes q}\sqrt{p_a}|a\rangle\sqrt{p_b}|b\rangle\frac{\sqrt{1-p_a-p_b}}{2^{2n}-2}\sum_{i=0,i\neq a,b}^{2^{2n}-2}|i\rangle\end{aligned}$$

Here, a and b are n-qubit quantum states representing positional information.

Step 5: Prepare the grayscale values of the quantum noise image. For the intermediate state $|\Psi_2\rangle$, all pixels have already appeared in the NEQR form. The grayscale value generation process can be divided into 2^{2n} sub-operations: $\Omega_{yx}^i : |0\rightarrow|0+C_{yx}^i$ In this way, if $C_{yx}^i=1$, then Ω_{yx}^i is a $2n$-*CONT* gate. Otherwise, it is an identity gate. Therefore, the quantum operation Ω_{yx} can be obtained through Ω_{yx}^i, in the form of $\Omega_{yx} = \bigotimes_{i=0}^{q-1}\Omega_{yx}^i$ Specifically,

$$\Omega_{yx}\left(|0\rangle^{\otimes q}\right) = \left(\bigotimes_{i=0}^{q-1}\Omega_{yx}^i\right)\left(|0\rangle^{\otimes q}\right) = \bigotimes_{i=0}^{q-1}\left(\Omega_{yx}^i|0\rangle\right) = \bigotimes_{i=0}^{q-1}\left|0+C_{yx}^i\right\rangle = \bigotimes_{i=0}^{q-1}\left|C_{yx}^i\right\rangle = |f(Y,X)\rangle$$

Design the unitary transformation U_3:

$$U_3 = \Omega_a\otimes|a\rangle\langle a| + \Omega_b\otimes|b\rangle\langle b| + \sum_{i=0,i\neq a,b}^{2^{2n}-2}\Omega_i\otimes|i\rangle\langle i|$$

$|\Psi_2\rangle$ achieves the purpose of preparing the grayscale values

$$U_3\left(|\Psi\rangle_2\right)=\left(\Omega_a\otimes|a\rangle\langle a|+\Omega_b\otimes|b\rangle\langle b|+\sum_{i=0,i\neq a,b}^{2^{2n-2}}\Omega_i\otimes|i\rangle\langle i|\right)$$
$$\left(|0\rangle^{\otimes q}\sqrt{p_a}|a\rangle\sqrt{p_b}|b\rangle\frac{\sqrt{1-p_a-p_a}}{2^{2n}-2}\sum_{i=0,i\neq a,b}^{2^{2n-2}}|i\rangle\right)$$
$$=|c_a\rangle\sqrt{p_a}|a\rangle+|c_b\rangle\sqrt{p_b}|b\rangle+\frac{\sqrt{1-p_a-p_b}}{2^{2n}-2}\sum_{i=0,i\neq a,b}^{2^{2n}-2}|c_i\rangle|i\rangle$$

The preparation of the quantum salt-and-pepper noise image can be completed through the above five steps. For an image of size $2^n\times 2^n$, the cost of the entire preparation process does not exceed $O\left(qn2^{2n}\right)$.

3.3 SUMMARY OF THIS CHAPTER

This chapter briefly studies the arithmetic operations of the quantum color image QIRHSI, namely inversion, complementation, addition, and subtraction operations. It is investigated how to use a quantum comparator to locate the same pixel positions in two images [21, 22], and then perform intensity addition and subtraction operations. Finally, the corresponding intensity values are output according to different value ranges. In addition, the complexity analysis and a simple 4 × 4 example demonstrate the performance of the proposed operations. Finally, a preparation method for quantum additive noise images based on the NEQR representation is presented.

REFERENCES

[1] Nielsen M A; Chuang I L. *Quantum Computation and Quantum Information*. Cambridge University Press, Cambridge, 2005: 57–89, 157–223.

[2] Barenco A; Bennett C H; Cleve R; et al. Elementary Gates for Quantum Computation. *Physical Review A*, 52(5), 1995: 1–31.

[3] Shende V V; Prasad A K; Markov I L; et al. Synthesis of Reversible Logic Circuits. *IEEE Transactions on Computer-Aided Design of Integrated Circuits and Systems*, 22(6), 2003: 710–722.

[4] Saeedi M; Markov I L. Synthesis and Optimization of Reversible Circuits – A Survey. *ACM Computing Surveys (CSUR)*, 45(2), 2013: 1–34.

[5] Chen G L; Song X H; Venegas-Andraca S E; et al. QIRHSI: Novel Quantum Image Representation Based on HSI Color Space Model. *Quantum Information Processing*, 21(1), 2022: 5.

[6] Yan F; Li N Q; Hirota K. QHSL: A Quantum Hue, Saturation, and Lightness Color Model. *Information Sciences*, 577(4), 2021: 196–213.

[7] Le P Q; Dong F Y; Hirota K. A Flexible Representation of Quantum Images for Polynomial Preparation, Image Compression, and Processing Operations. *Quantum Information Processing*, 10(1), 2011: 63–84.

[8] Li P C; Liu X D. Color Image Representation Model and Its Application Based on an Improved FRQI. *International Journal of Quantum Information*, 16(15), 2018: 1850005.

[9] Zhang Y; Lu K; Gao Y H; et al. NEQR: A Novel Enhanced Quantum Representation of Digital Images. *Quantum Information Processing*, 12(8), 2013: 2833–2860.

[10] Liu K; Zhang Y; Lu K; et al. An Optimized Quantum Representation for Color Digital Images. *International Journal of Theoretical Physics*, 57, 2018: 2938–2948.

[11] Sang J Z; Wang S; Li Q. A Novel Quantum Representation of Color Digital Images. *Quantum Information Processing*, 16(2), 2017: 42.

[12] Wang L; Ran Q W; Ma J; et al. QRCI: A New Quantum Representation Model of Color Digital Images. *Optics Communications*, 438, 2019: 147–158.

[13] Sun B; Le P Q; Iliyasu A M; et al. A Multi-Channel Representation for Images on Quantum Computers Using the RGBα Color Space. *IEEE International Symposium on Intelligent Signal Processing, Floriana, Malta. IEEE,* 2011: 1–6.

[14] Le P Q; Iliyasu A M; Dong F Y; et al. Efficient Color Transformations on Quantum Images. *Journal of Advanced Computational Intelligence and Intelligent Informatics*, 15(6), 2011: 698–706.

[15] Le P Q; Iliyasu A M; Dong F Y; et al. Fast Geometric Transformations on Quantum Images. *International Journal of Applied Mathematics*, 40(3), 2010: 113–123.

[16] Le P Q; Iliyasu A M; Dong F Y; et al. Strategies for Designing Geometric Transformations on Quantum Images. *Theoretical Computer Science*, 412(15), 2011: 1406–1418.

[17] Li H S; Fan P; Xia H Y; et al. Quantum Implementation Circuits of Quantum Signal Representation and Type Conversion. *IEEE Transactions on Circuits and Systems I: Regular Papers*, 66(1), 2018: 341–354.

[18] Yan F; Iliyasu A M; Venegas-Andraca S E. A Survey of Quantum Image Representations. *Quantum Information Processing*, 15, 2016: 1–35.

[19] Su J; Guo X C; Liu C Q; et al. A New Trend of Quantum Image Representations. *IEEE Access*, 8, 2020: 214520–214537.

[20] Venegas-Andraca S E; Ball J L. Processing Images in Entangled Quantum Systems. *Quantum Information Processing*, 9(1), 2010: 1–11.

[21] Zhu H H; Chen X B; Yang Y X. Image Preparations of Multi-Mode Quantum Image Representation and Their Application on Quantum Image Reproduction. *Optik*, 251, 2022: 168321.

[22] Chen G; Song X. Quantum Color Image Scaling on QIRHSI Model. *Data Science: ICPCSEE 2021. Springer*, Taiyuan, China, 2021: 453–467.

4 Geometric Operations of Quantum Images

Zhixin Zhang, Xianhua Song, Ahmed A. Abd El-Latif, Bassem Abd-El-Atty, and Basma Abd El-Rahiem

4.1 BASIC GEOMETRIC TRANSFORMATIONS OF QIRHSI[6]

4.1.1 Two-Point Exchange

Definition 4.1: The two-point exchange operator G_p based on the QIRHSI color image is defined as

$$\begin{aligned} G_p\left(\left|I(\theta)\right\rangle\right) &= \frac{1}{2^n}\sum_{k=0}^{2^{2n}-1}\left|H_k\right\rangle\left|S_k\right\rangle\left|I_k\right\rangle \otimes P\left(\left|k\right\rangle\right) \\ &= \frac{1}{2^n}\left\{\left|H_j\right\rangle\left|S_j\right\rangle\left|I_j\right\rangle \otimes \left|i\right\rangle + \left|H_i\right\rangle\left|S_i\right\rangle\left|I_i\right\rangle \otimes \left|j\right\rangle \right. \\ &\quad \left. + \sum_{k=0,k\neq i,j}^{2^{2n}-1}\left|H_k\right\rangle\left|S_k\right\rangle\left|I_k\right\rangle \otimes \left|k\right\rangle\right\} \end{aligned} \tag{4.1}$$

where $\left|I(\theta)\right\rangle$ represents a QIRHSI color image, as shown in Equation 2.1. The two-point exchange operator G_p can also be expressed as

$$G_p = I^{\otimes 2} \otimes I^{\otimes q} \otimes P = I^{\otimes 2} \otimes I^{\otimes q} \otimes \left\{\left|i\right\rangle\left\langle j\right| + \left|j\right\rangle\left\langle i\right|\right\} + \sum_{k=0,k\neq i,j}^{2^{2n}-1}\left|k\right\rangle\left\langle k\right| \tag{4.2}$$

It can be known from Equation 4.2 that $P\left(\left|k\right\rangle\right) = \left|k\right\rangle, k \neq i, j$, and $P\left(\left|i\right\rangle\right) = \left|j\right\rangle, P\left(\left|j\right\rangle\right) = \left|i\right\rangle$.

To design the quantum circuit of the two-point exchange operation G_p based on the quantum image QIRHSI, the assistance of Gray codes is required. Suppose for two different binary numbers $i = i_{2n-1}\cdots i_1 i_0$ and $j = j_{2n-1}\cdots j_1 j_0$, a set of Gray codes connecting i and j is a set of binary numbers that originate from i and end at j, and in this set of binary numbers, exactly two adjacent binary numbers differ from each other by 1 bit. For the $2n$-bit binary number $i = 0$ and the $2n$-bit binary number $j = 2^{2n} - 1$, their binary expansions are shown in Equation 4.3:

$$i = \underbrace{0\cdots0\cdots0}_{2n},\ j = \underbrace{1\cdots1\cdots1}_{2n} \tag{4.3}$$

DOI: 10.1201/9781003621423-5

There exists a set of Gray codes as shown in Equation 4.4:

$$\begin{matrix} 0 & 0 & \cdots & 0 & 0 & 0 & \cdots & 0 & 0 \\ 0 & 0 & \cdots & 0 & 0 & 0 & \cdots & 0 & 1 \\ \vdots & \vdots & & \vdots & \vdots & \vdots & & \vdots & \vdots \\ 0 & 0 & \cdots & 0 & 1 & 1 & \cdots & 1 & 1 \\ \vdots & \vdots & & \vdots & \vdots & \vdots & & \vdots & \vdots \\ 0 & 1 & \cdots & 1 & 1 & 1 & \cdots & 1 & 1 \\ 1 & 1 & \cdots & 1 & 1 & 1 & \cdots & 1 & 1 \end{matrix} \tag{4.4}$$

Denote g_1 to g_m as the elements of the Gray code connecting i and j, and $g_1 = i$ to $g_m = j$. A Gray code that satisfies the condition $m \le 2n+1$ can always be found, because i and j have at most $2n$ positions that are inconsistent.

Here, the implementation of the two-point exchange circuit of the quantum color image QIRHSI with a total of 2^{2n} pixels as shown in Equation 2.1 is taken as an example for illustration. Let us assume that the coordinates of the two pixels to be exchanged in the QIRHSI image are $|i\rangle = |0\rangle$ and $|j\rangle = |2^{2n}-1\rangle$, and the binary expansions of i and j are as shown in Equation 4.3. The Gray code connecting i and j is as shown in Equation 4.4. $|g_1\rangle = |i\rangle = |0\rangle$ to $|g_{2n+1}\rangle = |j\rangle = |2^{2n}-1\rangle$ are the elements of this Gray code. Here, it is only necessary to achieve the state transformation through a series of quantum gates:

$$|g_1\rangle \to |g_2\rangle \to \cdots \to |g_{2n}\rangle \tag{4.5}$$

Then perform the NOT gate operation on multiple control bits. The bit where $|g_{2n}\rangle$ and $|g_{2n+1}\rangle$ are exactly different is the target qubit. Then, restore the operations in the first stage and carry out the operations:

$$|g_{2n}\rangle \to |g_{2n+1}\rangle \to \cdots \to |g_1\rangle \tag{4.6}$$

Finally, the exchange of the pixels $|i\rangle = |0\rangle$ and $|j\rangle = |2^{2n}-1\rangle$ in the quantum color image QIRHSI is achieved. The quantum circuit for exchanging the pixels $|i\rangle = |0\rangle$ and $|j\rangle = |2^{2n}-1\rangle$ in the quantum color image QIRHSI is shown in Figure 4.1.

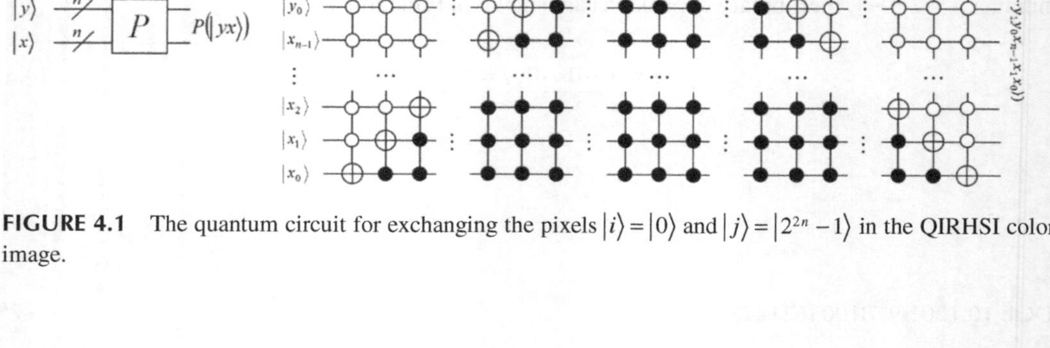

FIGURE 4.1 The quantum circuit for exchanging the pixels $|i\rangle = |0\rangle$ and $|j\rangle = |2^{2n}-1\rangle$ in the QIRHSI color image.

As there are usually multiple cases of Gray codes connecting i and j, the quantum circuit for exchanging pixels $|i\rangle$ and $|j\rangle$ is not fixed. Taking $i = 0010$ and $j = 1111$ as examples, two groups of Gray codes are listed as shown in Equations 4.7 and 4.8:

$$\begin{matrix} 0 & 0 & 1 & 0 \\ 0 & 0 & 1 & 1 \\ 0 & 1 & 1 & 1 \\ 1 & 1 & 1 & 1 \end{matrix} \tag{4.7}$$

and

$$\begin{matrix} 0 & 0 & 1 & 0 \\ 1 & 0 & 1 & 0 \\ 1 & 1 & 1 & 0 \\ 1 & 1 & 1 & 1 \end{matrix} \tag{4.8}$$

Theorem 4.1: The complexity of the quantum gates required for the two-point exchange operator G_p using Gray codes on the quantum image QIRHSI is $O(n^2)$.

Proof: Without loss of generality, assume that the pixel positions to be exchanged by the two-point exchange operator G_p in the quantum image QIRHSI are $|i\rangle$ and $|j\rangle$. Let the elements of a set of Gray codes connecting the $2n$-bit binary numbers i and j be $g_1, g_2, \cdots, g_{m-1}, g_m$, where $g_1 = 0$ and $g_m = 2^{2n} - 1$. A series of quantum gates are needed to perform the state transformation step by step.

$$|g_1\rangle \rightarrow |g_2\rangle \rightarrow \cdots \rightarrow |g_{m-1}\rangle$$

Then, perform the multi-controlled NOT gate operation, and then carry out the transformation:

$$|g_{m-1}\rangle \rightarrow |g_{m-2}\rangle \rightarrow \cdots \rightarrow |g_1\rangle$$

In this way, the G_p operator is completed, and the complexity of the G_p operator is $O(n^2)$. Next, we will analyze the complexity of this quantum circuit in detail.

First, swap the states of $|g_1\rangle$ and $|g_2\rangle$. If the values of $|g_1\rangle$ and $|g_2\rangle$ are different at the lth bit, then it is only necessary to flip the controlled bit of the lth qubit to complete the swap operation. The condition for this is that the qubits of the remaining bits of $|g_1\rangle$ and $|g_2\rangle$ are all the same. Next, use a controlled operation to swap $|g_2\rangle$ and $|g_3\rangle$, and perform such operations one by one until the swap between $|g_{m-2}\rangle$ and $|g_{m-1}\rangle$ is completed. The above $m-2$ operations are to complete the operations in Equation 4.9:

$$\begin{aligned} &|g_1\rangle \rightarrow |g_{m-1}\rangle \\ &|g_2\rangle \rightarrow |g_1\rangle \\ &|g_3\rangle \rightarrow |g_2\rangle \\ &\cdots\cdots \\ &|g_{m-1}\rangle \rightarrow |g_{m-2}\rangle \end{aligned} \tag{4.9}$$

It should be clearly explained that the entire other states of the computational basis remain unchanged during the above-mentioned series of operations. Secondly, assume that $|g_{m-1}\rangle$ and $|g_m\rangle$ are different at the wth bit. Under the condition that the remaining qubits are the same, perform the multi-controlled NOT gate operation on the target qubit at the wth bit. The multi-controlled NOT gate operation is completed by using swap operations: swap $|g_{m-1}\rangle$ and $|g_{m-2}\rangle$, then continue to swap $|g_{m-2}\rangle$ and $|g_{m-3}\rangle$, and so on, until swapping $|g_2\rangle$ and $|g_1\rangle$.

As there are at most $2n$ different positions between i and j, it is certain that a Gray code satisfying $m \leq 2n+1$ can be found. To achieve two-level unitary operations, no more than $2(2n-1)$ controlled-NOT gate operations are required to swap $|g_1\rangle$ to $|g_{m-1}\rangle$ and then swap back. Each such controlled operation can be implemented using $O(n)$ NOT gates and multi-controlled NOT gates. Implementing the multi-controlled NOT gate operation with the wth qubit as the target also requires $O(n)$ basic quantum gates. Therefore, achieving the transformation from $|g_1\rangle$ to $|g_{m-1}\rangle$ requires $O(n^2)$ basic quantum gates. Thus, the complexity of the quantum gates required for the quantum circuit to implement the G_p operator based on the quantum color image QIRHSI is $O(n^2)$.

□

4.1.2 Cyclic Translation

The cyclic translation of the QIRHSI image along the coordinate axes is as defined in Definition 4.2. Which gives examples of the cyclic translation of the image along the coordinate axes.

Definition 4.2: The operators T_{y+1} and T_{x+1} for the cyclic translation of the QIRHSI color image by l pixels along the y-axis and x-axis are defined as follows, respectively (Figure 4.2):

$$T_{y+1}\left(|I(\theta)\rangle\right) = \frac{1}{2^n}\sum_{y=0}^{2^n-1}\sum_{x=0}^{2^n-1}|H_{yx}\rangle|S_{yx}\rangle|I_{yx}\rangle \otimes |(y+1) \bmod 2^n, x\rangle \tag{4.10}$$

$$T_{x+1}\left(|I(\theta)\rangle\right) = \frac{1}{2^n}\sum_{y=0}^{2^n-1}\sum_{x=0}^{2^n-1}|H_{yx}\rangle|S_{yx}\rangle|I_{yx}\rangle \otimes |y, (x+1) \bmod 2^n\rangle \tag{4.11}$$

where $|I(\theta)\rangle$ represents a QIRHSI color image, as shown in Equation 2.1. $l = l_{n-1} \cdots l_1 l_0$ and $l \in (0, 2^n - 1)$, so $l_i\ (i = 0, 1, \cdots n-1)$ are not all zero. The cyclic shift operators T_{y+1} and T_{x+1} can be expressed as, respectively

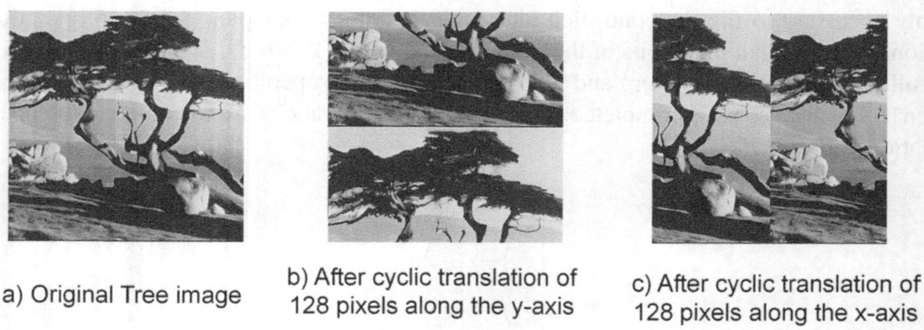

FIGURE 4.2 The image Tree is cyclically translated along the coordinate axes.

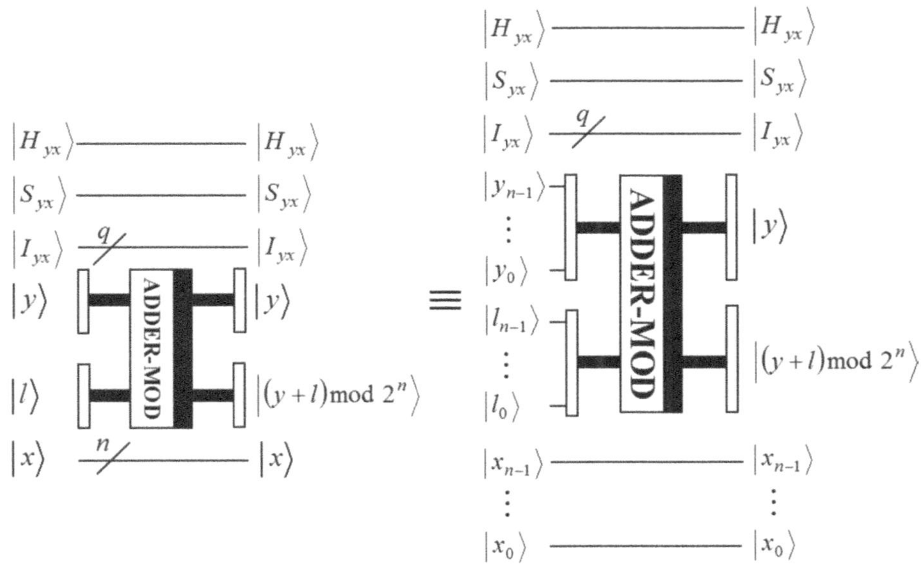

FIGURE 4.3 The quantum circuit for translating the QIRHSI image by l pixels along the *y*-axis.

$$\begin{aligned}T_{y+l} &= I^{\otimes 2} \otimes I^{\otimes q} \otimes \sum_{j=0}^{2^n-1} |(j+l)\text{mod } 2^n\rangle\langle j| \otimes I^{\otimes n} \\ &= I^{\otimes 2} \otimes I^{\otimes q} \otimes \left\{|(0+l)\text{mod } 2^n\rangle\langle 0| + \cdots + |(2^n-1+l)\text{mod } 2^n\rangle\langle 2^n-1|\right\} \otimes I^{\otimes n}\end{aligned} \tag{4.12}$$

$$\begin{aligned}T_{x+l} &= I^{\otimes 2} \otimes I^{\otimes q} \otimes I^{\otimes n} \otimes \sum_{j=0}^{2^n-1} |(j+l)\text{mod } 2^n\rangle\langle j| \\ &= I^{\otimes 2} \otimes I^{\otimes q} \otimes I^{\otimes n} \otimes \left\{|(0+l)\text{mod } 2^n\rangle\langle 0| + \cdots + |(2^n-1+l)\text{mod } 2^n\rangle\langle 2^n-1|\right\}\end{aligned} \tag{4.13}$$

The quantum circuits of the cyclic shift operators T_{y+1} and T_{x+1} are shown in Figures 4.3 and 4.4.

Theorem 4.2: The complexity of the quantum gates required for the cyclic shift operators T_{y+1} and T_{x+1} on the quantum image QIRHSI is $O(n)$.

Proof: To calculate the number of quantum gates required for the cyclic shift operators T_{y+1} and T_{x+1}, it is only necessary to calculate the number of basic quantum gates required for the modulo 2^n adder.

As a Toffoli gate is equivalent to 6 CNOT gates[1]. In the basic carry and sum operations (see Figure 1.5), the numbers of CNOT gates required are 13 and 2, respectively. Also, a common adder (see Figure 1.4) contains $2n-1$ carry operations, n sum operations and one CNOT gate. Therefore, the number of CNOT gates required for the common adder is $28n-12$, which has a linear relationship with the input n of the quantum circuit.

The modulo 2^n adder (see Figure 1.6) contains five common adders, 2 NOT gates, and at most no more than $2n-4$ CNOT gates. Therefore, the modulo 2^n adder requires 2 NOT gates and at most $142n-56$ CNOT gates, that is, the complexity of the quantum gates required for the modulo 2^n adder is $O(n)$. In other words, the complexity of the quantum gates required for the cyclic shift operators T_{y+1} and T_{x+1} is $O(n)$.

□

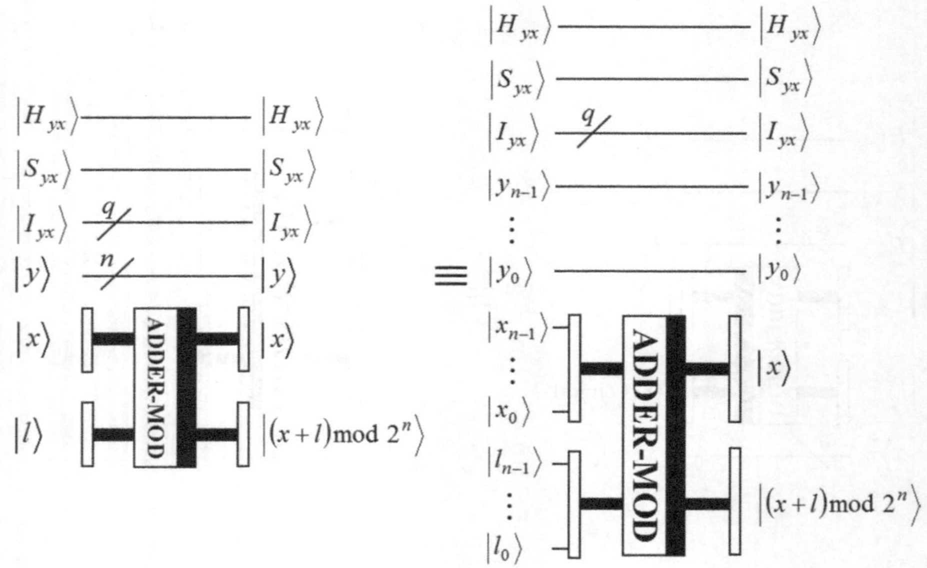

FIGURE 4.4 The quantum circuit for translating the QIRHSI image by l pixels along the x-axis.

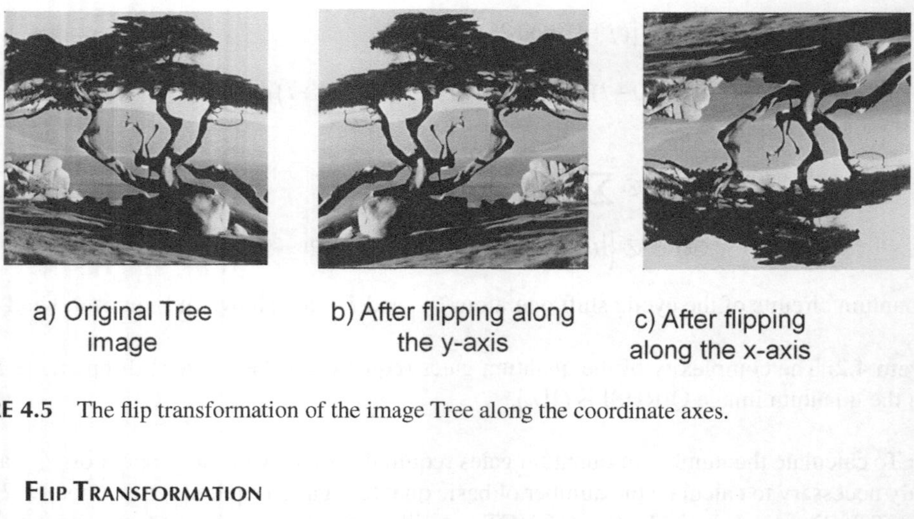

FIGURE 4.5 The flip transformation of the image Tree along the coordinate axes.

4.1.3 Flip Transformation

The flip transformation of the QIRHSI image along the coordinate axes is as defined in Definition 4.3. Figure 4.5 gives an example of the flip transformation of the image along the coordinate axes.

Definition 4.3: The flip transformation operators F_y and F_x for the QIRHSI color image along the y-axis and x-axis are defined as follows, respectively:

$$\begin{aligned} F_y\left(\left|I(\theta)\right\rangle\right) &= \frac{1}{2^n}\sum_{y=0}^{2^n-1}\sum_{x=0}^{2^n-1}\left|H_{yx}\right\rangle\left|S_{yx}\right\rangle\left|I_{yx}\right\rangle \otimes \left|y\bar{x}\right\rangle \\ &= \frac{1}{2^n}\sum_{y=0}^{2^n-1}\sum_{x=0}^{2^n-1}\left|H_{yx}\right\rangle\left|S_{yx}\right\rangle\left|I_{yx}\right\rangle \otimes \left|y, 2^n-1-x\right\rangle \end{aligned} \tag{4.14}$$

$$
\begin{aligned}
F_x\left(|I(\theta)\rangle\right) &= \frac{1}{2^n}\sum_{y=0}^{2^n-1}\sum_{x=0}^{2^n-1}|H_{yx}\rangle|S_{yx}\rangle|I_{yx}\rangle\otimes|y\bar{x}\rangle \\
&= \frac{1}{2^n}\sum_{y=0}^{2^n-1}\sum_{x=0}^{2^n-1}|H_{yx}\rangle|S_{yx}\rangle|I_{yx}\rangle\otimes|2^n-1-y,x\rangle
\end{aligned}
\tag{4.15}
$$

where $|I(\theta)\rangle$ represents a quantum color image QIRHSI, as shown in Equation 2.1. Also,

$$
\begin{gathered}
|y\rangle = |y_{n-1}\cdots y_1 y_0 \\
|x\rangle = |x_{n-1}\cdots x_1 x_0 \\
|\bar{y}\rangle = |\bar{y}_{n-1}\cdots \bar{y}_1\bar{y}_0 = |2^n-1-y\rangle \\
|\bar{x}\rangle = |\bar{x}_{n-1}\cdots \bar{x}_1\bar{x}_0 = |2^n-1-x\rangle \\
\bar{y}_i = 1-y_i, \bar{x}_i = 1-x_i \\
i = 0,1,\cdots,n-1
\end{gathered}
$$

The flip transformation operators F_y and F_x are expressed as, respectively,

$$F_y = I^{\otimes 2}\otimes I^{\otimes q}\otimes I^{\otimes n}\otimes X^{\otimes n} \tag{4.16}$$

$$F_x = I^{\otimes 2}\otimes I^{\otimes q}\otimes X^{\otimes n}\otimes I^{\otimes n} \tag{4.17}$$

The quantum circuits of the flip transformation operators F_y and F_x are shown in Figures 4.6 and 4.7, respectively.

Theorem 4.3: The complexity of the quantum gates required for the flip transformation operators F_y and F_x on the quantum image QIRHSI is $O(n)$.

Proof: As shown in Equations 4.16 and 4.17, the flip operators F_y and F_x both use n NOT gates. Therefore, the complexity of the quantum gates required to implement the flip operators F_y and F_x on the QIRHSI image is $O(n)$.

□

The flip transformations of the QIRHSI image along the axis $y = x$ and the axis $y = -x$ are as defined in Definition 4.4. Figure 4.8 shows examples of the flip transformations of the image along the axis $y = x$ and the axis $y = -x$.

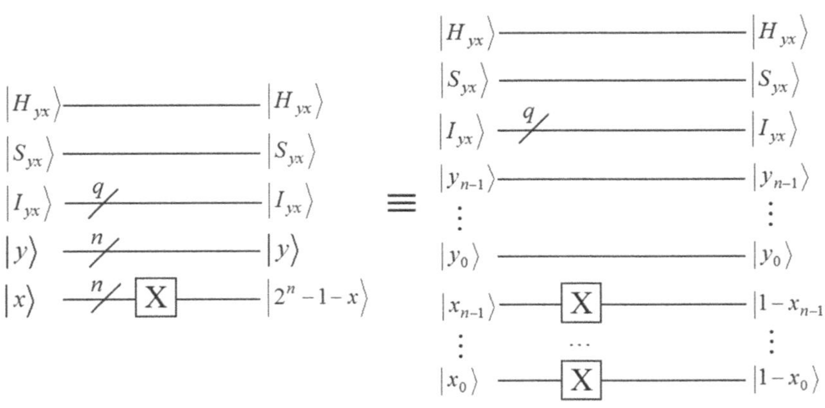

FIGURE 4.6 The quantum circuit for flipping the QIRHSI image along the y-axis.

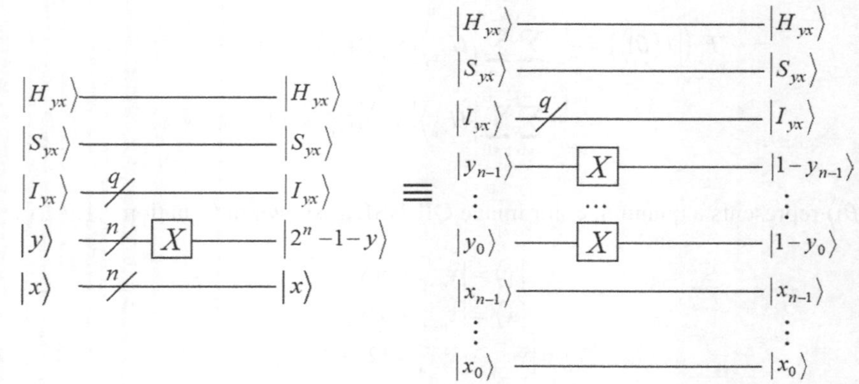

FIGURE 4.7 The quantum circuit for flipping the QIRHSI image along the x-axis.

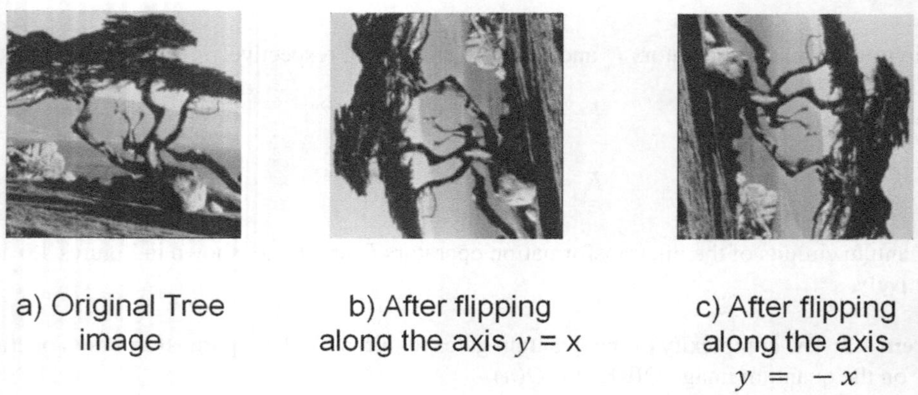

FIGURE 4.8 The flip transformations of the Tree image along the axis $y = x$ and the axis $y = -x$.

Definition 4.4: The flip operators $F_{y=x}$ and $F_{y=-x}$ for the QIRHSI color image along the axis $y = x$ and the axis $y = -x$ are defined as follows, respectively:

$$\begin{aligned} F_{y=x}\left(|I(\theta)\rangle\right) &= \frac{1}{2^n}\sum_{y=0}^{2^n-1}\sum_{x=0}^{2^n-1}\left|H_{yx}\right\rangle\left|S_{yx}\right\rangle\left|I_{yx}\right\rangle \otimes V\left(|yx\rangle\right) \\ &= \frac{1}{2^n}\sum_{y=0}^{2^n-1}\sum_{x=0}^{2^n-1}\left|H_{yx}\right\rangle\left|S_{yx}\right\rangle\left|I_{yx}\right\rangle \otimes |xy\rangle \end{aligned} \tag{4.18}$$

$$\begin{aligned} F_{y=-x}\left(|I(\theta)\rangle\right) &= \frac{1}{2^n}\sum_{y=0}^{2^n-1}\sum_{x=0}^{2^n-1}\left|H_{yx}\right\rangle\left|S_{yx}\right\rangle\left|I_{yx}\right\rangle \otimes \left|\overline{xy}\right\rangle \\ &= \frac{1}{2^n}\sum_{y=0}^{2^n-1}\sum_{x=0}^{2^n-1}\left|H_{yx}\right\rangle\left|S_{yx}\right\rangle\left|I_{yx}\right\rangle \otimes \left|2^n-1-x, 2^n-1-y\right\rangle \end{aligned} \tag{4.19}$$

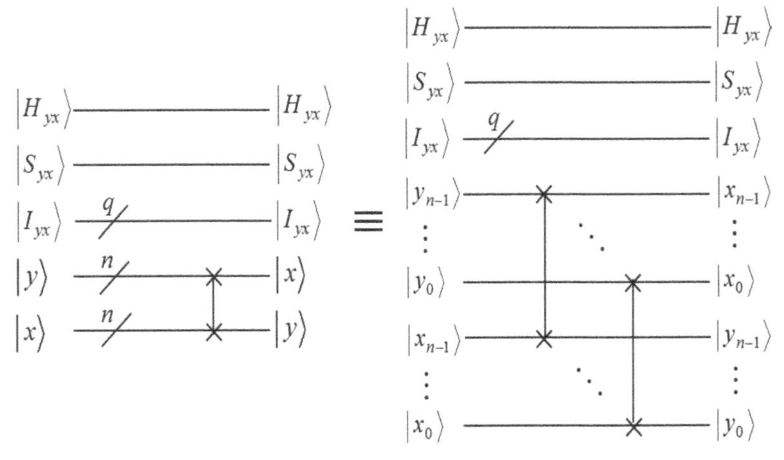

FIGURE 4.9 The quantum circuit for flipping the QIRHSI image along the axis $y = x$.

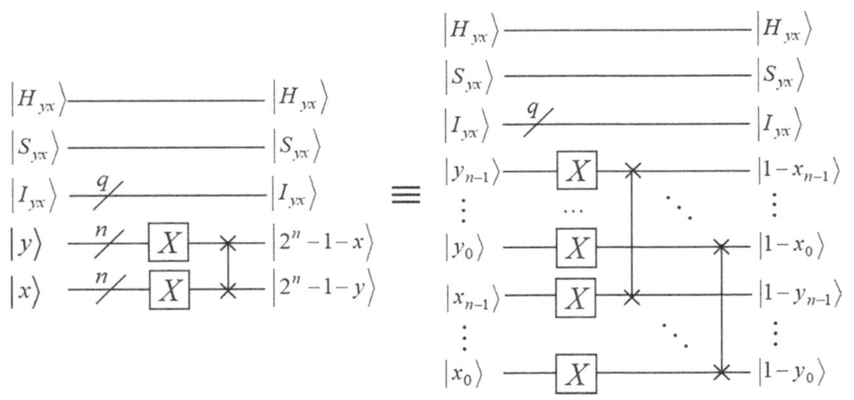

FIGURE 4.10 The quantum circuit for flipping the QIRHSI image along the axis $y = -x$.

where $|I(\theta)\rangle$ represents a QIRHSI color image, as shown in Equation 2.1. The flip transformation operators $F_{y=x}$ and $F_{y=-x}$ can be expressed as follows, respectively:

$$F_{y=x} = I^{\otimes 2} \otimes I^{\otimes q} \otimes V \tag{4.20}$$

$$F_{y=-x} = F_{y=x} F_y F_x \tag{4.21}$$

The quantum circuits of the flip transformation operators $F_{y=x}$ and $F_{y=-x}$ are shown in Figures 4.9 and 4.10, respectively. The quantum circuit of the operator $x_{n+1} = \eta x_n (1 - x_n)$ is shown in Figure 4.11.

Theorem 4.4: The complexity of the quantum gates required for the flip operators $F_{y=x}$ and $F_{y=-x}$ on the quantum image QIRHSI is $O(n)$.

Proof: As shown in Equations 4.20 and 4.21, the flip operator $F_{y=x}$ uses $F_{y=-x}$ swap gates, and the flip operator $F_{y=-x}$ uses $2n$ NOT gates and n swap gates. As 1 swap gate is equivalent to 3 CNOT gates,

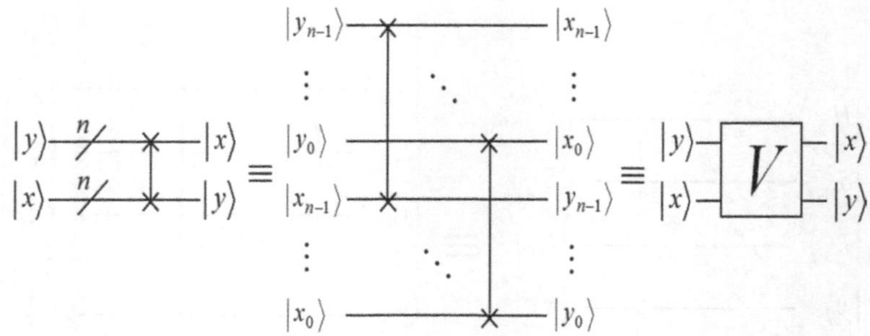

FIGURE 4.11 The quantum circuit of the operator V.

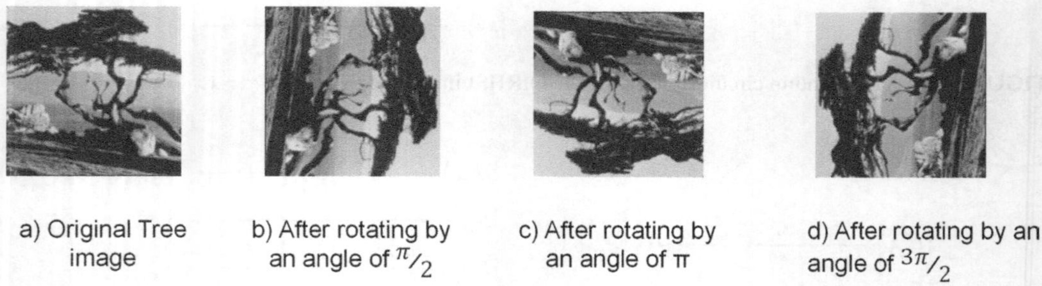

FIGURE 4.12 The image of the Tree after right-angle rotation.

the complexity of the quantum gates required to implement the flip operators $F_{y=x}$ and $F_{y=-x}$ on the QIRHSI image is $O(n)$.

□

4.1.4 Right-Angle Rotation

The transformations of the QIRHSI image with rotation angles of $\pi/2, \pi, 3\pi/2$ are as defined in Definition 4.5. Figure 4.12 shows examples of the image transformations with rotation angles of $\pi/2, \pi, 3\pi/2$.

Definition 4.5: The right-angle rotation operators $\pi/2, \pi, 3\pi/2$ for the QIRHSI color image with rotation angles of $R_{\pi/2}, R_{\pi}, R_{3\pi/2}$ are defined as follows, respectively:

$$\begin{aligned} R_{\pi/2}\left(|I(\theta)\rangle\right) &= \frac{1}{2^n}\sum_{y=0}^{2^n-1}\sum_{x=0}^{2^n-1}|H_{yx}\rangle|S_{yx}\rangle|I_{yx}\rangle\otimes|x\bar{y}\rangle \\ &= \frac{1}{2^n}\sum_{y=0}^{2^n-1}\sum_{x=0}^{2^n-1}|H_{yx}\rangle|S_{yx}\rangle|I_{yx}\rangle\otimes|x,2^n-1-y\rangle \end{aligned} \tag{4.22}$$

$$\begin{aligned} R_{\pi}\left(|I(\theta)\rangle\right) &= \frac{1}{2^n}\sum_{y=0}^{2^n-1}\sum_{x=0}^{2^n-1}|H_{yx}\rangle|S_{yx}\rangle|I_{yx}\rangle\otimes|\overline{yx}\rangle \\ &= \frac{1}{2^n}\sum_{y=0}^{2^n-1}\sum_{x=0}^{2^n-1}|H_{yx}\rangle|S_{yx}\rangle|I_{yx}\rangle\otimes|2^n-1-y,2^n-1-x\rangle \end{aligned} \tag{4.23}$$

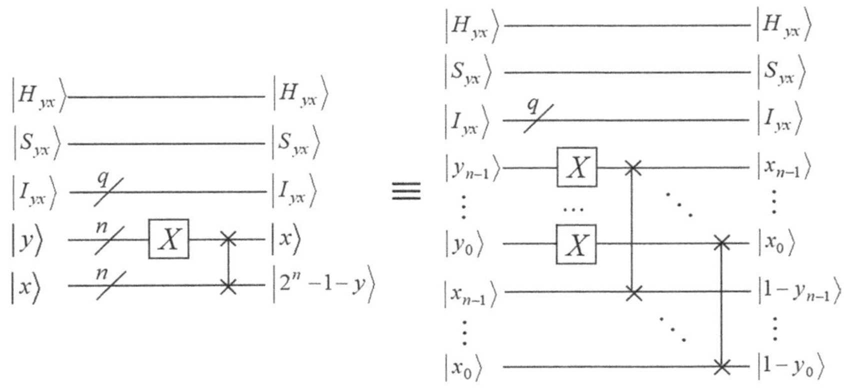

FIGURE 4.13 The quantum circuit for rotating the QIRHSI image by an angle of $\pi/2$.

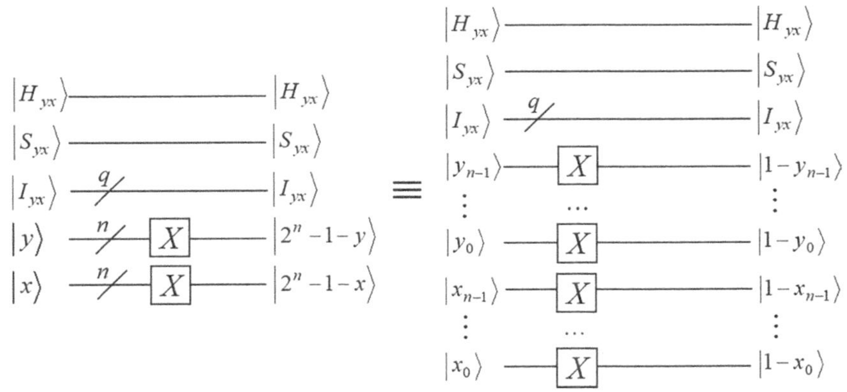

FIGURE 4.14 The quantum circuit for rotating the QIRHSI image by an angle of π.

$$
\begin{aligned}
R_{3\pi/2}\left(|I(\theta)\rangle\right) &= \frac{1}{2^n}\sum_{y=0}^{2^n-1}\sum_{x=0}^{2^n-1}|H_{yx}\rangle|S_{yx}\rangle|I_{yx}\rangle\otimes|\bar{x}y\rangle \\
&= \frac{1}{2^n}\sum_{y=0}^{2^n-1}\sum_{x=0}^{2^n-1}|H_{yx}\rangle|S_{yx}\rangle|I_{yx}\rangle\otimes|2^n-1-x,y\rangle
\end{aligned}
\tag{4.24}
$$

where $|I(\theta)\rangle$ represents a QIRHSI color image, as shown in Equation 2.1. The right-angle rotation operators $R_{\pi/2}$, R_{π}, and $R_{3\pi/2}$ can be expressed as follows, respectively:

$$R_{\pi/2} = F_{y=x}F_x \tag{4.25}$$

$$R_{\pi} = F_yF_x \tag{4.26}$$

$$R_{3\pi/2} = F_{y=x}F_y \tag{4.27}$$

The quantum circuits of the operators $R_{\pi/2}$, R_{π}, and $R_{3\pi/2}$ are shown in Figures 4.13–4.15, respectively.

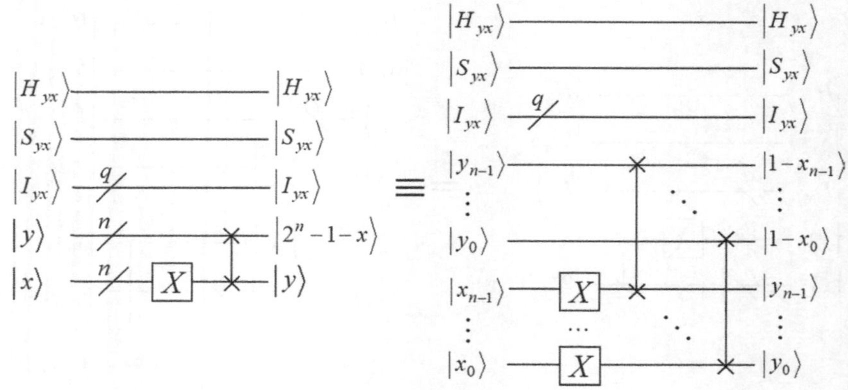

FIGURE 4.15 The quantum circuit for rotating the QIRHSI image by an angle of $3\pi/2$.

TABLE 4.1
Comparison of the Complexity of Quantum Gates Required for Geometric: Transformations under Different Quantum Image Models

Geometric Transformation	Based on the QIRHSI Representation	Based on the NASS Representation	Based on the FRQI Representation
Two-point exchange	$O(n^2)$	$O(n^2)$	$O(n^2)$
Cyclic translation	$O(n)$	$O(2^n n^2)$	–
Flip transformation	$O(n)$	$O(n)$	$O(n)$
Right-angle rotation	$O(n)$	$O(n)$	$O(n)$

Theorem 4.5: The complexity of the quantum gates required for the right-angle rotation operators $R_{\pi/2}$, R_{π}, and $R_{3\pi/2}$ on the quantum image QIRHSI is $O(n)$.

Proof: As shown in Equations 4.25–4.27, the right-angle rotation operators $R_{\pi/2}$, R_{π}, and $R_{3\pi/2}$ are considered. The operator $R_{\pi/2}$ uses n NOT gates and n swap gates, the operator R_{π} uses n swap gates, and the operator $R_{3\pi/2}$ uses n NOT gates and n swap gates. Therefore, the complexity of the quantum gates required to implement the right-angle rotation operations $R_{\pi/2}$, R_{π}, and $R_{3\pi/2}$ on the QIRHSI image is $O(n)$.

□

4.1.5 Complexity Analysis

Reference [2] has designed the geometric transformation of two-dimensional grayscale images in the FRQI model, and reference [3] has constructed the geometric transformation operations for multi-dimensional color images in the NASS model. This paper presents the geometric transformation operations for two-dimensional QIRHSI quantum color images. Table 4.1 compares the complexity of quantum gates required for geometric transformation operations based on the FRQI representation, the NASS representation, and the QIRHSI representation. It should be noted that the size of all the images is $2^n \times 2^n$.

By observing Table 4.1, it can be seen that the complexity of quantum gates required for the geometric transformations (two-point exchange, flip transformation, and right-angle rotation) based

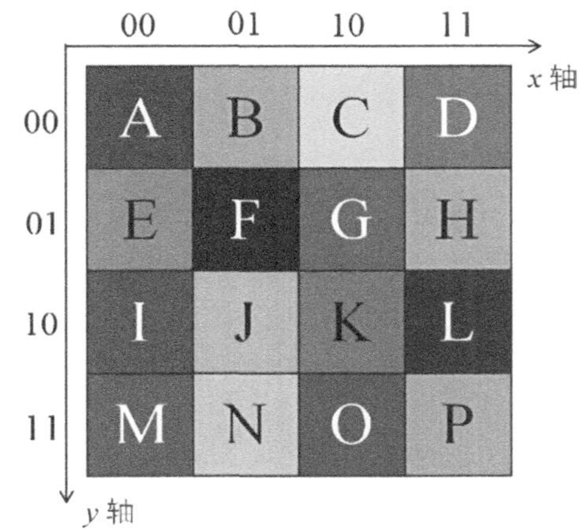

FIGURE 4.16 A 4 × 4 QIRHSI color image.

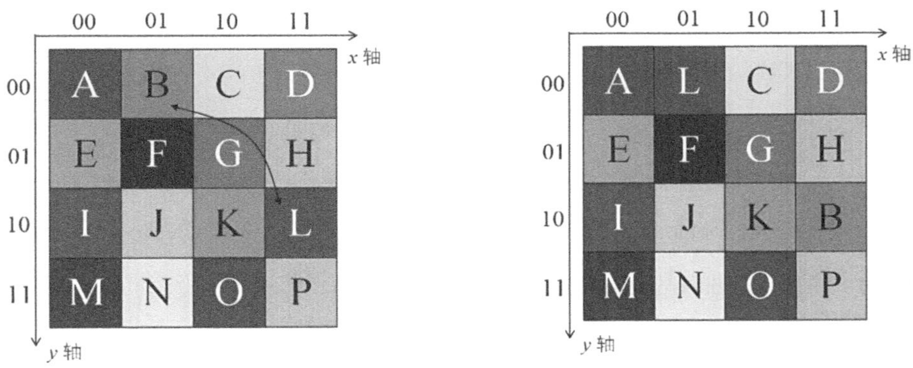

a) Original image

b) After Two-point Exchange

FIGURE 4.17 Two-point exchange.

on the QIRHSI representation is the same as that based on the FRQI representation and the NASS representation. Moreover, the complexity of quantum gates required for cyclic translation in the QIRHSI representation is much lower than that in the NASS representation.

4.1.6 Experimental Examples

To make the geometric transformation based on the QIRHSI image more intuitive, take the QIRHSI quantum color image with a size of 4 × 4 as shown in Figure 4.16 as an example, where $n = 2$ and $q = 8$. Figure 4.16 presents a 4 × 4 QIRHSI color image, and its representation is shown in Equation 4.28:

$$\left|I(\theta)\right\rangle = \frac{1}{2^2}\sum_{k=0}^{2^4-1}\left|H_k\right\rangle\left|S_k\right\rangle\left|I_k\right\rangle \otimes |k\rangle = \frac{1}{4}\sum_{y=0}^{2^2-1}\sum_{x=0}^{2^2-1}\left|H_{yx}\right\rangle\left|S_{yx}\right\rangle\left|I_{yx}\right\rangle \otimes |yx\rangle \tag{4.28}$$

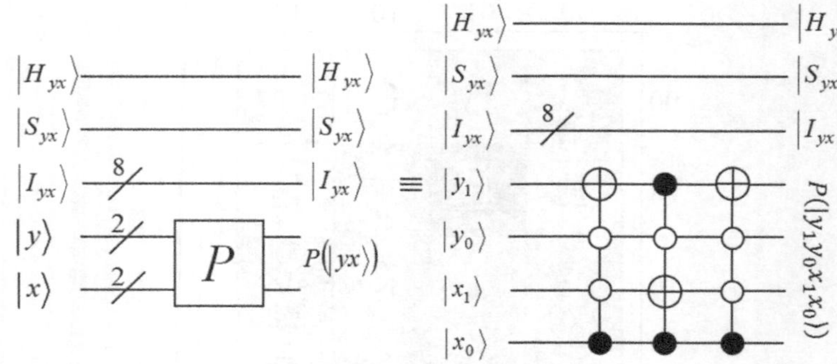

FIGURE 4.18 The quantum circuit for implementing the two-point exchange.

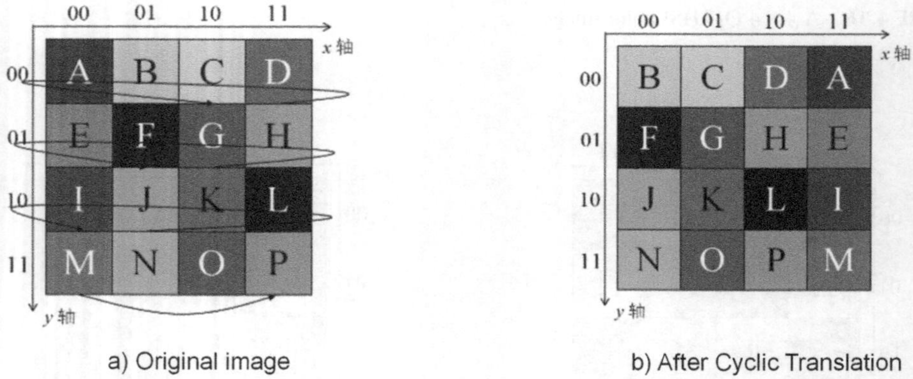

FIGURE 4.19 Translate three pixels along the x-axis.

1. *Two-Point Exchange:* Figure 4.17 shows an example of a QIRHSI color image where two pixels $|i\rangle = |1\rangle = |00\rangle|01\rangle$ and $|j\rangle = |11\rangle = |10\rangle|11\rangle$ are exchanged. Figure 4.18 shows the corresponding quantum circuit. The two-point exchange operator G_p is defined as

$$G_p = I^{\otimes 2} \otimes I^{\otimes 8} \otimes P = I^{\otimes 2} \otimes I^{\otimes 8} \left\{ |1\rangle\langle 11| + |11\rangle\langle 1| + \sum_{k=0,k\neq 1,11}^{15} |k\rangle\langle k| \right\}$$

Applying the operator G_p to the image $|I(\theta)\rangle$ yields

$$G_P |I(\theta)\rangle = \frac{1}{4}\sum_{k=0}^{15} |H_k\rangle|S_k\rangle|I_k\rangle \otimes P(|k\rangle)$$
$$= \frac{1}{4}\left\{ |H_{11}\rangle|S_{11}\rangle|I_{11}\rangle \otimes |1\rangle + |H_1\rangle|S_1\rangle|I_1\rangle \otimes |11\rangle + \sum_{k=0,k\neq 1,11}^{15} |H_k\rangle|S_k\rangle|I_k\rangle \otimes |k\rangle \right\}$$

where $P(|k\rangle) = |k\rangle, k \neq 1,11$, and $P(|1\rangle) = |11\rangle, P(|11\rangle) = |1\rangle$.

2. *Cyclic Translation:* Figure 4.19 shows an example of the cyclic translation of the QIRHSI color image along the x-axis, where $l = 3$. Figure 4.20 presents the corresponding quantum circuit. The cyclic translation operator T_{x+3} is defined as

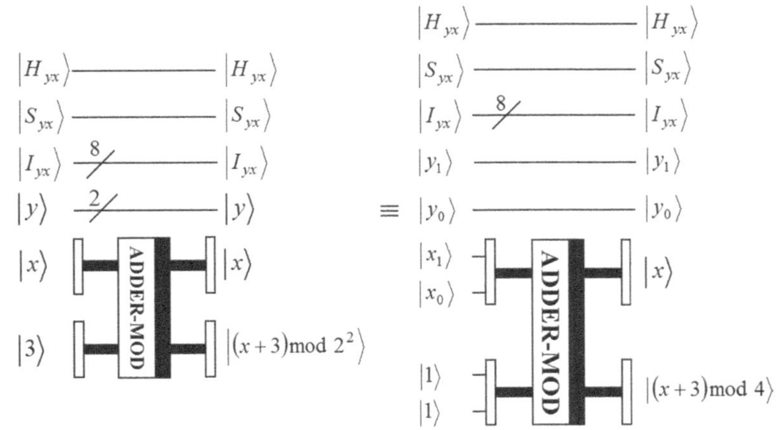

FIGURE 4.20 The quantum circuit for translating three pixels along the x-axis.

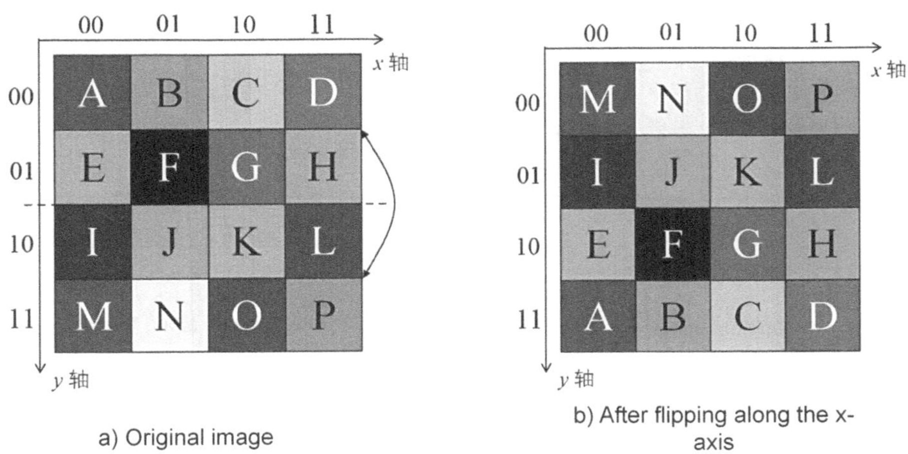

FIGURE 4.21 Flipping along the x-axis.

$$
\begin{aligned}
T_{x+3} &= I^{\otimes 2} \otimes I^{\otimes 8} \otimes I^{\otimes 2} \otimes \sum_{j=0}^{3} |(j+3)\mathrm{mod}4\rangle\langle j| \\
&= I^{\otimes 2} \otimes I^{\otimes 8} \otimes I^{\otimes 2} \otimes (|3\rangle\langle 0| + |0\rangle\langle 1| + |1\rangle\langle 2| + |2\rangle\langle 3|)
\end{aligned}
$$

Applying the operator T_{x+3} to the image $|I(\theta)$, we obtain

$$
T_{x+3}|I(\theta)\rangle = \frac{1}{4}\sum_{y=0}^{3}\sum_{x=0}^{3}|H_k\rangle|S_k\rangle|I_k\rangle \otimes |y,(x+3)\mathrm{mod}\,4\rangle
$$

3. *Flip Transformation:* Figure 4.21 shows an example of the flip of the QIRHSI color image along the x-axis. Figure 4.22 presents the corresponding quantum circuit. The flip transformation operator F_x is defined as

$$
F_x = I^{\otimes 2} \otimes I^{\otimes 8} \otimes X^{\otimes 2} \otimes I^{\otimes 2}
$$

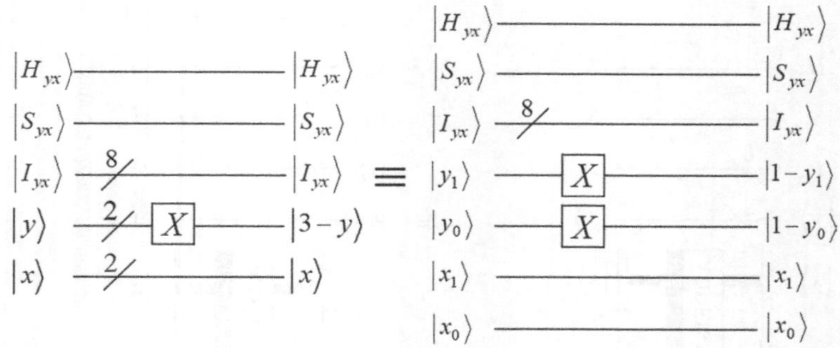

FIGURE 4.22 The quantum circuit for flipping along the x-axis.

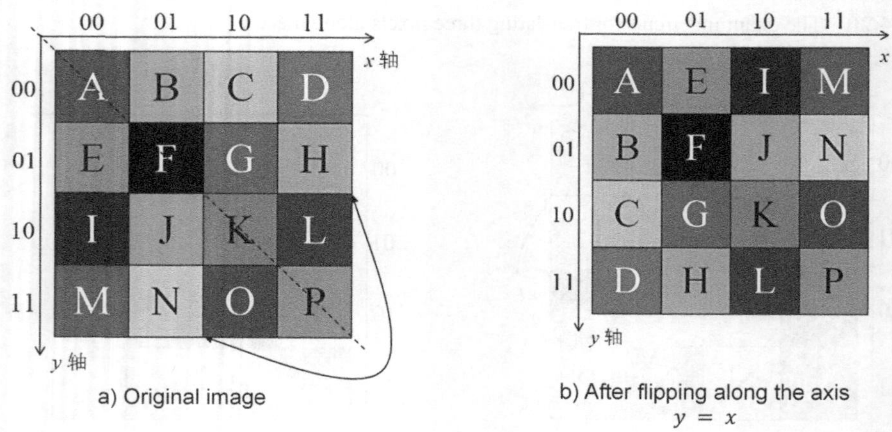

FIGURE 4.23 Flipping along the axis $y = x$.

Applying the flip transformation operator F_x to the image $|I(\theta)\rangle$ can obtain

$$F_x(|I(\theta)\rangle) = \frac{1}{4}\sum_{y=0}^{3}\sum_{x=0}^{3}|H_{yx}\rangle|S_{yx}\rangle|I_{yx}\rangle \otimes |3-y, x\rangle$$

4. *Flip Transformation:* Figure 4.23 shows an example of the flip transformation of the QIRHSI color image along the axis $y = x$. Figure 4.24 presents the corresponding quantum circuit. The flip transformation operator $F_{y=x}$ is defined as

$$F_{y=x} = I^{\otimes 2} \otimes I^{\otimes 8} \otimes V$$

Applying the flip transformation operator $F_{y=x}$ to the image $|I(\theta)\rangle$ can obtain

$$F_{y=x}(|I(\theta)\rangle) = \frac{1}{4}\sum_{y=0}^{3}\sum_{x=0}^{3}|H_{yx}\rangle|S_{yx}\rangle|I_{yx}\rangle \otimes |xy\rangle$$

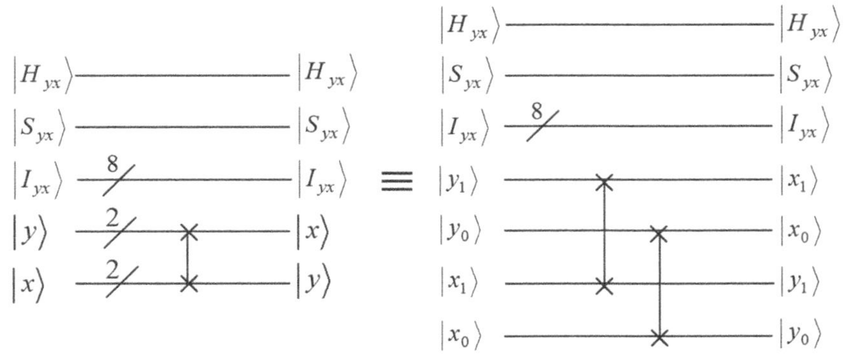

FIGURE 4.24 The quantum circuit of the operator *V*.

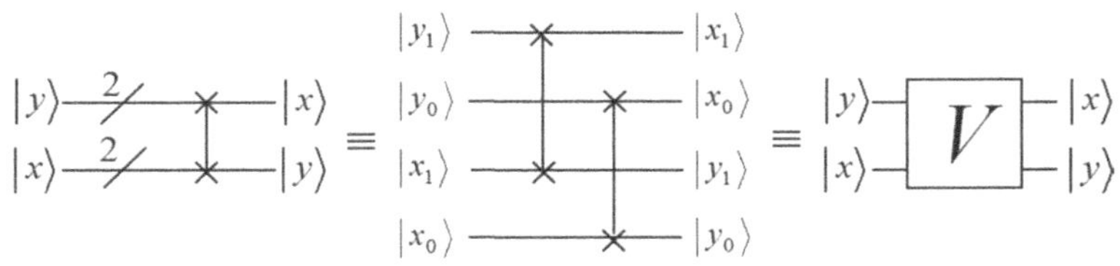

FIGURE 4.25 The quantum circuit for flipping along the axis $y = x$.

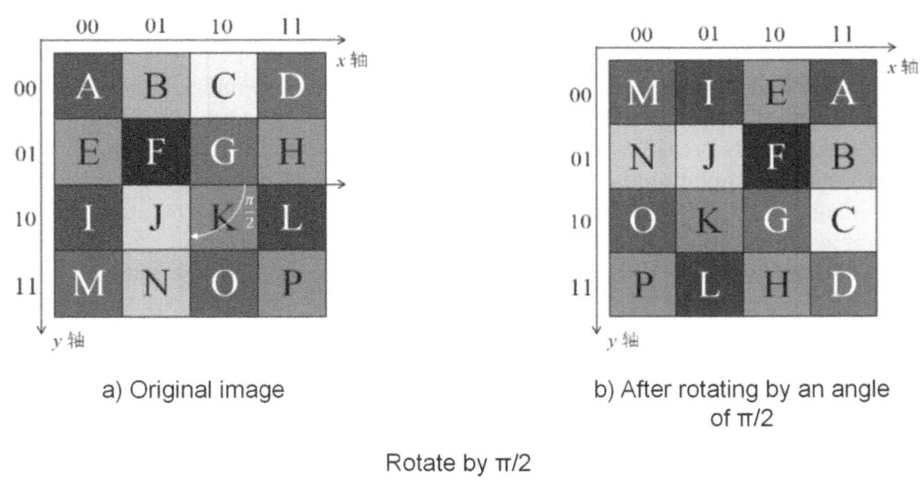

FIGURE 4.26 Rotate by π/2.

Among them, the operator V is shown in Figure 4.25.

5. *Right-Angle Rotation:* Figure 4.26 shows an example of the QIRHSI color image being rotated by an angle of $\pi/2$. Figure 4.27 presents the corresponding quantum circuit. The right-angle rotation operator $R_{\pi/2}$ is defined as

$$R_{\pi/2} = F_{y=x} F_x$$

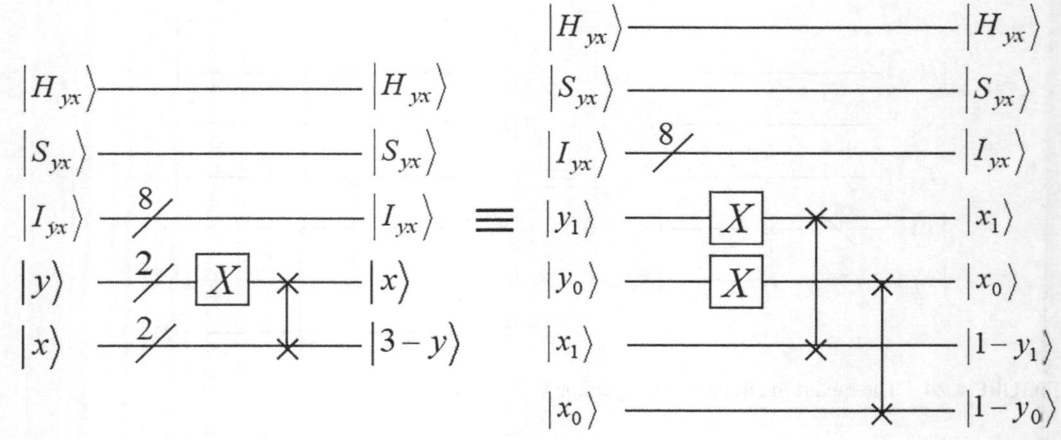

FIGURE 4.27 The quantum circuit for the rotation with an angle of π/2.

Applying the right-angle rotation operator $R_{\pi/2}$ to the image $|I(\theta)\rangle$ gives

$$R_{\pi/2}\left(|I(\theta)\rangle\right)=\frac{1}{4}\sum_{y=0}^{3}\sum_{x=0}^{3}|H_{yx}\rangle|S_{yx}\rangle|I_{yx}\rangle\otimes|x,3-y\rangle$$

4.2 SCALING OF QIRHSI

Based on the quantum image representation model, the research on related quantum computing methods is also booming, for example, feature extraction, image segmentation, digital watermarking, and quantum video. Image scaling is different from other geometric transformations, and there are two reasons for this phenomenon: first, the size of the image needs to be adjusted; second, interpolation methods need to be used to add new pixels or delete redundant pixels. As image scaling involves interpolation methods, different interpolation methods seek a balance between accuracy and computational complexity. The essence of interpolation is the process of estimating the positions of unknown pixels based on the known pixel data. The nearest-neighbor method, bilinear interpolation method, and cubic interpolation method are three of the most common interpolation methods.

The scaling of quantum images is still in its infancy. In this section, taking the nearest-neighbor interpolation as a tool, a method of scaling for quantum color images based on the improved QIRHSI (IQIRHSI) representation is proposed. Figure 4.28 shows the results of scaling the color image Lena using three interpolation methods. The sizes of Figure 4.28b–d are twice that of Figure 4.28a. The following introduces the scaling of quantum color images based on the QIRHSI representation.

4.2.1 Nearest-Neighbor Interpolation Method for Classical Image Scaling

Image scaling is one of the most widely used image processing operations, and its main purpose is to adjust the size of the image. Usually, with the help of interpolation methods, it adds new pixels or deletes redundant pixels. Nearest-neighbor, bilinear, and bicubic interpolation are three common interpolation methods. Relatively speaking, the interpolation method that is the easiest to implement and the most effective is the nearest-neighbor method.

FIGURE 4.28 The effects after scaling using different interpolation methods.

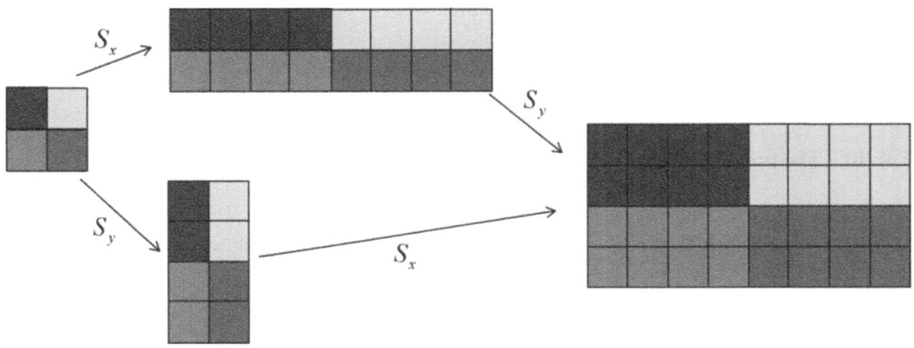

FIGURE 4.29 The decomposability and commutativity of S.

Generally, the nearest-neighbor interpolation method is decomposed and considered in two directions, horizontal and vertical, as shown in Equation 4.29:

$$I' = S\left(I,r_x,r_y\right)=S_x\left(S_y\left(I,r_y\right),r_y\right) = S_y\left(S_x\left(I,r_x\right),r_y\right) \tag{4.29}$$

where the scaling function is S, the original image is I, and the scaled image is denoted as I'. r_x, and r_y are the scaling ratios in the directions of the x-axis and y-axis, respectively. S_x and S_y are combined into a scaling function S. The scaling function along the horizontal direction is denoted as S_x, and the scaling function along the vertical direction is denoted as S_y. The order of the actions of the horizontal and vertical scaling functions does not affect the scaled image, so the order of S_x and S_y is interchangeable. In other words, r_x and r_y form the scaling ratio of the entire image. That is to say, the scaling of the entire color image is composed of the combination of horizontal and vertical scaling. The scaling is shown in Equation 4.29, where $r_x = 4$ and $r_y = 2$. Therefore, first study the scaling of the quantum color image in a single direction, then extend it to the scaling of the entire quantum color image, and finally design a complete quantum circuit through examples to explain and illustrate (Figure 4.29).

Suppose that $2^{n1} \times 2^{n2}$ is the size of the original image, and $2^{m1} \times 2^{m2}$ is the size of the scaled image. The scaling ratios are $r_y = 2^{m_1-n_1}$ and $r_x = 2^{m_2-n_2}$, where n_1,n_2,m_1,m_2 are all non-negative integers. Take the horizontal direction as an example to deeply explore the scaling of quantum color images.

- *Enlargement:* Repeat the pixel value r_x times. For example, for the original image "AB" and $r_x = 4$, the scaled image obtained is "AAAABBBB."

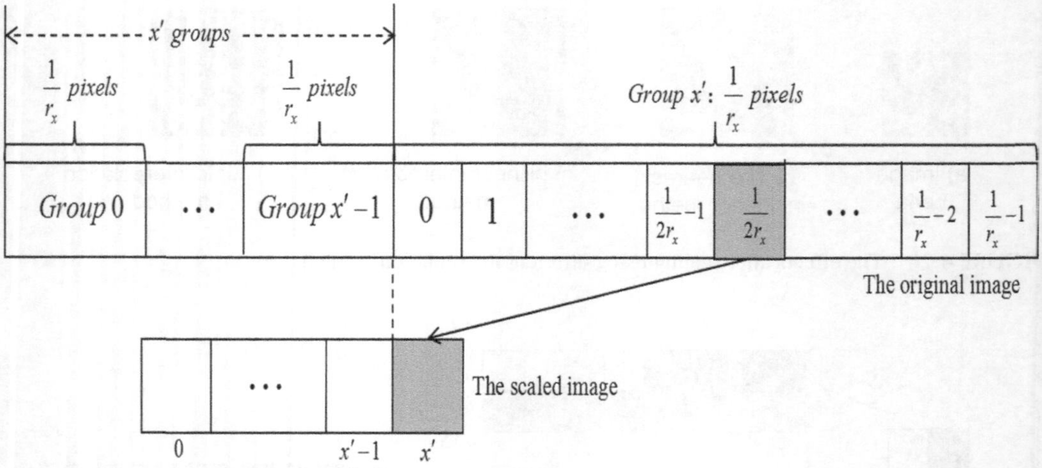

FIGURE 4.30 The principle of reduction. (Figure 3 in reference [4].)

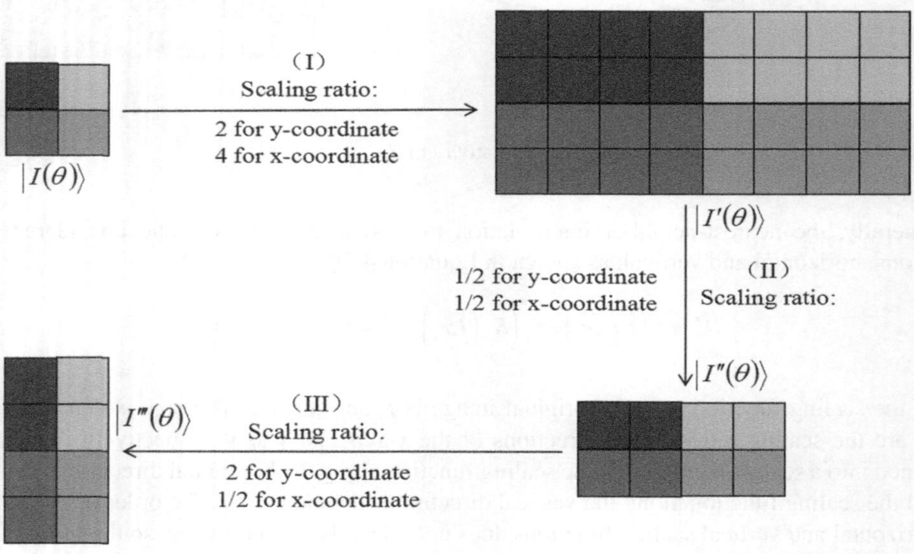

FIGURE 4.31 A simple example of image scaling.

- *Reduction:* Each $1/r_x$ pixel point in the original image will become one pixel after scaling. The color value of the new pixel position $x'$is equal to the color value of the first group of $1/2r_x$ pixels, that is, the pixel in the initial image $x'/r_x + 1/2r_x = x' \cdot 2^{n2-m2} + 2^{n2-m2-1}$, as shown in Figure 4.30. For example, "ABCDEFGH" is the initial image, $r_x = 1/4$ is the scaling ratio, and "BDFH" is the scaled image; if the scaling ratio is $r_x = 1/2$, the scaled image is "CG." Figure 4.31 describes a simple example of image scaling.

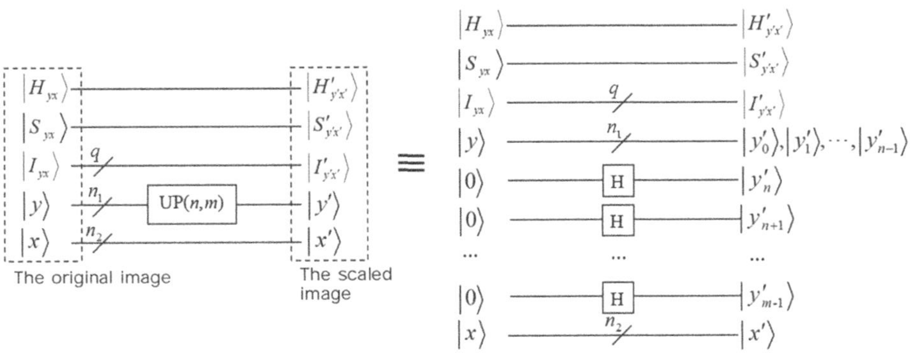

FIGURE 4.32 The quantum circuit of the color image enlarged in the direction of the y-axis.

4.2.2 Scaling of the Improved Quantum Color Image

Suppose that the size of the original color image is $2^{n1} \times 2^{n2}$, and the size of the scaled color image is $2^{m1} \times 2^{m2}$. The scaling ratio is $2^{m1-n1} \times 2^{m2-n2}$. Based on IQIRHSI, the color and position information of the original color image and the scaled color image are $C_{yx}(y,x)$ and $C_{y'x'}(y',x')$, respectively.

According to Equation 4.29, the overall scaling of the image is composed of the scaling in two directions. Therefore, in this section, only the scaling of the quantum color image in the direction of the *y*-axis is taken as an example. For the sake of convenience, the subscript 1 of n_1 and m_1 is omitted here.

4.2.2.1 Quantum Circuit for the Enlarging Operation of Color Images

The quantum circuit for the enlargement of a color image is obtained by using the quantum modular operation UP(n, m), where n and m represent the sizes of the color image from 2^n to 2^m, respectively. As the processing target in reference [4] is a grayscale image, the quantum circuit needs to be changed. The structure of the improved UP(n, m) is shown in Figure 4.32, where $m > n$.

- It is necessary to have $(m-n)$ state $|0\rangle$ to set the new position of the *y*-axis;
- In order to enlarge the coordinate axis, it is necessary to add $(m-n)$ Hadamard gates for expansion to obtain $y_n', y'_{n+1}, \cdots, y'_m$.

Obviously, adding $(m-n)$ Hadamard gates acts on the initial state, and the initial state appears as $|0\rangle$ and $|1\rangle$ with equal probability. All the possible values that appear are $\{|00\cdots00\rangle, |00\cdots01\rangle, \cdots |11\cdots10\rangle, |11\cdots11\rangle\}$. That is to say, each pixel $|y_0 y_1 \cdots y_{n+1}\rangle$ of the original image is enlarged by a factor of 2^{m-n}, and the corresponding pixel coordinates become

$$|y_0 y_1 \cdots y_{n+1} 0 \cdots 0\rangle, \cdots |y_0 y_1 \cdots y_{n-1} 1 \cdots 1\rangle$$

At the same time, the color values are kept unchanged. According to the above method, each pixel in the original color image is repeated r_y times. Figure 4.33 shows an example of the enlargement of the original image, where $m-n=1$, that is, $r_y = 2$.

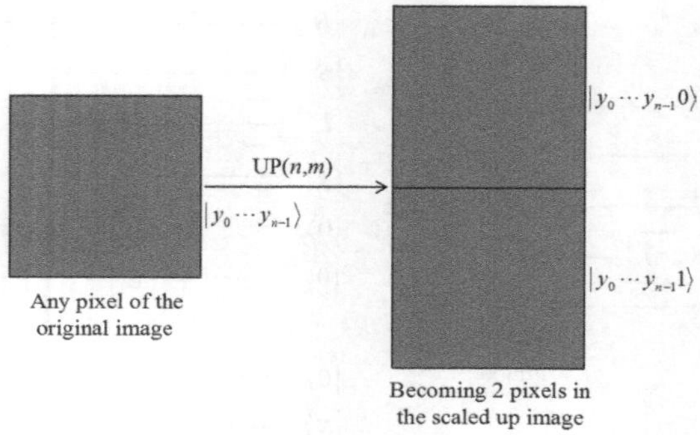

FIGURE 4.33 A brief example of UP(n, m).

4.2.2.2 Quantum Circuit for the Reducing Operation of Color Images

To design the circuit for reducing a quantum color image, by utilizing the reduction principle, it is necessary to conduct a further analysis with the help of Figure 4.30 in Section 4.2.1. The position information of the pixels in the original color image is $|y_0y_1\cdots y_{m-1}y_my_{m+1}\cdots y_{n-2}y_{n-1}\rangle$. The position information is divided into two parts: $|y_0y_1\cdots y_{m-2}y_{m-1}\rangle$ and $|y_my_{m+1}\cdots y_{n-2}y_{n-1}\rangle$, and it has the following characteristics:

1. All pixels belonging to the same group have the same $|y_0y_1\cdots y_{m-1}\rangle$, which is also the label of this group, that is, the position information of the pixels in the reduced color image generated by this group.
2. The color value of the pixel y' in the reduced image is equal to the color value of the pixel $|y_0'y_1'\cdots y'_{m-1}10\cdots 0\rangle$ in the original image, that is, $|y_my_{m+1}\cdots y_{n-1}\rangle = |10\cdots 0\rangle$.

The proofs of the above two characteristics are given in reference [4]. The quantum circuit for reduction in the same direction is represented by the modulo DOWN(n, m), where n and m represent the image being reduced from 2^n to 2^m. The structure of DOWN(n, m) is shown in Figure 4.34. The case where $m < n$ is presented as follows:

- Add $(q+2)|0\rangle$ as new color qubits and m $|0\rangle$ to set as the new position y of the reduced image;
- Add m Hadamard gates to generate $|y_0'\rangle, |y_1'\rangle, \cdots, |y'_{m-1}\rangle$;
- When $|y_my_{m+1}\cdots y_{n-1}\rangle = |10\cdots 0\rangle$, with the help of the CNOT gate, copy $|C_{yx}\rangle$ to $|C_{y'x'}\rangle$.
  - According to Feature (1), the control values of parts A and C are the same in Figure 4.34.
  - According to Feature (2), it can be concluded that the control value of part B in Figure 4.34 is $|10\cdots 0\rangle$.

In other words, the color value of the pixel $|y_0'y_1'\cdots y'_{m-1}10\cdots 0\rangle$ in the image before reduction is copied to the pixel $|y_0'y_1'\cdots y'_{m-1}\rangle$ in the reduced image. To more easily control the hue and saturation in the IQIRHSI model, the hue H and saturation S in the original image are in part A, while the hue H and saturation S in the reduced image are transferred to part B.

4.2.2.3 Complexity of the Quantum Network

Relatively speaking, the circuit for enlarging a quantum color image is simpler. If the scaling ratio is 2^{m-n}, only $(m-n)$ Hadamard gates are needed to solve the problem.

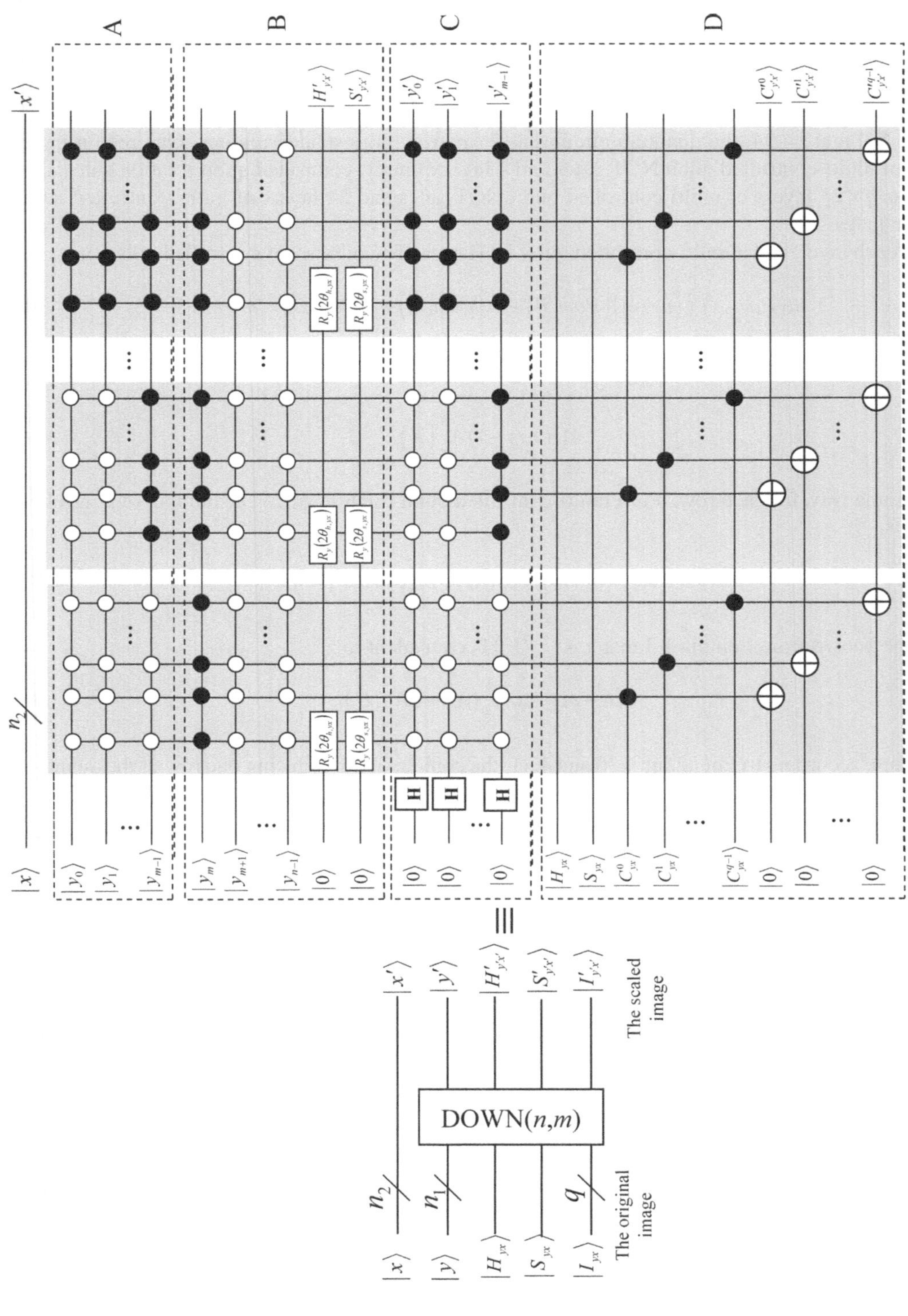

FIGURE 4.34 The quantum circuit of the color image reduced in the direction of the y-axis.

Next, we study the complexity of the quantum circuit for reducing a quantum color image. The number of basic quantum gates plays a crucial role in the complexity of the quantum circuit. In this paper, the CNOT gate and the controlled rotation gate are selected as the basic units.

Theorem 4.6: The complexity of the quantum gates required for preparing a quantum image reduction circuit in the IQIRHSI state with a size of $2^{n_1} \times 2^{n_2}$ is $O\left(3 \times 2^{m+3} \times (q+2)(n+m-1)\right)$.

Proof: In Figure 4.34, the quantum circuit has 2^m narrow gray shaded regions. One region has q layers of multi-controlled qubit NOT gates and 1 layer of multi-controlled qubit rotation gates, that is, it has $q \times 2^m$ layers of multi-controlled qubit NOT gates and 2^m layers of multi-controlled qubit rotation gates.

For each layer, it is a multi-controlled qubit NOT gate. The number of controlled qubits is

$$m(Part\ A)+(n-m)(Part\ B)+m(Part\ C)+1(Part\ D)=n+m+1 \tag{4.30}$$

With the help of Lemma 1.3, a $\Lambda_{n+m+1}(X)$ is equivalent to

$$4(n+m-1)\Lambda_2(X)$$

In the same way, for each row, it is a multi-controlled qubit rotation gate. The number of controlled qubits is

$$m(Part\ A)+(n-m)(Part\ B)+m(Part\ C)=n+m. \tag{4.31}$$

It can be known from Lemma 1.3 that a $\Lambda_{n+m}(U)$ is equivalent to

$$4(n+m)-10\Lambda_2(X)+2\Lambda_2(U)$$

Therefore, according to Equations 4.30 and 4.31, the complexity of reducing one row of the quantum image in Figure 4.34 is

$$8(n+m)-20\Lambda_2(X)+4\Lambda_2(U)+q\times 4(n+m-1)\Lambda_2(X)$$

Therefore, the quantum modulo DOWN(n, m) can be simulated as:

$$2^m \times \left[8(n+m)-20\Lambda_2(X)+4\Lambda_2(U)+q\times 4(n+m-1)\Lambda_2(X)\right]+(n+m-1)\text{NOT}$$

Reference [5] points out that a $\Lambda_2(X)$ gate can be simulated by six $\Lambda_1(X)$ gates. Through Lemma 1.2, the above equation can be simplified as

$$3\times 2^{m+3}\times\left((q+2)\cdot(n+m-1)-\frac{5}{3}\right)\Lambda_1(X)+2^{m+5}\Lambda_0+(n+m-1)\text{NOT}$$

That is to say, the complexity of the quantum gates required in the quantum circuit is

$$O\left(3\times 2^{m+3}\times(q+2)(n+m-1)\right)$$

□

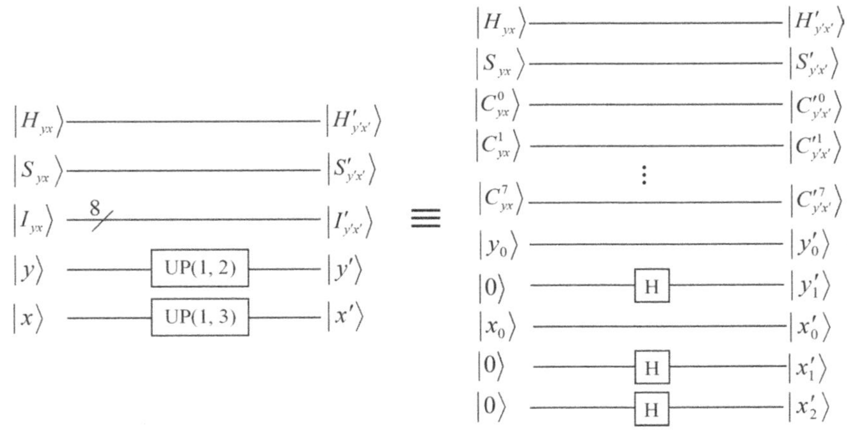

FIGURE 4.35 The circuit design in Step I of Figure 4.31.

4.2.3 Experimental Examples

Select the image scaling transformation in Figure 4.31 as an example to illustrate some details of the quantum scaling circuit. The original image is

$$|I(\theta)\rangle = \frac{1}{2}\sum_{y=0}^{1}\sum_{x=0}^{1}|H_{yx}\rangle|S_{yx}\rangle|I_{yx}\rangle \otimes |y\rangle|x\rangle$$

where $n_1 = n_2 = 1$, $q = 8$.

1. $|I(\theta)\rangle \to |I'(\theta)\rangle$. Here $r_y = 2$, $r_x = 4$, and $m_1 = 2$, $m_2 = 3$. Figure 4.35 depicts the scaling circuit.
2. $|I'(\theta)\rangle \to |I''(\theta)\rangle$, where $r_y = r_x = 1/2$, $m_1' = 1$, and $m_2' = 2$. Figure 4.36 shows the scaling circuit.
3. $|I''(\theta)\rangle \to |I'''(\theta)\rangle$, where $r_y = 2$, $r_x = 1/2$, $m_1'' = 2$, and $m_2'' = 1$. Figure 4.37 shows the scaling circuit.

4.3 SUMMARY OF THIS CHAPTER

In this chapter, firstly, the basic geometric transformation operations of the QIRHSI quantum color image are introduced, such as two-point exchange, cyclic translation, folding transformation, and right-angle rotation. The unitary operators and quantum circuits for realizing the above geometric transformations are designed, and the complexity of the quantum gates required for the quantum circuits is analyzed. Then, the scaling of the QIRHSI color image based on the nearest-neighbor interpolation is studied. The scaling quantum circuit based on the IQIRHSI image is designed, and the complexity of the quantum gates required for the quantum circuit is analyzed.

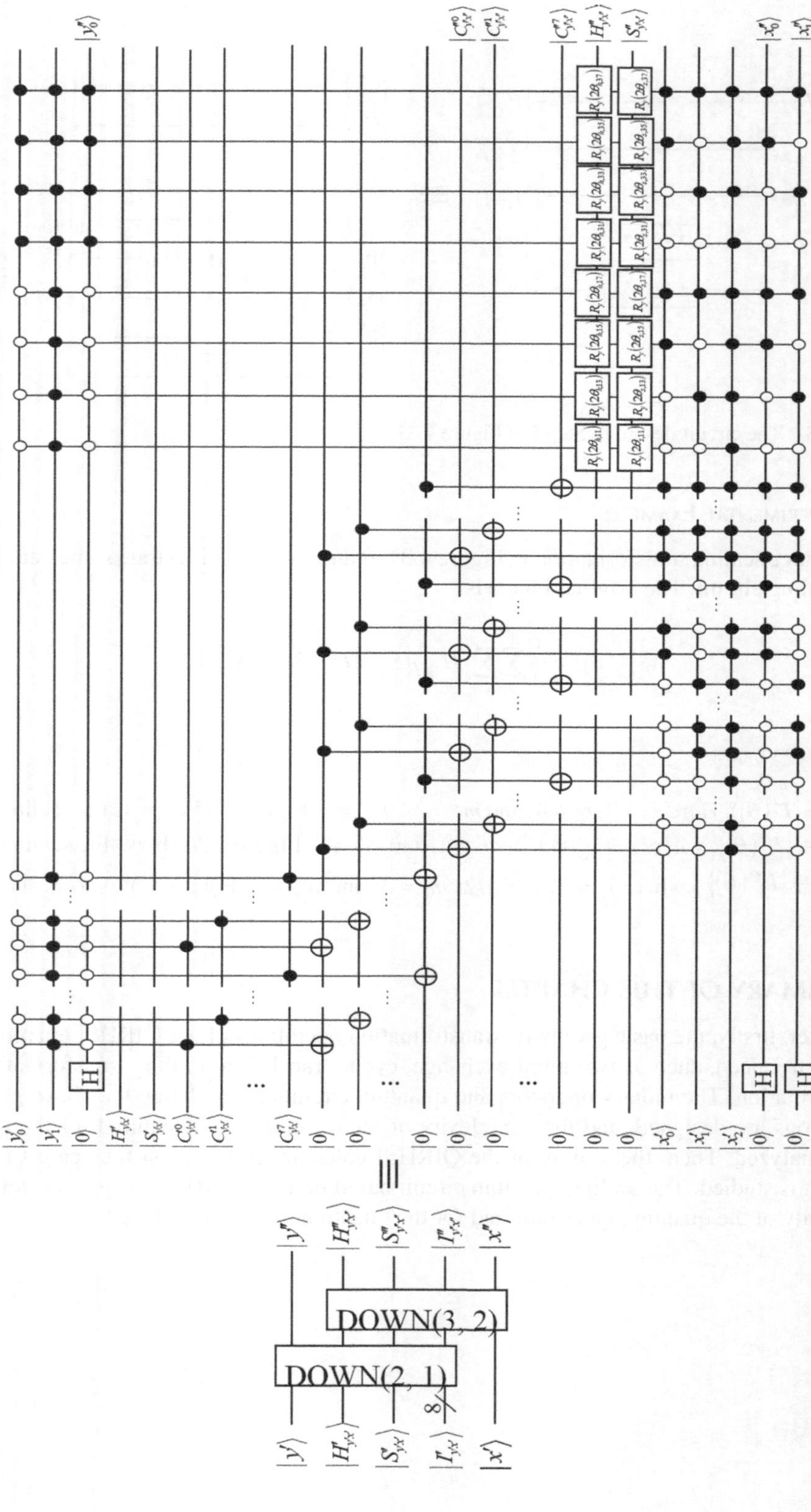

FIGURE 4.36 The circuit design in Step II of Figure 4.31.

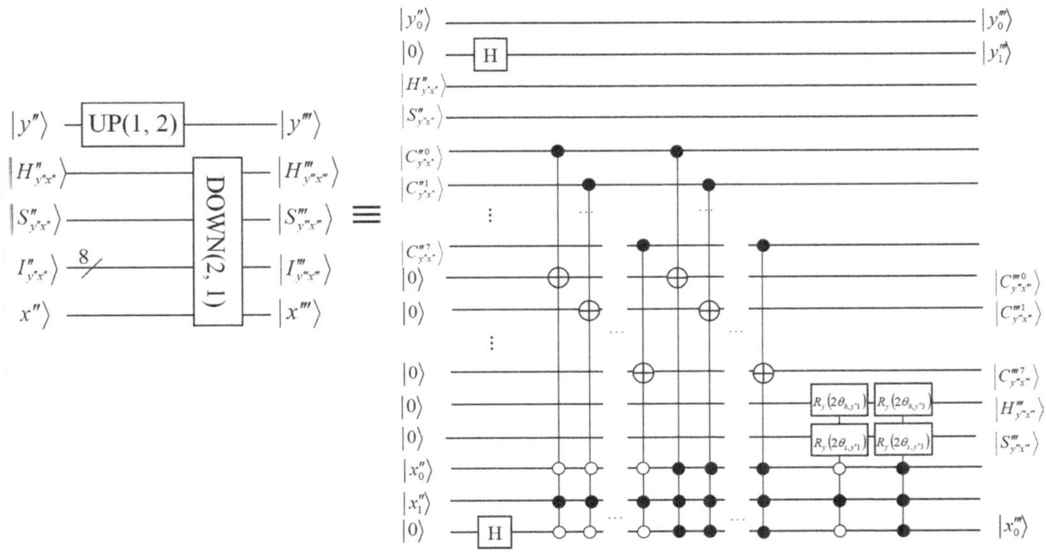

FIGURE 4.37 The circuit design in Step III of Figure 4.31.

REFERENCES

[1] J. I. Latorre, Image compression and entanglement, *Computer Science*, 2005, 1–4.

[2] P. Q. Le, A. M. Iliyasu, and F. Y. Dong, et al., Fast geometric transformations on quantum images, *Int. J. Appl. Math.*, 2010, 40, 3, 113–123.

[3] P. Fan, R. G. Zhou, and N. H. Jing, et al., Geometric transformations of multidimensional color images based on NASS, *Inf. Sci.*, 2016, 340, 191–208.

[4] N. Jiang, X. W. Lu, and H. Hu, et al., A novel quantum image compression method based on JPEG, *Int. J. Theor. Phys.*, 2018, 57, 3, 611–636.

[5] V. Vedral, A. Barenco, and A. Ekert, et al., Quantum networks for elementary arithmetic operations, *Quantum Phys.*, 1995, 54, 1, 1–12.

[6] X. Song, G. Chen, A. A. Abd El-Latif, M. A. Wani, and B. Abd-El-Atty, Quantum geometric transformation based on QIRHSI quantum color images, *IEEE Access*, 2023, 11, 21883–21899.

[7] P. Q. Le, A. M. Iliyasu, F. Dong, and K. Hirota, Fast geometric transformations on quantum images, *IAENG Int. J. Appl. Math.*, 2010, 40, 3, 1–2.

[8] P. Q. Le, A. M. Iliyasu, F. Dong, and K. Hirota, Strategies for designing geometric transformations on quantum images, *Theor. Comput. Sci.*, 2011, 412, 1406–1418.

[9] P. Fan, R.-G. Zhou, N. Jing, and H.-S. Li, Geometric transformations of multidimensional color images based on NASS, *Inf. Sci.*, 2016, 340–341, 191–208.

[10] F. Yan, K. Chen, S. E. Venegas-Andraca, and J. Zhao, Quantum image rotation by an arbitrary angle, *Quantum Inf. Process.*, 2017, 16, 11, 282.

[11] J. Wang, N. Jiang, and L. Wang, Quantum image translation, *Quantum Inf. Process.*, 2015, 14, 5, 1589–1604.

[12] R.-G. Zhou, C. Tan, and H. Ian, Global and local translation designs of quantum image based on FRQI. *Int. J. Theor. Phys.*, 2017, 56, 4, 1382–1398.

[13] N. Jiang and L. Wang, Quantum image scaling using nearest neighbor interpolation, *Quantum Inf. Process.*, 2015, 14, 5, 1559–1571.

[14] J. Sang, S. Wang, and X. Niu, Quantum realization of the nearest neighbor interpolation method for FRQI and NEQR, *Quantum Inf. Process.*, 2016, 15, 1, 37–64.

[15] P. Li and X. Liu, Bilinear interpolation method for quantum images based on quantum Fourier transform, *Int. J. Quantum Inf.*, 2018, 16, 4, 1850031.

[16] R.-G. Zhou, Y. Cheng, X. Qi, H. Yu, and N. Jiang, Asymmetric scaling scheme over the two dimensions of a quantum image, *Quantum Inf. Process.*, 2020, 19, 9, 343.
[17] P. Q. Le, A. M. Iliyasu, F. Dong, and K. Hirota, Efficient color transformations on quantum images, *J. Adv. Comput. Intell. Intell. Inform.*, 2011, 15, 6, 698–706.
[18] N. Jiang, W.-Y. Wu, and L. Wang, The quantum realization of Arnold and Fibonacci image scrambling, *Quantum Inf. Process.*, 2014, 13, 5, 1223–1236.
[19] N. Jiang and L. Wang, Analysis and improvement of the quantum Arnold image scrambling, *Quantum Inf. Process.*, 2014, 13, 7, 1545–1551.
[20] N. Jiang, L. Wang, and W.-Y. Wu, Quantum Hilbert image scrambling, *Int. J. Theor. Phys.*, 2014, 53, 7, 2463–2484.
[21] R.-G. Zhou, Y.-J. Sun, and P. Fan, Quantum image gray-code and bitplane scrambling, *Quantum Inf. Process.*, 2015, 14, 5, 1717–1734.
[22] S. Caraiman and V. I. Manta, Image segmentation on a quantum computer, *Quantum Inf. Process.*, 2015, 14, 5, 1693–1715.
[23] P. Li, T. Shi, Y. Zhao, and A. Lu, Design of threshold segmentation method for quantum image, *Int. J. Theor. Phys.*, 2020, 59, 2, 514–538.
[24] S. Yuan, C. Wen, B. Hang, and Y. Gong, The dual-threshold quantum image segmentation algorithm and its simulation, *Quantum Inf. Process.*, 2020, 19, 12, 425.
[25] Y. Zhang, K. Lu, K. Xu, Y. Gao, and R. Wilson, Local feature point extraction for quantum images, *Quantum Inf. Process.*, 2015, 14, 5, 1573–1588.
[26] N. Jiang, Z. Ji, H. Li, and J. Wang, Quantum image interest point extraction, *Mod. Phys. Lett. A*, 2021, 36, 9, 2150063.

5 Quantum Image Compression

Zhixin Zhang, Xianhua Song, Ahmed A. Abd El-Latif, and Basma Abd El-Rahiem

5.1 COMPRESSION OF QUANTUM IMAGE QIRHSI

The time consumption for preparing the quantum color image Quantum Image Representation based on HSI (QIRHSI)[1–4]mainly concentrates on steps B and R. The time complexity of the entire process depends on the number of controlled NOT gates and controlled rotation gates used in steps B and R. To simplify the optimization process of the resources for preparing the quantum color image[5], only the case where all pixels have the same hue and saturation is considered here. Therefore, this section focuses on how to optimize these controlled NOT gates to simplify the preparation work in step B.

First of all, recall Equation 2.8. For the convenience of the following discussion, this formula is given below:

$$B_m = I^{\otimes q+2} \otimes \sum_{k=0,k\neq m}^{2^{2n}-1} |k\rangle\langle k| + I^{\otimes 2} \otimes \Lambda_m \otimes |m\rangle\langle m|$$

Equation 2.8 represents the suboperation B_m in operation B corresponding to each pixel m in step B.

In step B, the original preparation[6] method of QIRHSI model is based on pixel positions, and it is necessary to traverse all the pixels. That is, the value of m ranges from 0 to $2^{2n}-1$. Therefore, there are 2^{2n} operations of B_m. In other words, there are 2^{2n} operations of Λ_m. As the number of controlled-NOT gates Λ_m at each pixel position depends on the number of 1s in the binary sequence of the intensity at that pixel position[7], if there exists a pixel position with an intensity of 0, then all the Λ_m at that pixel position are identity gates. The designed optimized preparation method of the QIRHSI model is carried out according to the q layers of bit planes[8]. The number of controlled-NOT gates in each group of bit planes is closely related to the number of 1s in that layer. In addition, the number of controlled-NOT gates in each layer of bit planes is determined by minimizing the Boolean expressions of the pixel positions with a value of 1 in that layer. If the values in some bit planes are all 0, then the quantum operation gates in that layer of bit planes are all identity gates.

To more clearly demonstrate the above method for optimizing the resources for preparing quantum images, take the 2×2 image in Figure 5.1a as an example. This image has four pixels, namely 00, 01, 10, and 11. Figure 5.1b presents the original quantum operations Λ_{00}, Λ_{01}, Λ_{10}, and Λ_{11}. As the value of q is 8 (q represents the range of intensity values), the intensity is divided into eight layers of bit planes. By minimizing the pixel positions with a value of 1 in each layer through Boolean expressions, respectively[9], the optimized quantum operations $C'^0, C'^1, \cdots, C'^7$ are obtained, as shown in Figure 5.1c.

DOI: 10.1201/9781003621423-6

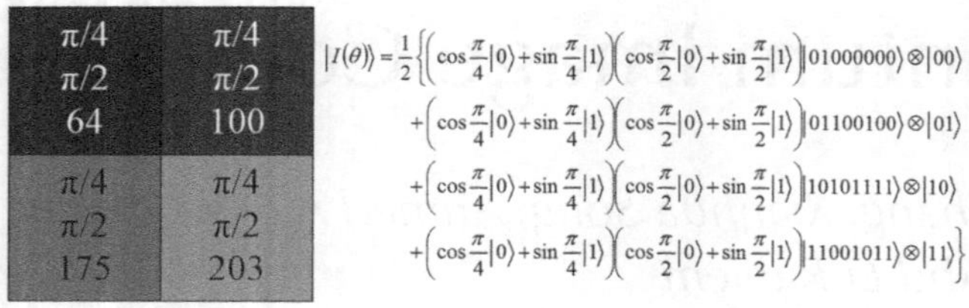

a) An example image and its QIRHSI representation

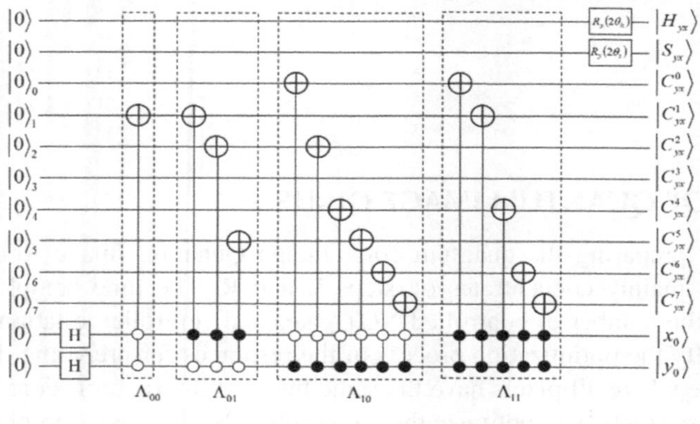

b) The image preparation resources required before optimization (15 controlled NOT gates)

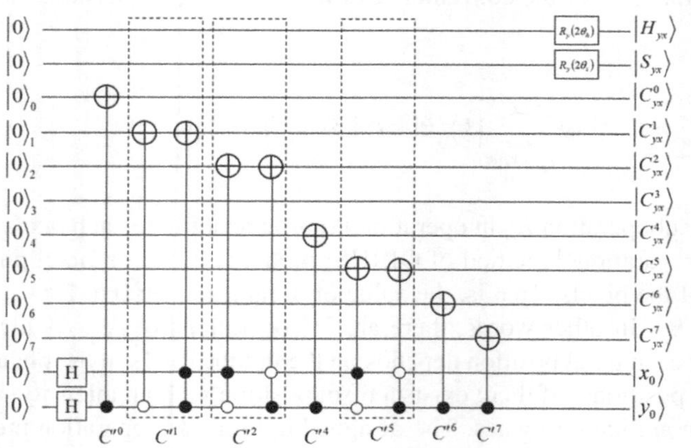

c) The image preparation resources required after optimization (10 controlled NOT gates)

FIGURE 5.1 Representation, preparation and optimization of QIRHSI.

Take the binary string corresponding to each pixel position as a Boolean variable, and then convert it into a Boolean minterm. If x is a Boolean variable at a certain position and its value is 1, then x is used in the Boolean minterm; otherwise, $\overline{x}$ is used. For example, the binary strings 00 and 10 are equivalent to $\overline{x}_1\overline{x}_0$ and $x_1\overline{x}_0$, respectively. Using this method, there is a one-to-one correspondence between the set of all binary strings of length n and all the Boolean minterms generated by n Boolean variables.

First, divide the intensity of the quantum image into q layers of bit planes, and then optimize the control information with a value of 1 in each bit plane. To this end, the binary strings

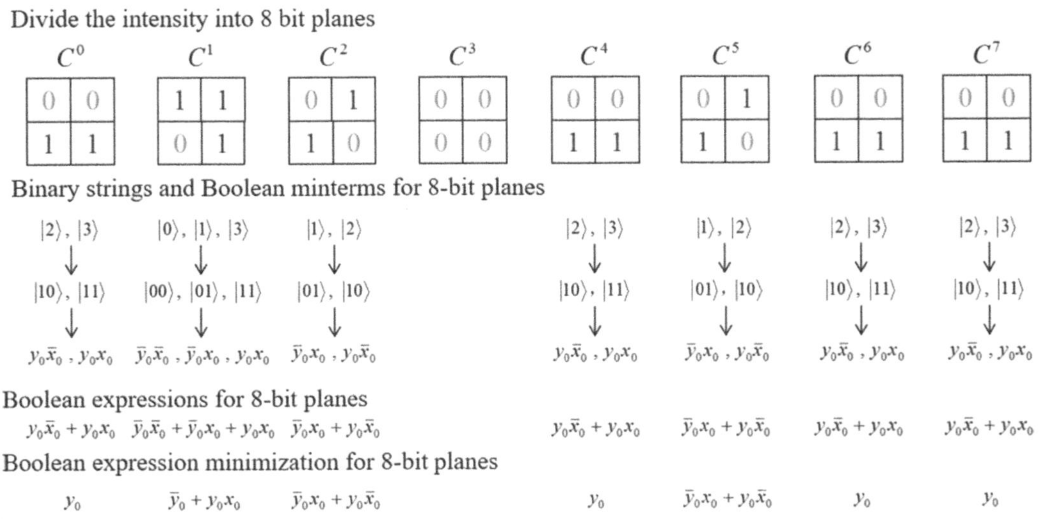

FIGURE 5.2 Boolean expressions of the eight layers of bit planes and the corresponding minimized expressions.

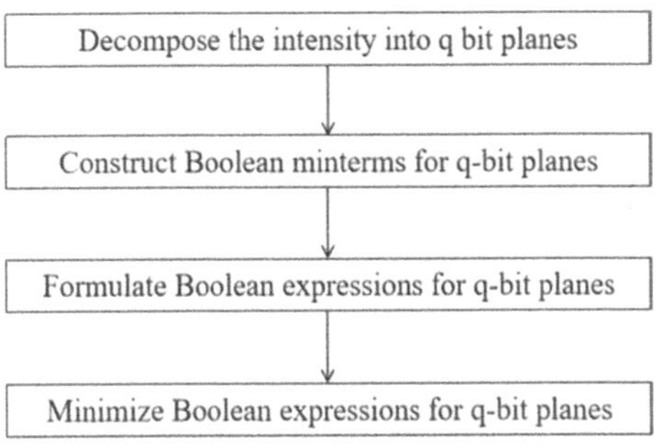

FIGURE 5.3 Flowchart of optimizing the preparation resources of quantum images.

corresponding to the positions with a value of 1 in each bit plane play a pivotal role. Each binary string corresponds to a Boolean minterm. Next, combine the Boolean minterms of each bit plane. Finally, minimize the Boolean minterms of the combination of each bit plane. Taking a 2×2 image as an example, Figure 5.2 shows the detailed process of optimizing the preparation resources of the quantum image.

Figure 5.3 shows an algorithm for optimizing the preparation resources of quantum images based on the minimization of Boolean expressions to reduce the controlled NOT gates in step B. This process first divides the intensity channel into q layers of bit planes, then constructs Boolean minterms, constructs Boolean expressions, and finally minimizes the Boolean expressions.

5.2 COMPARISON OF COMPRESSION RATES OF QUANTUM COLOR IMAGES

General simulation results show that the representation methods of QIRHSI and Novel Quantum representation of Color digital Images (NCQI)[12,13] can optimize more resources for image preparation than the Multi-Channel representation for Quantum Images model (MCQI) model[14–16]. When all pixel positions in the entire image have the same hue and saturation, the preparation resources of the MCQI quantum image can only be optimized through the minimization of Boolean expressions when adjacent pixels in a certain area have the same intensity value. This is because it requires three qubits to store the color information of the MCQI image. The NCQI quantum image uses 24 qubits to store color information, and optimizes each qubit[17] on the R, G, and B channels[18] through the method of Boolean expression minimization. In the QIRHSI model, the hue and saturation information of each pixel is stored using 2 qubits, and the intensity information is stored using the basic state of an 8-qubit sequence. Each qubit in the sequence does not affect each other, and each qubit for preparing the intensity information can be simplified through the minimization of Boolean expressions. Therefore, in terms of the optimization effect, compared with the optimization rate[10,11] of MCQI, NCQI and QIRHSI can obtain a better quantum image optimization rate by using the method of Boolean expression minimization.

A simple example is introduced to illustrate the similarities and differences among the optimizations of the preparation resources of MCQI, NCQI, and QIRHSI quantum images. Figure 5.4a presents an 8×8 color image with a hue of 0.71, a saturation of 1, and the intensity values ranging from 0 to 255 and Figure 5.4b shows the detailed intensity values of all pixels. Figure 5.4c shows detailed *R*, *G*, and *B* component values of all pixels corresponding to those shown in Figure 5.4b. If the MCQI is used to store the image, 64 controlled rotation gates need to be prepared before the optimization. As the hues and saturations of all pixels in the entire image are the same, and the intensity values are also distributed according to a certain pattern, the number of controlled rotation gates in the MCQI model is reduced to 5, as shown in Equation 5.1:

$$\begin{aligned} &column\ 0,1,2,3: \overline{x} = 0, R_{64} \\ &column\ 4: x_0\overline{x}_1\overline{x}_2 = 0, R_{100} \\ &column\ 5: x_0\overline{x}_1\overline{x}_2 = 0, R_{175} \\ &column\ 6: x_0\overline{x}_1\overline{x}_2 = 0, R_{203} \\ &column\ 7: x_0\overline{x}_1\overline{x}_2 = 0, R_{245} \end{aligned} \tag{5.1}$$

where R_x indicates that the intensity value set in the rotation operation is x.

Figure 5.4 shows 8 × 8 color image, intensity values, the corresponding *R*, *G*, and *B* component values, and eight layers of bit planes.

As the range of the intensity values varies from 0 to 255, when QIRHSI stores an image, 8 qubits are required to store the intensity information. Figure 5.4d shows all the information of the quantum operations of the eight layers of bit planes during the image preparation process. It should be noted that 192 controlled NOT gates are needed to prepare the quantum image before optimization. The Boolean expression minimization method can be used to simplify the quantum operations of each intensity qubit. After optimization, the number of controlled NOT gates required for the quantum operations of the eight layers of bit planes is 2, 3, 2, 1, 2, 2, 2, and 2, respectively. The quantum operations of the first group after reconstruction are shown in Equation 5.2:

$$\begin{aligned} &column\ 5,7: x_0x_2 = 11, 2-\text{CNOT} \\ &column\ 6: x_0\overline{x}_1\overline{x}_2 = 110, 3-\text{CNOT} \end{aligned} \tag{5.2}$$

For the color image given in Figure 5.4a, the three channels of the NCQI model require 24 qubits to store the image. Similarly, as shown in Figure 5.4c, the RGB color image can also be divided into 24 groups[19]. The Boolean expression minimization method can be used to simplify the quantum

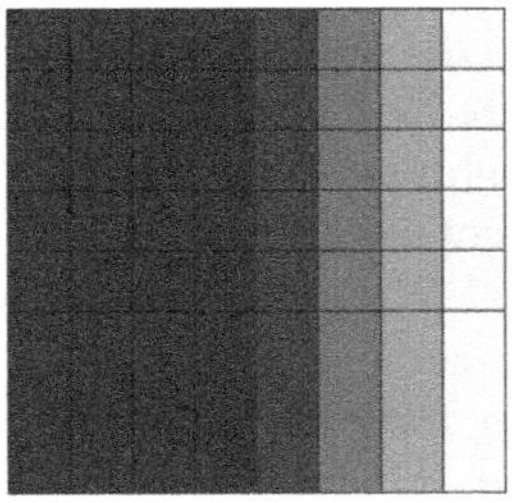

a) A color image with the size of 8 × 8

64	64	64	64	100	175	203	245
64	64	64	64	100	175	203	245
64	64	64	64	100	175	203	245
64	64	64	64	100	175	203	245
64	64	64	64	100	175	203	245
64	64	64	64	100	175	203	245
64	64	64	64	100	175	203	245
64	64	64	64	100	175	203	245

b) The intensity values of image a)

R

30	30	30	30	42	129	173	240
30	30	30	30	42	129	173	240
30	30	30	30	42	129	173	240
30	30	30	30	42	129	173	240
30	30	30	30	42	129	173	240
30	30	30	30	42	129	173	240
30	30	30	30	42	129	173	240
30	30	30	30	42	129	173	240

G

0	0	0	0	0	95	151	235
0	0	0	0	0	95	151	235
0	0	0	0	0	95	151	235
0	0	0	0	0	95	151	235
0	0	0	0	0	95	151	235
0	0	0	0	0	95	151	235
0	0	0	0	0	95	151	235
0	0	0	0	0	95	151	235

B

128	128	128	128	200	255	255	255
128	128	128	128	200	255	255	255
128	128	128	128	200	255	255	255
128	128	128	128	200	255	255	255
128	128	128	128	200	255	255	255
128	128	128	128	200	255	255	255
128	128	128	128	200	255	255	255
128	128	128	128	200	255	255	255

c) The R, G and B components corresponding to the color image a)

C^0

0	0	0	0	0	1	1	1
0	0	0	0	0	1	1	1
0	0	0	0	0	1	1	1
0	0	0	0	0	1	1	1
0	0	0	0	0	1	1	1
0	0	0	0	0	1	1	1
0	0	0	0	0	1	1	1
0	0	0	0	0	1	1	1

C^1

1	1	1	1	1	0	1	1
1	1	1	1	1	0	1	1
1	1	1	1	1	0	1	1
1	1	1	1	1	0	1	1
1	1	1	1	1	0	1	1
1	1	1	1	1	0	1	1
1	1	1	1	1	0	1	1
1	1	1	1	1	0	1	1

C^2

0	0	0	0	1	1	0	1
0	0	0	0	1	1	0	1
0	0	0	0	1	1	0	1
0	0	0	0	1	1	0	1
0	0	0	0	1	1	0	1
0	0	0	0	1	1	0	1
0	0	0	0	1	1	0	1
0	0	0	0	1	1	0	1

C^3

0	0	0	0	0	0	0	1
0	0	0	0	0	0	0	1
0	0	0	0	0	0	0	1
0	0	0	0	0	0	0	1
0	0	0	0	0	0	0	1
0	0	0	0	0	0	0	1
0	0	0	0	0	0	0	1
0	0	0	0	0	0	0	1

C^4

0	0	0	0	0	1	1	0
0	0	0	0	0	1	1	0
0	0	0	0	0	1	1	0
0	0	0	0	0	1	1	0
0	0	0	0	0	1	1	0
0	0	0	0	0	1	1	0
0	0	0	0	0	1	1	0
0	0	0	0	0	1	1	0

C^5

0	0	0	0	1	1	0	1
0	0	0	0	1	1	0	1
0	0	0	0	1	1	0	1
0	0	0	0	1	1	0	1
0	0	0	0	1	1	0	1
0	0	0	0	1	1	0	1
0	0	0	0	1	1	0	1
0	0	0	0	1	1	0	1

C^6

0	0	0	0	0	1	1	0
0	0	0	0	0	1	1	0
0	0	0	0	0	1	1	0
0	0	0	0	0	1	1	0
0	0	0	0	0	1	1	0
0	0	0	0	0	1	1	0
0	0	0	0	0	1	1	0
0	0	0	0	0	1	1	0

C^7

0	0	0	0	0	1	1	1
0	0	0	0	0	1	1	1
0	0	0	0	0	1	1	1
0	0	0	0	0	1	1	1
0	0	0	0	0	1	1	1
0	0	0	0	0	1	1	1
0	0	0	0	0	1	1	1
0	0	0	0	0	1	1	1

d) The intensity values of the eight layers of bit planes corresponding to the color image a) in b)

FIGURE 5.4 An 8 × 8 color image, intensity values, the corresponding R, G, and B component values, and eight layers of bit planes.

operations of the three quantum channels[20]. After optimization, the numbers of controlled NOT gates required for the eight quantum operations of the R channel are 2, 1, 2, 2, 2, 2, 2, and 2, respectively; the numbers of controlled NOT gates required for the eight quantum operations of the G channel are 1, 1, 1, 2, 1, 2, 2, and 2, respectively; and the numbers of controlled NOT gates required for the eight quantum operations of the B channel are 1, 1, 2, 2, 1, 2, 2, and 2, respectively. The subsequent calculation method is consistent with that of the QIRHSI model.

The Boolean expression minimization step in the algorithm for optimizing the preparation resources of quantum color images is completed by the Logic Friday software. According to Equation 5.3, the optimization rate of QIRHSI is 91.67%, the optimization rate of NCQI is 93.59%, and the optimization rate of MCQI is 92.19%:

$$R = \frac{N_{before} - N_{after}}{N_{before}} \times 100\% \tag{5.3}$$

The optimization rate in Equation 5.3 is approximately calculated where R is the optimization rate, N_{before} is the number of operations before optimization, N_{new} is the number of operations after optimization. Essentially, the optimization rate depends on the number of operations before and after the preparation of image resources. Obviously, for QIRHSI in Figure 5.4, the number of operations required before optimization is 5-CNOT gates, and after optimization, it becomes a basic quantum operation gate, including 2-CNOT gates and 3-CNOT gates. The quantum operation gates required for these quantum operations are also different, but the simplification of the quantum operation gates has no impact on the qualitative comparison results of MCQI, NCQI, and QIRHSI.

To compare different performances of MCQI, NCQI, and QIRHSI in optimizing the preparation resources of images, Figure 5.5 shows six test images. Among these images, the intensities of the first four images are distributed according to a certain pattern, while the last two images are randomly obtained.

Table 5.1 presents different optimization rates of six images in three quantum image models. In QIRHSI, the average optimization rate achieved across the six images is 80.72%, which is 1.34 times that of MCQI (59.90%) and is nearly equal to that of NCQI (82.75%). When the intensity of the image is distributed according to a certain pattern, all three models have a good optimization ratio (e.g., in images (A) (a), (B) (b), and (C) (c)). When the intensity information of the image is "messily" distributed, the optimization ratios of NCQI and QIRHSI become more pronounced. Especially when the image is "messily" distributed, in other words, when almost all adjacent pixels in the image have different intensity values, the Boolean expression minimization method has almost no effect on the optimization of MCQI. However, for "messy" images, NCQI and QRHSI can still achieve an optimization rate of 50–70%. These experimental results support the following analytical conclusion: using the Boolean expression minimization method, NCQI and QIRHSI can optimize the preparation resources of quantum images better than MCQI.

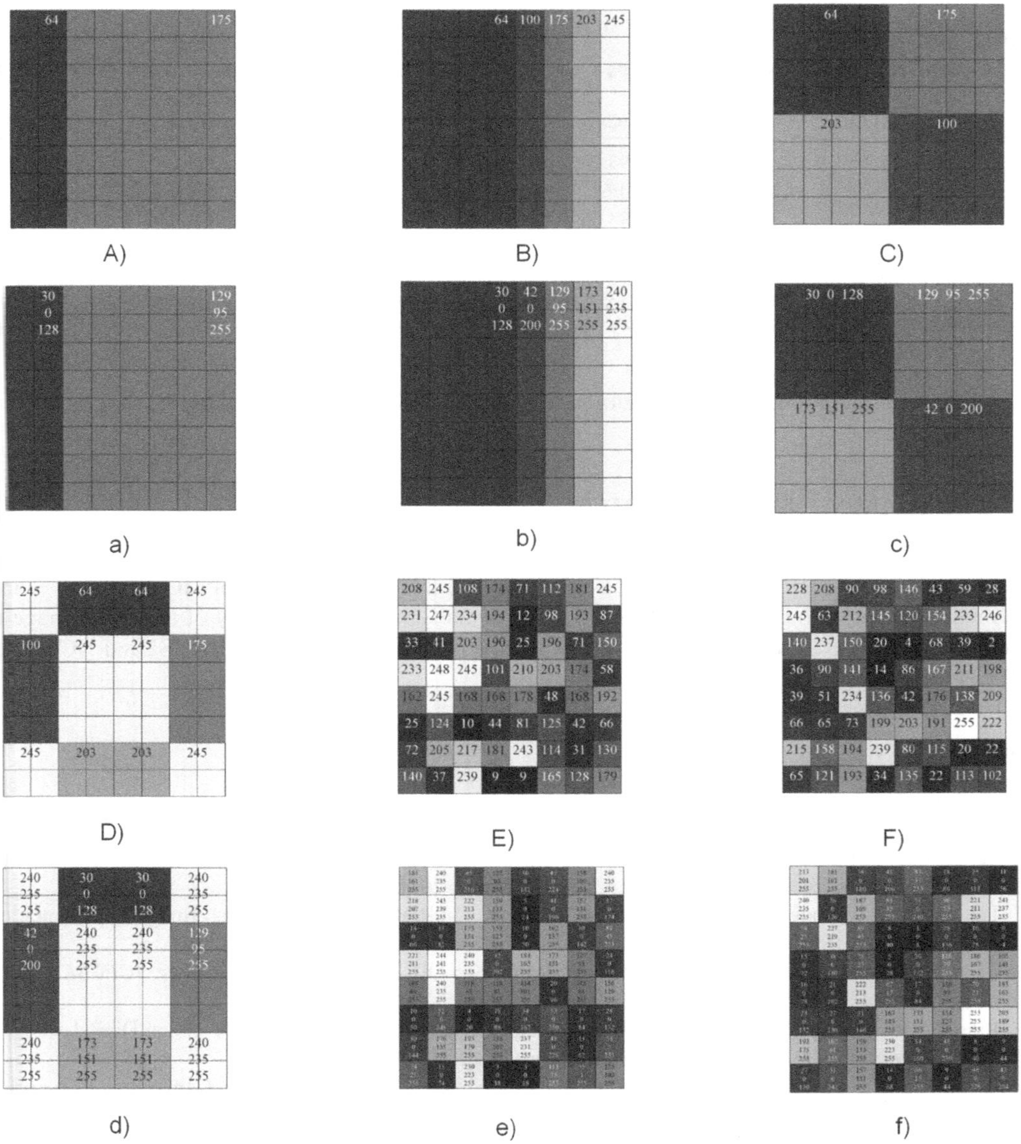

FIGURE 5.5 The I component diagrams of six 8 × 8 images and their corresponding RGB component diagrams.

TABLE 5.1
The Optimization Rates of Six Images Corresponding to the Three Quantum Image Models

	(A) (a)	(B) (b)	(C) (c)	(D) (d)	(E) (e)	(F) (f)	AVG (%)
MCQI	95.31% 3/64	92.19% 5/64	93.75% 4/64	75% 16/64	3.13% 62/64	0% 64/64	59.90%
NCQI	95.87% 35/848	93.59% 40/624	94.86% 37/720	84.83% 142/936	65.15% 276/792	62.21% 274/725	82.75%
QIRHSI	95.72% 13/304	91.67% 16/192	95% 12/240	86.54% 42/312	59.52% 102/252	55.87% 109/247	80.72%

①For each operation, the optimization rate and the number of operations before and after optimization are given.

5.3 SUMMARY OF THIS CHAPTER

In this chapter, by using the optimization method of Boolean expression minimization, the average optimization rate of QIRHSI is obtained as 80.72%, which is 1.34 times that of MCQI (59.90%) and is almost on a par with that of NCQI (82.75%). Especially when the intensity of the color image is "messily" distributed, the optimization rate of MCQI is almost zero, while QIRHSI and NCQI can still reach 50%-70%.

REFERENCES

[1] Yan F, Iliyasu A M, Venegas-Andraca S E. A Survey of Quantum Image Representations. *Quantum Information Processing*, 2016, 15: 1–35.

[2] Li H S. *Research on Key Technologies of Quantum Image Processing*. University of Electronic Science and Technology of China, Chengdu, China. 2014.

[3] Chen G L, Song X H, Venegas-Andraca S E, et al. QIRHSI: Novel Quantum Image Representation Based on HSI Color Space Model. *Quantum Information Processing*, 2022, 21(1): 5.

[4] Chen G, Song X. Quantum Color Image Scaling on QIRHSI Model. *Data Science: ICPCSEE 2021*. Taiyuan, China: Springer, 2021: 453–467.

[5] Liu K, Zhang Y, Lu K, et al. An Optimized Quantum Representation for Color Digital Images. *International Journal of Theoretical Physics*, 2018, 57: 2938–2948.

[6] Nielsen M A, Chuang I L. *Quantum Computation and Quantum Information*. Cambridge: Cambridge University Press, 2005, 57–89, 157–223.

[7] Zhang Y, Lu K, Gao Y H, et al. NEQR: A Novel Enhanced Quantum Representation of Digital Images. *Quantum Information Processing*, 2013, 12(8): 2833–2860.

[8] Zhou R G, Sun Y J, Fan P. Quantum Image Gray-Code and Bit-Plane Scrambling. *Quantum Information Processing*, 2015, 14(5): 1717–1734.

[9] Saeedi M, Markov I L. Synthesis and Optimization of Reversible Circuits – A Survey. *ACM Computing Surveys*, 2013, 45(2): 1–34.

[10] Jiang N, Lu X W, Hu H, et al. A Novel Quantum Image Compression Method Based on JPEG. *International Journal of Theoretical Physics*, 2018, 57(3): 611–636.

[11] Amankwah M G, Camps D, Bethel E W, et al. Quantum Pixel Representations and Compression for N-dimensional Images. *Scientific Reports*, 2022, 12(1): 7712.

[12] Sang J Z, Wang S, Li Q. A Novel Quantum Representation of Color Digital Images. *Quantum Information Processing*, 2017, 16(2): 42.

[13] Su J, Guo X, Liu C, et al. A New Trend of Quantum Image Representations. *IEEE Access*, 2020, 8: 214520–214537.

[14] Sun B, Iliyasu A M, Yan F, et al. An RGB Multi-Channel Representation for Images on Quantum Computers. *Journal of Advanced Computational Intelligence & Intelligent Informatics*, 2013, 17(3): 404–417.
[15] Sun B, Le P Q, Iliyasu A M, et al. A Multi-Channel Representation for Images on Quantum Computers Using the RGBα Color Space. *IEEE International Symposium on Intelligent Signal Processing, Floriana, Malta*, 2011: 1–6. Piscataway, NJ,USA: IEEE.
[16] Li H S, Zhu Q X, Zhou R G, et al. Multi-Dimensional Color Image Storage and Retrieval for a Normal Arbitrary Quantum Superposition State. *Quantum Information Processing*, 2014, 13(4): 991–1011.
[17] Jiang N, Zhao N, Wang L. LSB Based Quantum Image Steganography Algorithm. *International Journal of Theoretical Physics*, 2016, 55(1): 107–123.
[18] Li P, Liu X, Xiao H.A Quantum Image Representation Based on Bit-Plane Separation and Its Applications. Proceedings of the 2018 IEEE International Conference on Systems, Man, and Cybernetics (SMC 2018). Miyazaki, Japan,2018: 1262-1267. Piscataway, NJ, USA: IEEE.
[19] Şahin E, Yilmaz I. QRMW: Quantum Representation of Multi Wavelength Images. *Turkish Journal of Electrical Engineering & Computer Sciences*, 2018, 26(2): 768–779.
[20] Jiang N, Wang L, Wu W Y. Quantum Hilbert Image Scrambling. *International Journal of Theoretical Physics*, 2014, 53(7): 2463–2484.

6 Quantum Video Encryption

Zhiheng Zhao, Xianhua Song, Ahmed A. Abd El-Latif, Bassem Abd-El-Atty, and Basma Abd El-Rahiem

6.1 IMPROVED LOGISTIC MAP[1]

The Logistic map is a chaotic system[2–5] that utilizes a simple iterative formula to generate complex dynamic behaviors. It is particularly suitable for image and video encryption applications. Mathematically, the chaotic map is defined as:

$$x_{n+1} = \eta x_n (1 - x_n) \tag{6.1}$$

Here, $x_0 \in (0,1)$ represents the initial value, $n = 0,1,\ldots$, $0 \leq \eta \leq 4$. For $3.57 < \eta \leq 4$, the mapping is referred to as chaotic. The chaotic mapping utilized in this chapter is:

$$\begin{cases} x_{n+1} = L(\eta, x_n)G(k) - floor(L(\eta, x_n)G(k)) \\ L(\eta, x_n) = \eta x_n (1 - x_n) \\ G(k) = 2^k, k \in Z^+, k \geq 8 \end{cases} \tag{6.2}$$

The modified Logistic map explores the chaotic behavior when $0 \leq \eta \leq 4$. Figure 6.1 illustrates a comparison of the Lyapunov exponents between the standard Logistic map and the modified logistic map. The Lyapunov exponent is a metric used to quantify the sensitivity of chaotic mappings to initial conditions and control parameters.

From Figure 6.1a, it can be observed that the standard Logistic map exhibits a relatively large negative value region for certain parameter ranges. Conversely, Figure 6.1b indicates that the Lyapunov exponent of the modified Logistic map has no negative value region when $0 \leq \eta \leq 4$, thus demonstrating that the modified Logistic map possesses enhanced chaotic performance.

To more intuitively demonstrate the advantages of the modified Logistic map, a comparison of the histograms of data generated by the standard Logistic map and the modified Logistic map is presented in Figure 6.2. It is evident that the histogram of the modified Logistic map exhibits a more uniform distribution.

Considering these advantages, the designed quantum video encryption strategy will utilize the modified Logistic map for key generation.

6.2 QUANTUM VIDEO SCRAMBLING

The quantum video scrambling mechanism can be categorized into two types: inter-frame permutation and intra-frame pixel position scrambling.

 DOI: 10.1201/9781003621423-7

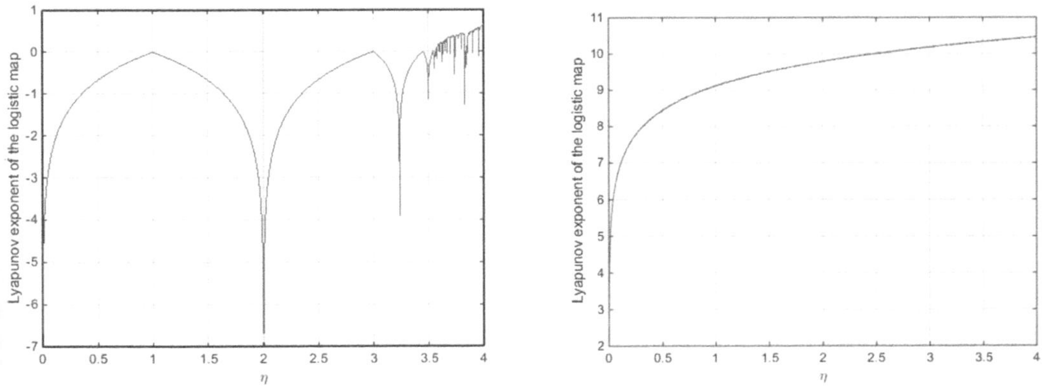

FIGURE 6.1 Lyapunov exponents of the Logistic map: (a) Standard Logistic map, (b) Modified Logistic map.

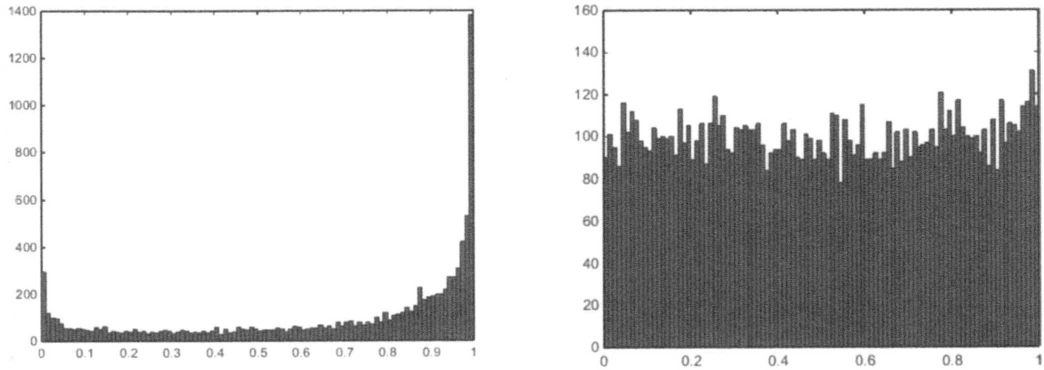

FIGURE 6.2 Histograms of the Logistic map: (a) Standard Logistic map, (b) Modified Logistic map.

6.2.1 Inter-Frame Quantum Video Frame Scrambling

For quantum video

$$\begin{aligned}|V\rangle &= \frac{1}{2^{m/2}}\sum_{j=0}^{2^m-1}\left|I_j\right\rangle\otimes|j\rangle \\ &= \frac{1}{2^{m/2+n}}\sum_{j=0}^{2^m-1}\sum_{i=0}^{2^{2n}-1}\left|c_{j,i}\right\rangle\otimes|i\rangle\otimes|j\rangle\end{aligned}$$

Inter-frame scrambling refers to

$$\begin{aligned}G_{inter} &= I^{q+n}\otimes S \\ &= I^{q+n}\otimes\left(|k\rangle\langle l|+|l\rangle\langle k|+\sum_{j\neq k,l}|j\rangle\langle j|\right)\end{aligned} \tag{6.3}$$

Here, I represents the identity transformation, and S denotes the transformation described in Equation 6.4:

$$\begin{aligned} G_{inter}(|V\rangle) &= G_{inter}\left(\frac{1}{2^{m/2+n}}\sum_{j=0}^{2^m-1}\left|I_j\right\rangle \otimes |j\rangle\right) \\ &= \frac{1}{2^{m/2+n}}\sum_{j=0}^{2^m-1}\sum_{i=0}^{2^{2n}-1}\left|c_{j,i}\right\rangle \otimes |i\rangle \otimes S(|j\rangle) \end{aligned} \tag{6.4}$$

6.2.2 Intra-Frame Video Frame Position Scrambling

If it is desired to exchange the positions of two pixels within a quantum video frame, the transformation can be carried out using the following formula (Equation 6.5):

$$\begin{aligned} G_{intra} &= I^q \otimes S \otimes I^m \\ &= I^q \otimes (|p\rangle\langle q| + |q\rangle\langle p| + \sum_{i\neq p,q}|i\rangle\langle i|) \otimes I^m \end{aligned} \tag{6.5}$$

Therefore,

$$\begin{aligned} G_{intra}(|V\rangle) &= G_{intra}\left(\frac{1}{2^{m/2+n}}\sum_{j=0}^{2^m-1}\left|I_j\right\rangle \otimes |j\rangle\right) \\ &= \frac{1}{2^{m/2+n}}\sum_{j=0}^{2^m-1}\sum_{i=0}^{2^{2n}-1}\left|c_{j,i}\right\rangle \otimes S(|i\rangle) \otimes |j\rangle \end{aligned} \tag{6.6}$$

6.3 QUANTUM VIDEO-CONTROLLED EXCLUSIVE OR OPERATION

The basic logical exclusive OR (XOR) operation [6, 7] outputs **true** if and only if the corresponding bits of the two 4-bit inputs differ – that is, for each bit position, if 1 bit is 1 and the other is 0, the output bit is 1. In quantum computing, for the input quantum states $|0\rangle, |1\rangle$, the quantum XOR operation can be implemented using a controlled-NOT (CNOT) gate. Figure 6.3 illustrates the quantum XOR circuit diagram for two quantum bit inputs.

The matrix representation of XOR is:

$$XOR = \begin{bmatrix} 1 & 0 & 0 & 0 \\ 0 & 1 & 0 & 0 \\ 0 & 0 & 0 & 1 \\ 0 & 0 & 1 & 0 \end{bmatrix}$$

Therefore, we have:

$$\text{XOR}(|ab\rangle) = |a\rangle|a \oplus b\rangle$$

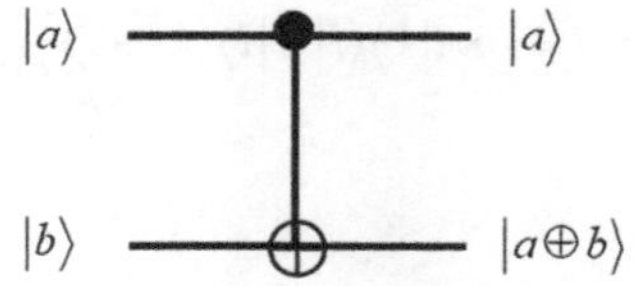

FIGURE 6.3 Quantum bit XOR circuit diagram.

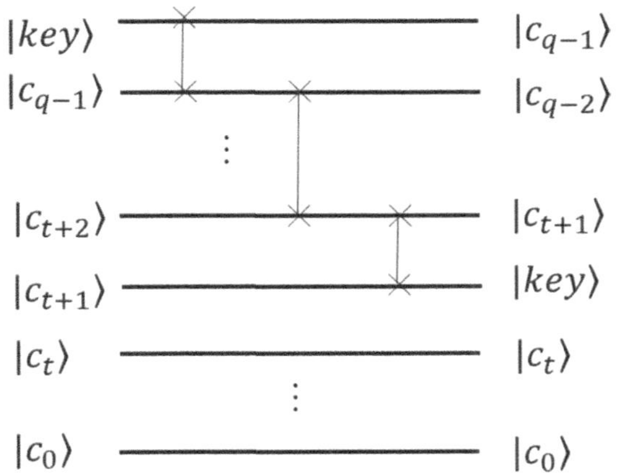

FIGURE 6.4 Quantum multi-step swap circuit (MS).

For the quantum video superposition state, if an XOR operation is to be performed between a quantum bit representing the key and a quantum bit in the color bit plane representing a pixel of a quantum video frame, the following unitary transformation is constructed: First, a multi-step swap operation is used to move the key bit to a position adjacent to the target quantum bit, as shown in Figure 6.4, where the target quantum bit $c_t^{j,i}$ is brought close to the key quantum bit.

$$\begin{aligned} |V_s\rangle &= MS(|V\rangle) \\ &= \frac{1}{2^{(\frac{m}{2}+n)}} \sum_{j=0}^{2^m-1} \sum_{i=0}^{2^{2n}-1} \left|c_{q-1}^{j,i} c_{q-2}^{j,i} \ldots c_{t+1}^{j,i}\right\rangle |key\rangle \left|c_t^{j,i} c_{t-1}^{j,i} \ldots c_0^{j,i}\right\rangle \otimes |ij\rangle \end{aligned} \tag{6.7}$$

Here, the purpose of the 2-qubit swap operation S is to exchange the positions of the 2 quantum bits, that is,

$$S = \begin{bmatrix} 1 & 0 & 0 & 0 \\ 0 & 0 & 1 & 0 \\ 0 & 1 & 0 & 0 \\ 0 & 0 & 1 & 0 \end{bmatrix}$$

Therefore, $S(|a,b\rangle) = |b,a\rangle$.

Then, the controlled quantum XOR transformation, as shown in the following equation, is used to implement the XOR operation for quantum video.

$$\begin{aligned} &|V_{k,l}\rangle \\ &= CXOR_{k,l}(|V_s\rangle) \\ &= I^{\otimes q-t-1} \otimes XOR \otimes I^{t-1} \otimes |kl\rangle\langle kl| \\ &\left(\frac{1}{2^{(\frac{m}{2}+n)}} \sum_{j=0}^{2^m-1} \sum_{i=0}^{2^{2n}-1} \left|c_{q-1}^{j,i} c_{q-2}^{j,i} \ldots c_{t+1}^{j,i}\right\rangle |key\rangle \left|c_t^{j,i} c_{t-1}^{j,i} \ldots c_0^{j,i}\right\rangle \otimes |ij\rangle \right) \\ &+ I^{\otimes q+1} \otimes \sum_{j=0, j\neq l}^{2^m-1} \sum_{i=0, i\neq k}^{2^{2n}-1} |ij\rangle\langle ij| \end{aligned}$$

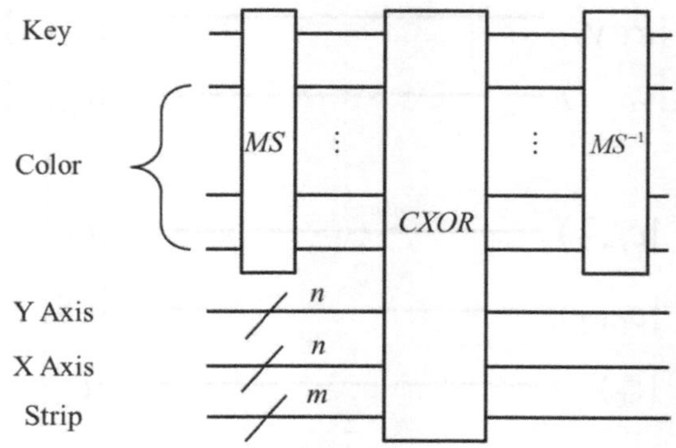

FIGURE 6.5 Controlled XOR operation for quantum video.

$$\left(\frac{1}{2^{\left(\frac{m}{2}+n\right)}}\sum_{j=0}^{2^m-1}\sum_{i=0}^{2^{2n}-1}\left|c_{q-1}^{j,i}c_{q-2}^{j,i}\ldots c_{t+1}^{j,i}\right\rangle|key\rangle\left|c_t^{j,i}c_{t-1}^{j,i}\ldots c_0^{j,i}\right\rangle\otimes|ij\rangle\right)$$
$$=\frac{1}{2^{\left(\frac{m}{2}+n\right)}}\left|c_{q-1}^{j,i}c_{q-2}^{j,i}\ldots c_{t+1}^{j,i}\right\rangle XOR(|key\rangle|c_t^{j,i}\rangle)\left|c_{t-1}^{j,i}\ldots c_0^{j,i}\right\rangle\otimes|kl\rangle$$
$$+\frac{1}{2^{\left(\frac{m}{2}+n\right)}}\sum_{j=0}^{2^m-1}\sum_{i=0}^{2^{2n}-1}\left|c_{q-1}^{j,i}c_{q-2}^{j,i}\ldots c_{t+1}^{j,i}\right\rangle|key\rangle\left|c_t^{j,i}c_{t-1}^{j,i}\ldots c_0^{j,i}\right\rangle\otimes|ij\rangle$$

By repeatedly performing the above operation, the controlled XOR operation on the lth quantum bit plane of the k-th pixel in the t-th frame of the quantum video can be achieved, that is,

$$|CV\rangle=\prod_{l=0}^{2^m-1}\prod_{k=0}^{2^{2n}-1}CXOR_{k,l}(|V_s\rangle)$$

Finally, the inverse multi-step operation of the process shown in Figure 6.4 is used to restore the position of the *key* bit:

$$|V_{CXOR}\rangle=MS^{-1}(|CV\rangle)$$

The entire controlled XOR transformation circuit is shown in Figure 6.5.

6.4 QUANTUM VIDEO ENCRYPTION AND DECRYPTION[8,9]

6.4.1 Quantum Video Encryption Algorithm[10–14]

As illustrated in the encryption flowchart depicted in Figure 6.6, the proposed quantum video encryption scheme[15–19] is implemented in three steps: scrambling of frame spatial positions, scrambling of intra-frame positions, and scrambling of quantum bit-plane positions within frames. In the frame spatial position scrambling step, the spatial positions of quantum video frames are transformed and re-encoded based on the scrambling of frame positions. During this process, the decision on which two frame images to exchange is determined by an improved Logistic mapping. Subsequently, the scrambling of each quantum video pixel position within a frame is determined by the chaotic sequences generated by the improved Logistic mapping. Finally, the controlled XOR operation on

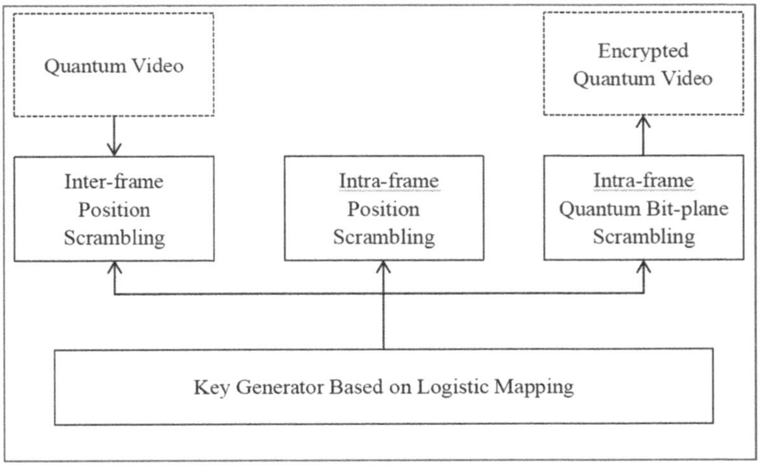

FIGURE 6.6 Quantum video encryption flowchart.

the quantum video bit-planes and the encryption *Key* generated by the Logistic mapping are used to diffuse the high 4 quantum bit-planes representing color information, thereby achieving encryption. Taking a grayscale quantum video as an example, the specific encryption process is as follows:

Step 1: Input the 8-qubit grayscale quantum video $|V\rangle$ to be encrypted. Assume the video consists of $M = 2^m$ frames, with each frame having a resolution of $N \times N = 2^n \times 2^n$. Set $lp = N^2$.

Note: If the number of quantum video frames is not an integer power of 2, additional blank image frames can be added to ensure that the video satisfies the size requirement of an integer power of 2.

Step 2: Inter-Frame Position Scrambling

1. Input the key $\{\mu_1, x_1\}$ and use the improved Logistic map to generate a sequence $K_1 = \{k_1, k_2, \ldots, k_{2^m}\}$ of length 2^m. Then, rearrange the elements of K_1 based on their values to obtain a new sequence $K_{1'} = \{k_{1'}, k_{2'}, \ldots, k_{2^{m'}}\}$ and a position sequence $P_1 = \{p_1, p_2, \ldots, p_{2^m}\}$. The size of P_1 is the same as K_1, and P_1 represents the indices of the elements in K_1 in the new sequence $K_{1'}$.
2. Utilize the position sequence P_1 and perform inter-frame position scrambling operations to rearrange the positions of the video frames, thereby obtaining the quantum video $|V_1\rangle$.

$$\begin{aligned} G_{inter_i} &= I^{q+n} \otimes S \\ &= I^{q+n} \otimes (|p_i\rangle\langle p_{i+1}| + |p_{i+1}\rangle\langle p_i| + \sum\nolimits_{j \neq p_i, p_{i+1}} |j\rangle\langle j|) \\ i &= 1, 3, 5, \cdots, 2^m - 1 \end{aligned} \tag{6.8}$$

Step 3: Intra-frame Pixel Position Geometric Transformation

For each frame in the quantum video $|V_1\rangle$:

1. Input the key $\{\mu_2, x_2\}$ as the initial values and execute the improved Logistic map lp times to generate the key sequence K_2. Rearrange the elements of K_2 based on their values to obtain a new sequence set $K_{2'}$ and the index sequence $P_2 = \{p_{1'}, p_{2'}, \ldots, p_{lp'}\}$. Clearly, P_2 and K_2 have the same size.

2. Utilize the index sequence P_2 and Equation 6.6 to scramble the pixel positions, thereby obtaining the quantum video $|V_2\rangle$:

$$G_{intra_i} = I^q \otimes (|p_{i'}\rangle\langle p_{i+1'}| + |p_{i+1'}\rangle\langle p_{i'}| + \sum_{i \neq p_{i'}, p_{i+1'}} |i\rangle\langle i|) \otimes I^m \quad i = 1,3,\ldots,lp-1 \tag{6.9}$$

$$|V_2\rangle = \prod_{i=1,3,\ldots,lp-1} G_{intra_i}(|V_1\rangle)$$

Step 4: Intra-Frame Quantum Qubit Plane Scrambling

1. Input the key $\{\mu_3, x_3\}$ as the initial values and execute the improved Logistic map $lp \times 2^m$ times to generate the key sequence K_3. Note that the decimal digits in positions 10–12 after the decimal point of each element in K_3 are extracted to form a new sequence $K_{3'}$, where all elements are integers. Then, calculate $K_{3'} \bmod 16$ to generate a random integer sequence K_4, with values ranging from 0 to 15. Finally, transform K_4 and apply it to the key sequence K_5 to generate a 4-bit binary key stream.
2. For the quantum video $|V_2\rangle$, utilize K_5 to perform controlled quantum XOR operations on the higher 4-qubit planes, thereby obtaining the encrypted quantum video $|V_3\rangle$.

$$|V_3\rangle = \prod_{l=0}^{2^m-1} \prod_{k=0}^{lp-1} CXOR_{k,l} \otimes MS(|V_2\rangle)$$

Note: If the quantum video to be encrypted is a colored QVNCQI, only Step 4 of the above encryption algorithm needs to be modified. For instance, if the colored image uses the RGB color model, the operation shown in Step 4 should be performed separately for each color channel, resulting in a total of three executions.

6.4.2 Quantum Video Decryption

In the context of quantum computing[20], the inverse process of encryption can be easily obtained, enabling the plaintext quantum video to be perfectly restored without any loss. The decryption algorithm is described as follows:

Step 1: Input the encrypted quantum video $|V_3\rangle$.
Step 2: Perform inverse intra-frame quantum qubit plane scrambling.

1. Input the key $\{\mu_3, x_3\}$ as the initial values and execute the improved Logistic map $lp \times 2^m$ times to generate the key sequence K_3. Then, use the method described in Step 4(1) of the encryption algorithm to obtain a new sequence $K_{3'}$. By calculating $K_{3'} \bmod 16$, generate a random integer sequence K_4 with values ranging from 0 to 15. Finally, transform K_4 into a 4-bit binary key stream K_5.
2. For the higher 4-qubit planes of $|V_3\rangle$, utilize the key K_5 to perform inverse controlled XOR operations, thereby obtaining the quantum video $|V_2\rangle$ after decryption:

$$|V_2\rangle = \prod_{l=0}^{2^m-1} \prod_{k=0}^{lp-1} CXOR_{k,l}^{-1} \otimes MS^{-1}(|V_3\rangle)$$

Step 3: Perform inverse intra-frame position scrambling.

For each frame in the quantum video $|V_2\rangle$, perform the following operations:

1. Input the key $\{\mu_2, x_2\}$ as the initial values and execute the improved Logistic map lp times to generate the key sequence K_2. Based on Step 3(1) of the encryption algorithm, the position sequence $P_2 = \{p_{1'}, p_{2'}, \ldots, p_{lp'}\}$ can be generated.
2. Utilize the position sequence P_2 and Equation 6.5 to reverse the scrambling of pixel positions, thereby obtaining the quantum video $|V_1\rangle$.

$$|V_1\rangle = \prod_{i=1,3,\ldots,lp-1} G_{intra_i}(|V_2\rangle)$$

Step 4: Perform inverse inter-frame position scrambling.

1. Input the key $\{\mu_1, x_1\}$ and use the improved Logistic map to iterate 2^m times, generating the sequence $K_1 = \{k_1, k_2, \ldots, k_{2^m}\}$ and the position sequence$P_1 = \{p_1, p_2, \ldots, p_{2^m}\}$. This step corresponds to the inverse process of Step 2(1) in the encryption procedure.
2. Utilize the position sequence P_1 and Equation 6.3 to reverse the scrambling of the video frame sequence, thereby restoring the original quantum video $|V\rangle$.

$$|V\rangle = \prod_{i=1,3,\ldots,2^m-1} G_{inter_i}(|V_1\rangle)$$

6.5 RESULT ANALYSIS

All simulations in this section are performed on a classical computer to represent and process quantum videos, analogous to quantum superposition states and unitary transformations in a quantum computer.

6.5.1 Quantum Video Encryption Demonstration

To demonstrate the effectiveness, two experimental demonstrations were conducted. One focuses on a single video frame, that is, a single quantum image. A grayscale image with a resolution of 256×256 was prepared and encrypted using Steps 3 and 4 of the encryption algorithm, which only involves intra-frame encryption. The experimental results are shown in Figure 6.7. The other experiment targets a video with 16 frames provided by MATLAB, where each frame has a resolution of 120×160. The simulation results are presented in Figure 6.8.

6.5.2 Security Analysis

For quantum video encryption, the cost of cracking the encryption key exceeds that of directly purchasing the video copyright. The encryption system proposed in this chapter is secure, with the encryption strength of the algorithm divided into three levels. First, inter-frame scrambling encryption establishes the foundational security of the video data. Second, intra-frame pixel position scrambling further enhances the security of the video. Finally, intra-frame pixel color quantum bit-plane encryption makes the quantum video encryption even more effective. As shown in Figures 6.7 and 6.8, the encrypted video is difficult to recognize. Additionally, the use of the Logistic map throughout the encryption process further improves the algorithm's security.

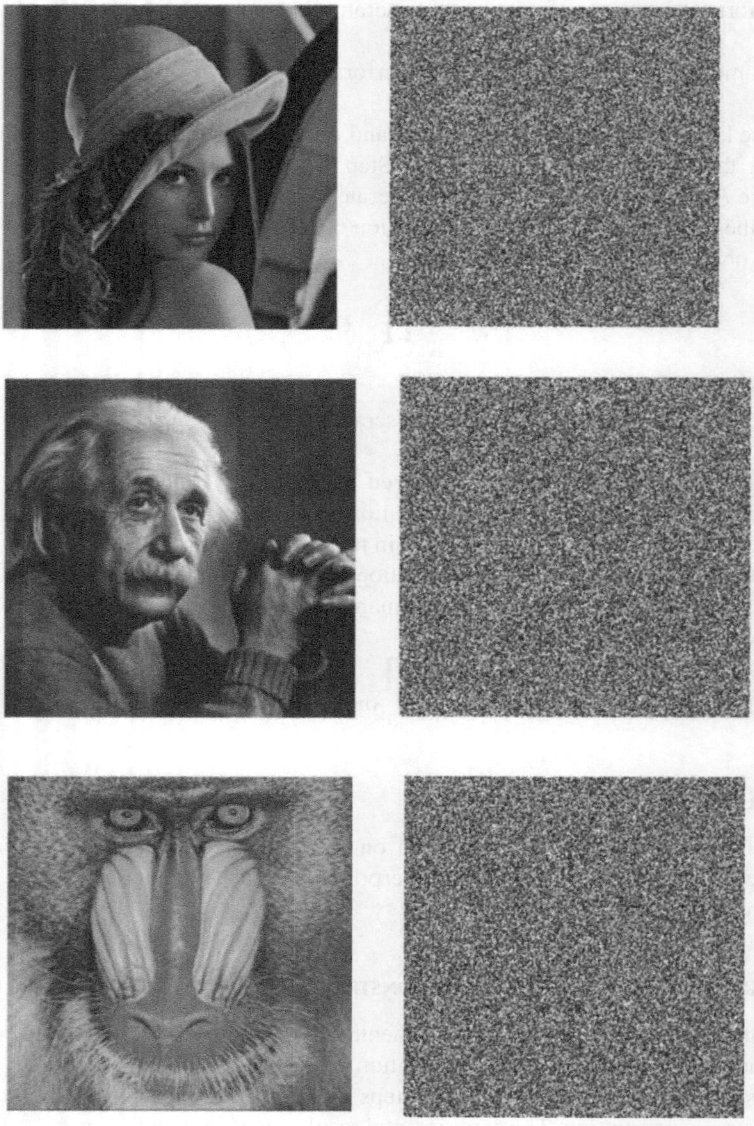

FIGURE 6.7 Quantum image encryption demonstration: (a) original image and (b) encrypted image.

6.5.3 Statistical Analysis

6.5.3.1 Histogram Analysis

The histogram of ciphertext is often used to estimate the degree of closeness between the information and a uniform distribution. The histograms of the encrypted images and video frames are shown in Figures 6.9 and 6.10.

Figure 6.9 illustrates the histogram comparison of three images – Lena, Einstein, and Baboon – before and after encryption. From the comparison, it is evident that the histograms of the encrypted quantum images and video frames are closer to a uniform distribution, thereby ensuring higher security. Figure 6.10 presents the histogram comparison of video frames before and after encryption across the three color channels. As the video is in color, the histograms of 16 video frames are

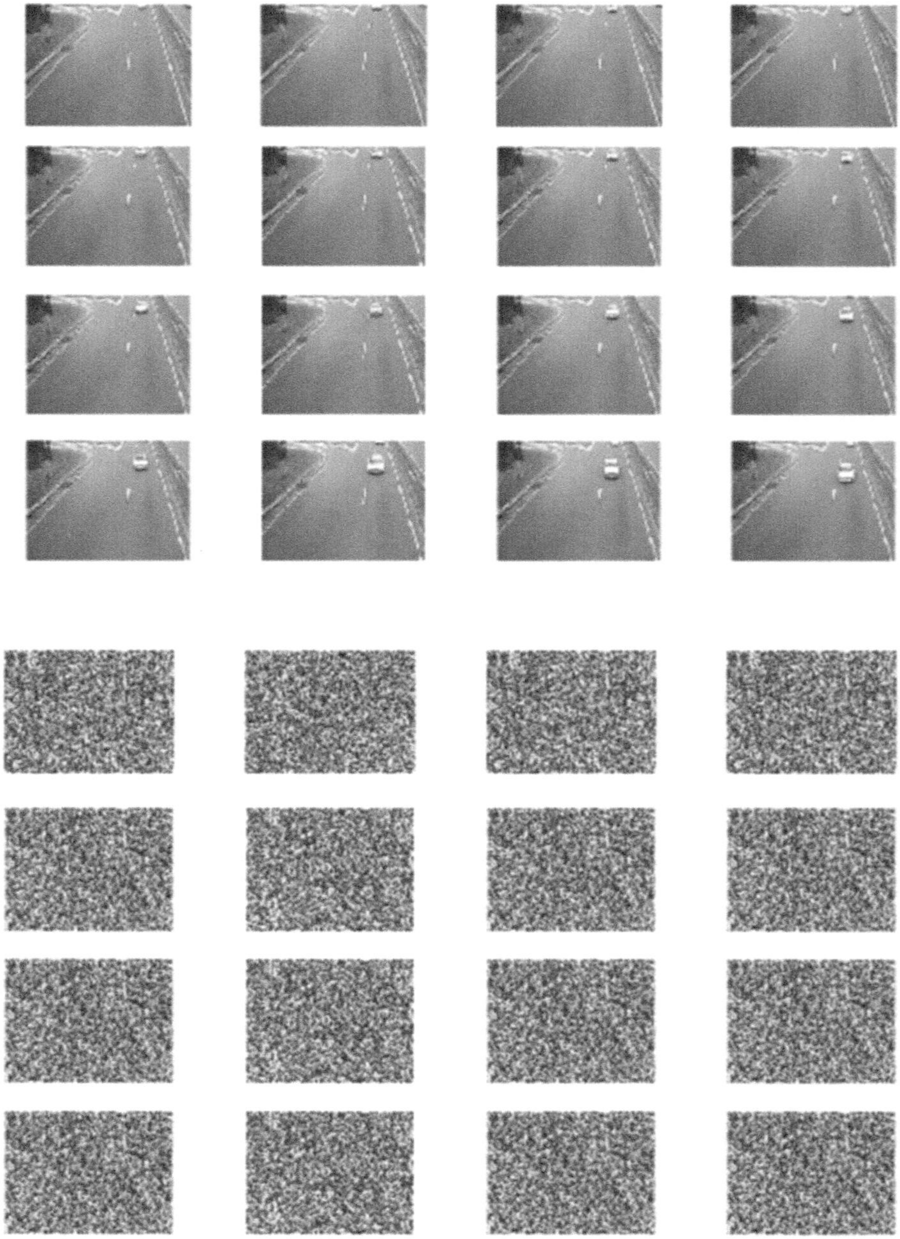

FIGURE 6.8 Quantum video encryption demonstration: (a) original video frames and (b) encrypted video frames.

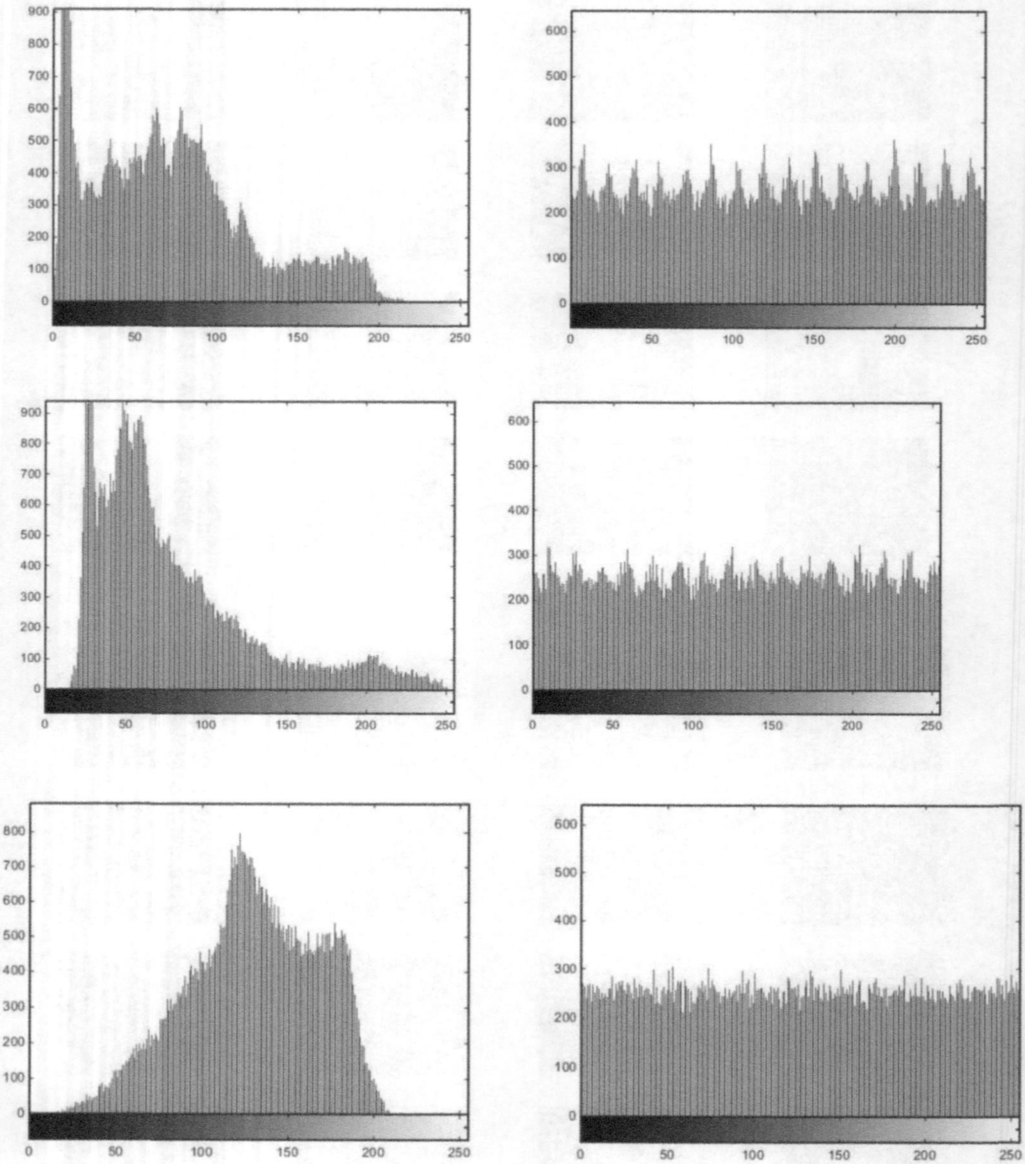

FIGURE 6.9 Histogram comparison: (a) histogram of the original image and (b) histogram of the encrypted image.

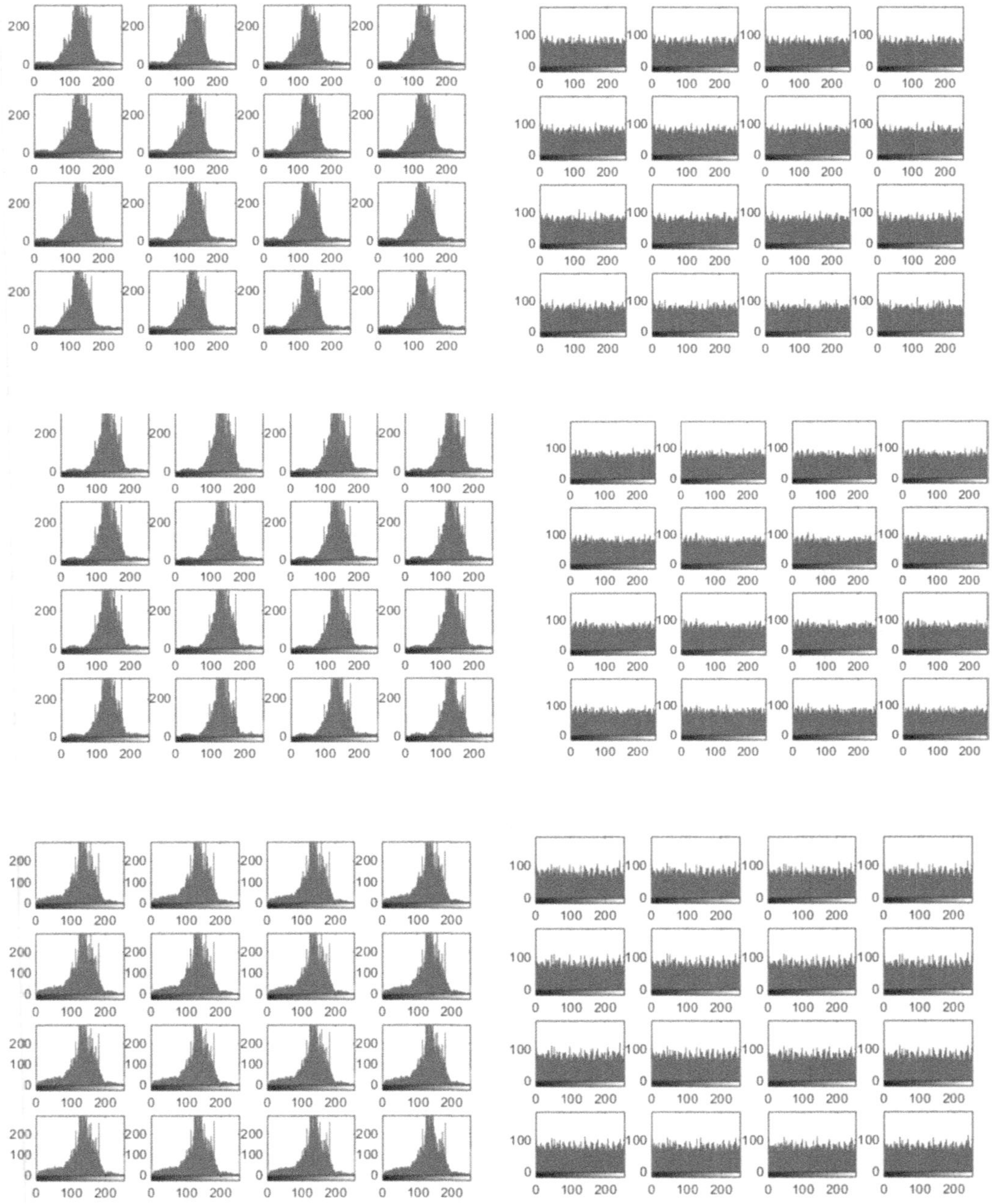

FIGURE 6.10 Histograms of video frames in R, G, and B channels: (a) histogram of the original video frame; and (b) histogram of the encrypted video frame.

TABLE 6.1
Information Entropy of Plaintext and Ciphertext Images

Test Media	Plaintext	Ciphertext
Lena	7.3275	7.9878
Einstein	7.2520	7.9934
Baboon	7.2269	7.9969
Video car	6.9417	7.8708

plotted separately for the R, G, and B channels. By comparing the two sets of histograms, it is clear that the encrypted quantum video frames exhibit greater randomness and histograms that are closer to a uniform distribution.

6.5.3.2 Information Entropy Analysis

Entropy is a quantitative measure of the average amount of information contained in an information source and can be used to evaluate the degree of disorder in a system. The information entropy of plaintext and ciphertext for quantum images and quantum video frames was calculated. Details are provided in Table 6.1.

As shown in Table 6.1, the information entropy of the encrypted images introduced in this chapter is 7.9878, 7.9934, and 7.9969, which are very close to the theoretical value of 8 bits. For quantum videos, the average information entropy of plaintext and ciphertext is 6.9417 and 7.8708, respectively. This indicates that the information leakage during the encryption process described in this chapter is negligible, further demonstrating the algorithm's robustness against entropy-based attacks.

6.5.3.3 Correlation Analysis

The correlation experiments for the quantum video encryption strategy proposed in this chapter are as follows: For each plaintext image (video frame) and ciphertext image (video frame), $P = 3000$ pairs of adjacent pixels (in the vertical, horizontal, and diagonal directions) are randomly selected for calculation. Figure 6.11 illustrates the correlation distribution of the quantum image Lena in three directions, while Figure 6.12 shows the correlation distribution of the quantum video frames in three directions. The corresponding correlation coefficients are detailed in Table 6.2. These values demonstrate the degree of decorrelation in the encrypted images and video frames across the three directions.

6.5.4 Key Space Analysis

The key space refers to the range of possible values that the encryption keys can take. A large key space ensures that the encryption algorithm is resistant to brute-force attacks. In the quantum video encryption algorithm described in this chapter, the keys are generated using three independent Logistic maps, with parameters $\{\mu_1, \mu_2, \mu_3\}$ and initial values $\{x_1, x_2, x_3\}$. Assuming the computer storage precision is 15 decimal places, the key space for each Logistic map is $2^{15} = 32768$. Therefore, the total key space for the three keys used in the encryption algorithm is $32768^6 = 1.23794003928538 \times 10^{27}$, which is sufficiently large to resist exhaustive search and brute-force attacks.

6.5.5 Complexity and Real-Time Performance Analysis

As video data encryption and decryption require real-time transmission, the encryption algorithm must not introduce significant delays to the video decoding process. This means that the

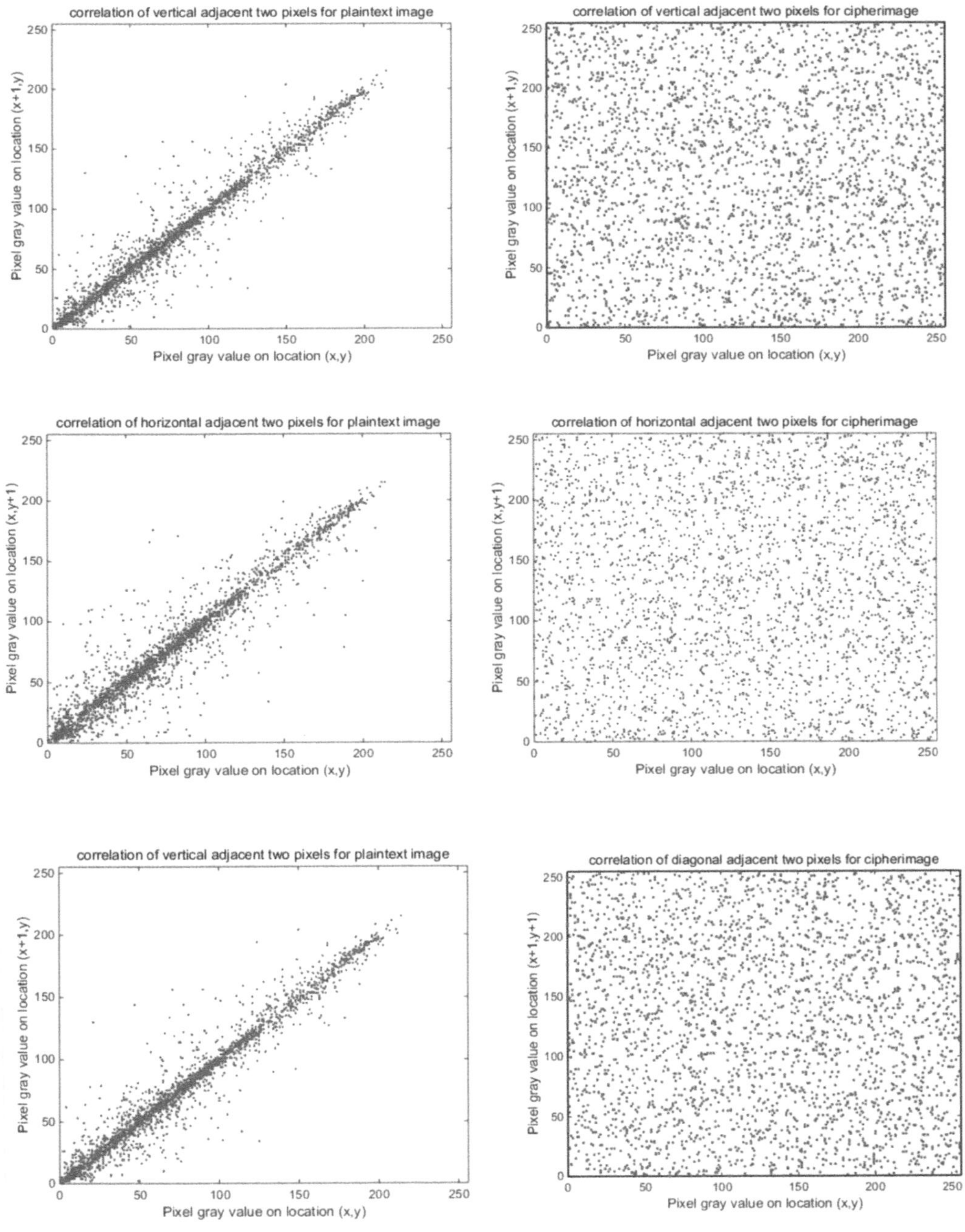

FIGURE 6.11 Correlation of adjacent pixels in the Lena image: (a) original image and (b) encrypted image.

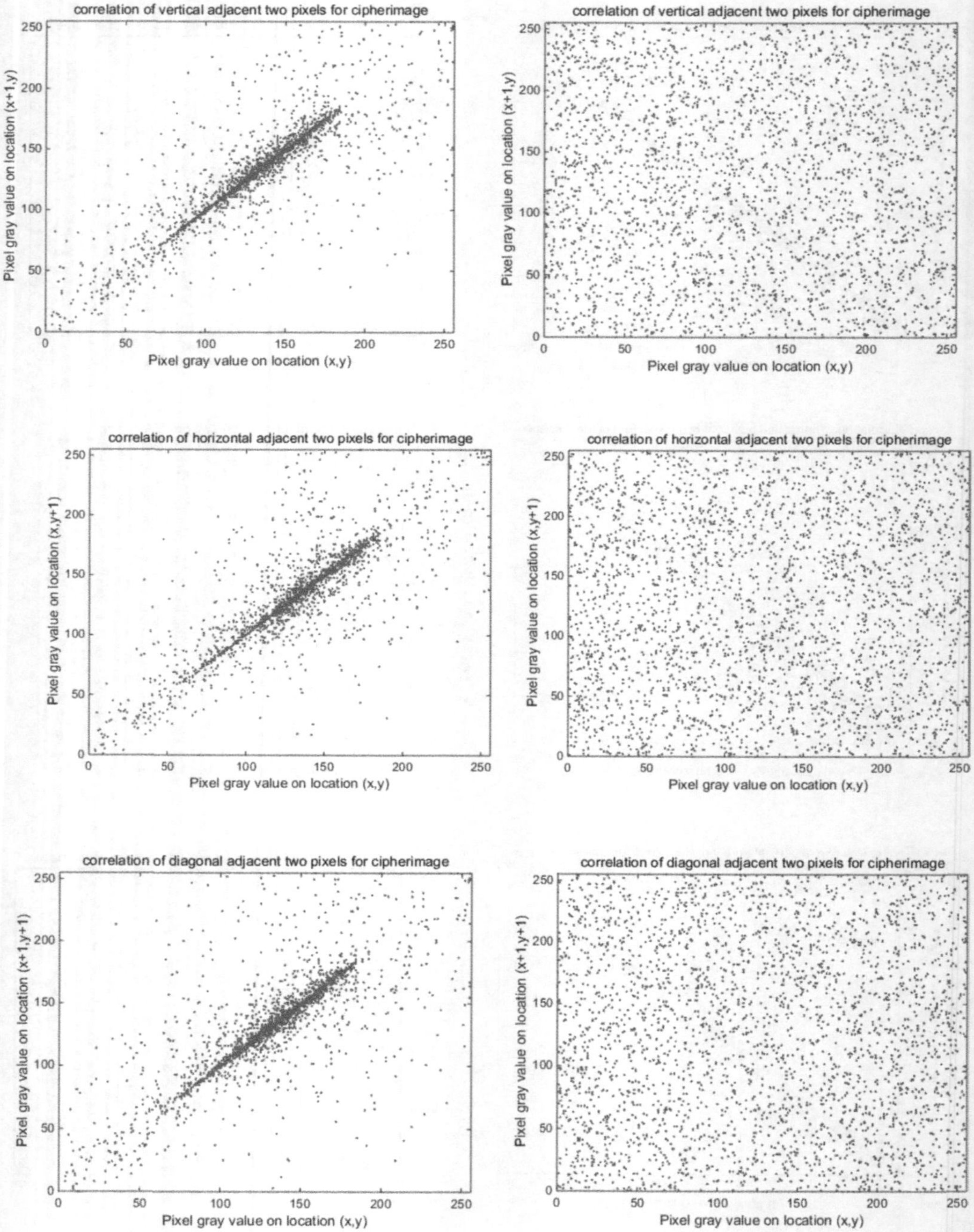

FIGURE 6.12 Correlation of adjacent pixels in video frames: (a) original video frame and (b) encrypted video frame.

TABLE 6.2
Correlation Coefficients of Adjacent Pixels in Plaintext and Ciphertext Images

Test Media	Image Type	Vertical	Horizontal	Diagonal
Lena	Plaintext	0.9754	0.9634	0.9433
	Ciphertext	0.0112	-0.0084	0.0044
Einstein	Plaintext	0.9739	0.9712	0.9565
	Ciphertext	-0.0033	0.0012	-0.0028
Baboon	Plaintext	0.8298	0.8822	0.8021
	Ciphertext	0.0116	-0.0473	-0.0005
Video car	Plaintext	0.8387	0.8257	0.7577
	Ciphertext	-0.0245	-0.0135	-0.0143

computational complexity of the encryption and decryption algorithms must be sufficiently low to meet the real-time requirements of video applications. In the encryption algorithm presented in this chapter, the complexity is primarily determined by the number of basic quantum gates used during the encryption process. For a quantum video of size $2^m \times 2^n \times 2^n$, the complexity is calculated as follows: 2^{m-1} inter-frame position scrambling operations, with each frame transformation G_{inter_i}, require $O\left(\frac{n^2}{2}\right)$ basic quantum gates. $2^m \times 2^{2n-1}$ intra-frame position scrambling operations, with each transformation G_{intra_i}, require $O(n^2)$ basic quantum gates.2^m multi-swap circuits MS, with each MS requiring six basic quantum gates. $4 \times 2^m \times 2^{2n}$ controlled XOR operations $CXOR_{k,l}$, with each $CXOR_{k,l}$ requiring $O(n^2)$ basic quantum gates. Thus, the total computational complexity of the quantum video encryption algorithm introduced in this chapter is:

$$\begin{aligned}&2^{m-1} \times O(\frac{n^2}{2}) + 2^m \times 2^{2n-1} \times O(n^2) + 6 \times 2^m + 4 \times 2^m \times 2^{2n} \times O(n^2)\\&= 9 \times 2^{m+2n-1} \times O(n^2) + m^2 \times O(2^{m-2}) + 6 \times O(2^m)\end{aligned}$$

For a quantum video of size $2^m \times 2^n \times 2^n$, the computational complexity is within an acceptable range, satisfying the real-time performance requirements.

6.5.6 Compression Ratio Analysis

An algorithm is referred to as a compression ratio-invariant algorithm if the data size remains unchanged before and after encryption. Compression ratio-invariant encryption algorithms do not alter the storage space required during storage and maintain consistent transmission speeds during data transfer. Therefore, an ideal video encryption algorithm should be compression ratio-invariant. The encryption and decryption algorithm introduced in this chapter ensures that the data size of the video remains unchanged both before and after encryption, thereby having no impact on the compression ratio.

6.6 SUMMARY OF THIS CHAPTER

This chapter proposes a quantum video encryption algorithm based on quantum controlled XOR operations and an improved Logistic map. The objective of video encryption research is to leverage advancements in the field of perceptual visual studies to determine the permissible range of visual perception for videos and to focus on encrypting perceptually significant regions under maximum security conditions. In the original domain of quantum videos, random numbers generated by the Logistic map are utilized to control the position scrambling and color diffusion operations in the quantum video encryption process. The proposed encryption algorithm combines chaos theory and quantum XOR operations to achieve secure and efficient quantum encryption of quantum video bitstreams.

REFERENCES

[1] Liu X B, Xiao D, Liu C. Three-Level Quantum Image Encryption Based on Arnold Transform and Logistic Map. *Quantum Information Processing*, 2021, 20(1): 23.

[2] Guitouni Z, Zayani M H. Quantum Chaotic Techniques for Medical Image Encryption in Next-Generation IoMT Applications. *The Journal of Supercomputing*, 2025, 81(9): 1070.

[3] Lalili Y, Bouden T, Grimes M, et al. FPGA Realization of a Controllable 4-D Hyperchaotic System for Quantum-Inspired Medical Image Encryption. *Physica Scripta*, 2025, 100(7): 075262.

[4] Abdelfatah I R, Elsobky M R, Khamis A S. Ultra-Secure Quantum Protection for e-Healthcare Images: Hybrid Chaotic One-time Pad with Cipher Chaining Encryption Framework. *Journal of King Saud University Computer and Information Sciences*, 2025, 37(6): 158.

[5] Zhang X, Wen H, Feng W, et al. Weighted Color Image Encryption Algorithm Based on RNA Extended Dynamic Coding and Quantum Chaotic System. *Entropy*, 2025, 27(8): 852.

[6] Gong L H, He X T, Cheng S, et al. Quantum Image Encryption Algorithm Based on Quantum Image XOR Operations. *International Journal of Theoretical Physics*, 2016, 55(7): 3234–3250.

[7] Song X H, Wang H Q, Venegas-Andraca S E, et al. Quantum Video Encryption Based on Qubit-Planes Controlled-XOR Operations and Improved Logistic Map. *Physica A: Statistical Mechanics and Its Applications*, 2019, 537: 122660.

[8] Yang Y G, Xia J, Jia X, et al. Novel Image Encryption/Decryption Based on Quantum Fourier Transform and Double Phase Encoding. *Quantum Information Processing*, 2013, 12(11): 3477–3493.

[9] Yan F, Iliyasu A M, Venegas-Andraca S E, et al. Video Encryption and Decryption on Quantum Computers. *International Journal of Theoretical Physics*, 2015, 54(8): 2893–2904.

[10] Wang H Q, Song X H, Chen L L, et al. A Secret Sharing Scheme for Quantum Gray and Color Images Based on Encryption. *International Journal of Theoretical Physics*, 2019, 58(5): 1626–1650.

[11] Wang H, Wang J, Geng Y C, et al. Quantum Image Encryption Based on Iterative Framework of Frequency-Spatial Domain Transforms. *International Journal of Theoretical Physics*, 2017, 56(8): 1–21.

[12] Ran Q W, Wang L, Ma J, et al. A Quantum Color Image Encryption Scheme Based on Coupled Hyper-Chaotic Lorenz System with Three Impulse Injections. *Quantum Information Processing*, 2018, 17(8): 1–30.

[13] Khan M, Rasheed A. Permutation-Based Special Linear Transforms with Application in Quantum Image Encryption Algorithm. *Quantum Information Processing*, 2019, 18(10): 1–21.

[14] Abd El-Latif A A, Abd-El-Atty B, Venegas-Andraca S E. Controlled Alternate Quantum Walk-Based Pseudo-Random Number Generator and Its Application to Quantum Color Image Encryption. *Physica A: Statistical Mechanics and its Applications*, 2019, 547: 123869.

[15] Li M, Song X, Zhao Y, et al. Space-Frequency-Based Multichannel Dual Encryption for Quantum Color Images Using Chaotic System and Quantum Walks. *Quantum Information Processing*, 2025, 24(9): 266.

[16] Yan F, Iliyasu A M, Venegas-Andraca S E, et al. Video Encryption and Decryption on Quantum Computers. *International Journal of Theoretical Physics*, 2015, 54: 2893–2904.

[17] Song X H, Wang S, Abd El-Latif A A, et al. Quantum Image Encryption Based on Restricted Geometric and Color Transformations. *Quantum Information Processing*, 2014, 13(8): 1765–1787.

[18] Zhou N R, Hua T X, Gong L H, et al. Quantum Image Encryption Based on Generalized Arnold Transform and Double Random-Phase Encoding. *Quantum Information Processing*, 2015, 14(4): 1193–1213.

[19] Song X, Chen G, Abd El-Latif A A. Quantum Color Image Encryption Scheme based on Geometric Transformation and Intensity Channel Diffusion. *Mathematics*, 2022, 10(17): 3038.

[20] Barenco A, Bennett C H, Cleve R, et al. Elementary Gates for Quantum Computation. *Physical Review A*, 1995, 52(5): 1–31.

7 Quantum Video Steganography

Zhiheng Zhao, Xianhua Song, Ahmed A. Abd El-Latif, and Basma Abd El-Rahiem

7.1 QUANTUM VIDEO FRAME MOTION DETECTION[1,2]

Quantum video motion can be determined by calculating the changes in corresponding blocks between two adjacent frames of the video^0. This process involves three steps: quantum video frame blocking, frame block comparison, and residual calculation.

7.1.1 Quantum Video Frame Blocking

For a quantum video frame of size $2^n \times 2^n$, the quantum pixels are encoded along the x-axis and y-axis using n quantum bits, respectively. Assuming the block size is 8×8, each frame can be divided into 2^{2n-6} image blocks. These 2^{2n-6} image blocks are then encoded, requiring $2n-6$ quantum bits to represent their positions. Specifically, $n-3$quantum bits are required for each of the x-axis and y-axis, and the binary representation of the block sequence number corresponds one-to-one with the encoding of these blocks.

To complete the frame blocking process, the quantum state $\left|F_j\right\rangle$ is encoded as follows:

$$
\begin{aligned}
\left|F_j\right\rangle &= \frac{1}{2^n}\sum_{i=0}^{2^{2n-1}}\left|c_{j,i}\right\rangle \otimes |i\rangle = \frac{1}{2^n}\sum_{y=0}^{2^n-1}\sum_{x=0}^{2^n-1}\left|c_{q-1}^{j,i}c_{q-2}^{j,i}\cdots c_0^{j,i}\right\rangle |y\rangle |x\rangle \\
&= \frac{1}{2^n}\sum_{y_i\in\{0,1\},x_i\in\{0,1\}}\left|c_{q-1}^{j,i}c_{q-2}^{j,i}\cdots c_0^{j,i}\right\rangle \left|\underbrace{000\cdots 0}_{n-3}\right\rangle \left|y_2y_1y_0\right\rangle \left|\underbrace{000\cdots 0}_{n-3}\right\rangle \left|x_2x_1x_0\right\rangle \\
&+ \frac{1}{2^n}\sum_{y_i\in\{0,1\},x_i\in\{0,1\}}\left|c_{q-1}^{j,i}c_{q-2}^{j,i}\cdots c_0^{j,i}\right\rangle \left|\underbrace{000\cdots 0}_{n-3}\right\rangle \left|y_2y_1y_0\right\rangle \left|\underbrace{000\cdots 1}_{n-3}\right\rangle \left|x_2x_1x_0\right\rangle + \cdots \\
&+ \frac{1}{2^n}\sum_{y_i\in\{0,1\},x_i\in\{0,1\}}\left|c_{q-1}^{j,i}c_{q-2}^{j,i}\cdots c_0^{j,i}\right\rangle \left|\underbrace{111\cdots 1}_{n-3}\right\rangle \left|y_2y_1y_0\right\rangle \left|\underbrace{111\cdots 0}_{n-3}\right\rangle \left|x_2x_1x_0\right\rangle \\
&+ \frac{1}{2^n}\sum_{y_i\in\{0,1\},x_i\in\{0,1\}}\left|c_{q-1}^{j,i}c_{q-2}^{j,i}\cdots c_0^{j,i}\right\rangle \left|\underbrace{111\cdots 1}_{n-3}\right\rangle \left|y_2y_1y_0\right\rangle \left|\underbrace{111\cdots 1}_{n-3}\right\rangle \left|x_2x_1x_0\right\rangle
\end{aligned}
$$

It can be observed that each term in the result represents an 8×8 image block and satisfies the formal structure of a quantum state.

DOI: 10.1201/9781003621423-8

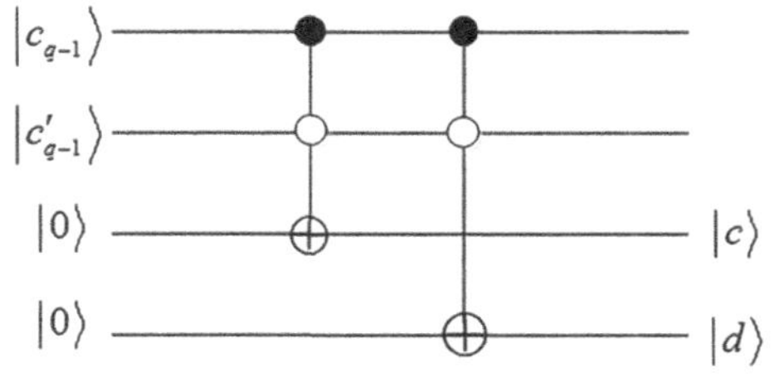

FIGURE 7.1 Quantum comparator.

7.1.2 Frame Block Comparison

The purpose of frame block comparison is to detect similar blocks between adjacent frames. Denote the image block sets of video frames $|F_1\rangle$ and $|F_2\rangle$ as $\{|B_i^1\rangle, i = 1,2,\cdots,2^{2n-6}\}$ and $\{|B_i^2\rangle, i = 1,2,\cdots,2^{2n-6}\}$, respectively. The pixel values $|p_s^{u,v}\rangle$, where $s = 1,2,\cdots,64$, $u = 0,1,\cdots,2^m - 1$ represent the sth pixel value in the v-th block of the uth frame.

The comparison of blocks can be performed using a quantum comparator (Figure 7.1). The comparison starts from the most significant bit and proceeds sequentially to compare each bit. Based on the results, the maximum range of the difference between two pixels $|p_s^{u,v_1}\rangle$ and $|p_s^{u+1,v_2}\rangle$ can be calculated: $2^{q-1} + 2^{q-2} + \cdots + 1 = 2^q - 1$.

The estimation of pixel differences can be calculated using Equation 7.1:

$$\begin{aligned}&\left|c'_{q-1} - c_{q-1}\right| \cdot 2^{q-1} + \left|c'_{q-2} - c_{q-2}\right| \cdot 2^{q-2} + \cdots + \left|c'_0 - c_0\right| \cdot 2^0 \\ &\le 2^{q-1} + 2^{q-2} + \ldots + 1 \\ &= 2^q - 1\end{aligned} \tag{7.1}$$

The computational complexity of the quantum comparator is $O(T)$. Therefore, the computational complexity of the video frame block comparison algorithm is $2^{2n-6} \cdot 64 \cdot 2O(T) = 2 \cdot 2^{2n} O(T)$, which falls within an acceptable range of complexity.

7.1.3 Residual Calculation

Let the adjacent video frames be denoted as $|F_1\rangle$ and $|F_2\rangle$. The tth block of the video frame $|F_k\rangle$ $(k = 1,2)$ is represented as $|B_t^k\rangle$, as shown in Equation 7.2.

$$|B_t^k\rangle = \frac{1}{2^n} \sum_{y_i \in \{0,1\}, x_i \in \{0,1\}} \left|c_{q-1}^{k,t} c_{q-2}^{k,t} \cdots c_0^{k,t}\right\rangle \left|t_{b_y}\right\rangle \left|y_2 y_1 y_0\right\rangle \left|t_{b_x}\right\rangle \left|x_2 x_1 x_0\right\rangle \tag{7.2}$$

Here, $|t_{b_y}\rangle$ and $|t_{b_x}\rangle$ represent the quantum states corresponding to the binary encoding of t required for the y-axis and x-axis, respectively.

Thus, the residual representation of the video frame blocks $|B_{t_1}^1\rangle$ and $|B_{t_1}^1\rangle$ is expressed as shown in Equation 7.3:

$$|\Delta B\rangle = \frac{1}{2^n} \sum_{y_i \in \{0,1\}, x_i \in \{0,1\}} \left|C^{1,t_1} - C^{2,t_2}\right\rangle \left|t_{b_y}\right\rangle \left|y_2 y_1 y_0\right\rangle \left|t_{b_x}\right\rangle \left|x_2 x_1 x_0\right\rangle \tag{7.3}$$

where: $|C^{1,t_1}\rangle = \left|c_{q-1}^{2,t_1} c_{q-2}^{2,t_1} \cdots c_0^{2,t_1}\right\rangle$, $|C^{2,t_2}\rangle = \left|c_{q-1}^{2,t_2} c_{q-2}^{2,t_2} \cdots c_0^{2,t_2}\right\rangle$.

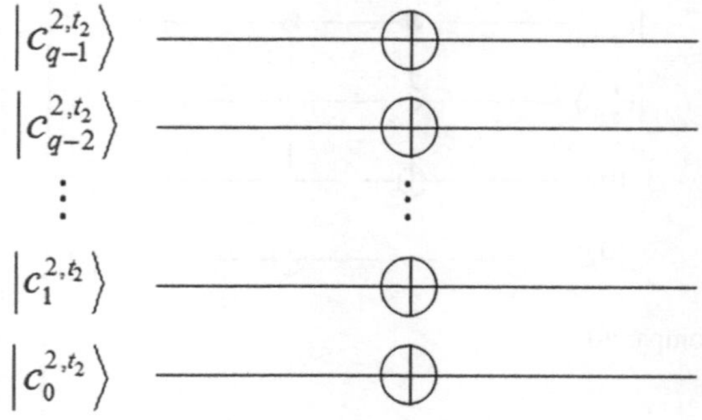

FIGURE 7.2 Quantum complement circuit.

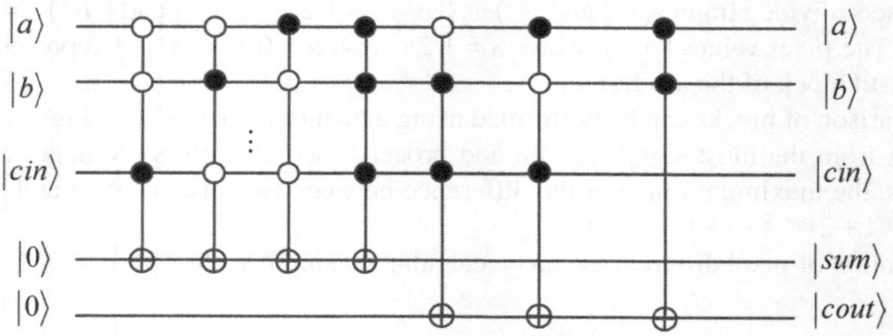

FIGURE 7.3 1-ADD quantum circuit.

The design of the quantum circuit for calculating the residual $|\Delta B\rangle$ is completed through the following two steps.

1. Design a quantum complement circuit to perform the complement operation on the quantum state $|C^{2,t_2}\rangle$. The circuit design is shown in Figure 7.2, resulting in the state $|\overline{C^{2,t_2}}\rangle$.
2. Utilize the q-ADD quantum circuit to compute the sum of $|C^{2,t_1}\rangle$ and $|\overline{C^{2,t_2}}\rangle$. The circuit design is shown in Figures 7.3 and 7.4.

The q-ADD circuit can implement the operation $|C\rangle = |A{+}B\rangle$, where $|A\rangle = |C^{2,t_1}\rangle$ and $|B\rangle = |\overline{C^{2,t_2}}\rangle$. This completes the calculation of the residual.

7.2 LEAST SIGNIFICANT QUANTUM BIT[3,4]

In Section 7.1, two quantum comparator circuits (Figure 7.1) were designed. The quantum circuit utilizes two quantum comparators controlled by a Toffoli gate to achieve the goal of determining whether two quantum states are identical. In Figure 7.1, $|a\rangle, |b\rangle$ represent two input quantum states, while $|c\rangle, |d\rangle$ represent the corresponding output states.

If $|c\rangle|d\rangle = |1\rangle|0\rangle$ or $|c\rangle|d\rangle = |0\rangle|1\rangle$, then the quantum state $|a\rangle$ is not equal to the quantum state $|b\rangle$. If $|c\rangle|d\rangle = |0\rangle|0\rangle$, then the quantum state $|a\rangle$ is equal to the quantum state $|b\rangle$. Clearly, the comparator can be used to compare the secret quantum state features with the carrier quantum image's features to determine whether the last quantum feature is identical.

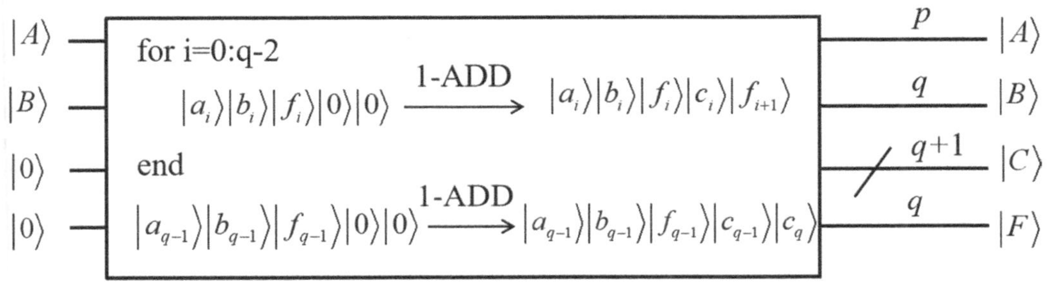

FIGURE 7.4 q-ADD quantum circuit.

The specific LSQb quantum image steganography algorithm consists of two steps: (1) quantum bit comparison and (2) LSQb steganography. In the first step, the quantum comparator shown in Figure 7.1 is used to compare the least significant quantum bit (LSQb) of the carrier image with the quantum bit of the secret information. Based on the output of the comparator, a unitary transformation is applied to the quantum states as needed. In the second step, if the output of the quantum comparator indicates equality, no changes are made to the color information of the carrier image. Otherwise, a unitary transformation can be defined and applied to the quantum carrier image as follows:

$$U_i = I^{\otimes q-1} \otimes U \otimes |i\rangle\langle i| + I^{\otimes q} \otimes \left(\sum_{j=0, j\neq i}^{2^{2n}-1} |j\rangle\langle j| \right) \tag{7.4}$$

where:

$$I = \begin{pmatrix} 1 & 0 \\ 0 & 1 \end{pmatrix}, \quad U = \begin{pmatrix} 0 & 1 \\ 1 & 0 \end{pmatrix},$$

Clearly, this unitary transformation is a controlled-NOT (CNOT) gate. The secret information needs to be embedded into all positions of the carrier image. If certain positions return identical values, the following unitary transformation is applied:

$$U_j = I^{\otimes q} \otimes \left(\sum_{j=0}^{2^{2n}-1} |j\rangle\langle j| \right) \tag{7.5}$$

From the above process, it can be observed that the quantum operation $\prod_{i=0}^{2^{2n-1}} U_i$ only modifies the LSQb of the color information corresponding to the pixels of the carrier image $|I\rangle$. If this operation is applied to the carrier image, it achieves the purpose of embedding the secret information into the quantum state of the carrier image:

$$|I'\rangle = \frac{1}{2^n} \sum_{i=0}^{2^{2n}-1} \left| c_{q-1}^i c_{q-2}^i \cdots c_1^i \right\rangle \left| \tilde{c}_0^i \right\rangle |i\rangle \tag{7.6}$$

$$\left| \tilde{c}_0^i \right\rangle = \begin{cases} \left| \bar{c}_0^i \right\rangle, & \left| c_{m0} \right\rangle \neq \left| c_0 \right\rangle \\ \left| c_0^i \right\rangle, & \left| c_{m0} \right\rangle = \left| c_0 \right\rangle \end{cases}$$

where $\left| \bar{c}_0^i \right\rangle$ represents the complement of $\left| c_0^i \right\rangle$.

7.3 STEGANOGRAPHIC ALGORITHM[5,6]

Due to the characteristics of quantum video superposition states, the first quantum qubit $|c_0\rangle$ and the quantum qubit $|c_{m0}\rangle$ containing the secret information cannot directly serve as the input to a quantum comparator. As a superposition state, it is not possible to directly process certain quantum qubits without considering other quantum qubits. Therefore, before extending the LSQb algorithm for quantum images, it is necessary to construct a controlled comparison (CC) operation over the entire quantum video superposition state to compare the least significant quantum qubit $|c_0\rangle$ of the carrier and the secret information encoded in $|c_{m0}\rangle$.

7.3.1 Quantum Video Controlled Comparator[7–9]

The QVNEQR quantum video can be reformulated as follows:

$$
\begin{aligned}
|V\rangle &= \frac{1}{2^{(\frac{m}{2}+n)}} \sum_{j=0}^{2^m-1} \sum_{i=0}^{2^{2n}-1} |c_{j,i}\rangle \otimes |ij\rangle \\
&= \frac{1}{2^{(\frac{m}{2}+n)}} \sum_{j=0}^{2^m-1} \sum_{i=0}^{2^{2n}-1} |c_{q-1}^{j,i} c_{q-2}^{j,i} \ldots c_0^{j,i}\rangle \otimes |ij\rangle
\end{aligned}
\tag{7.7}
$$

Here, $|ij\rangle$ represents the ith pixel of the jth frame in the video. In this way, the quantum circuit of the quantum video can be represented as shown in Figure 7.5. The first q qubits represent the color encoding, the next $2n$ qubits represent the position encoding, and the final m qubits represent the frame number 2^m.

Assuming the length of the secret information is equal to the minimum number of effective quantum qubits in the carrier video, if the secret information is too long, it can be divided into segments to match the qubit length. Conversely, if it is shorter, it can be padded with identical quantum qubits. Under this assumption, to compare the secret quantum state $|c_{m0}\rangle$ with the quantum state $|c_0\rangle$ of the first frame of the carrier video, a new quantum circuit is designed, as shown in Figure 7.6. In Figure 7.6, the three newly added quantum qubits are used to represent one qubit of secret quantum information and two auxiliary quantum qubits for storing the comparison results.

Note that the quantum secret information qubit $|0\rangle$ and the two auxiliary quantum qubits are not adjacent to the lowest effective qubit $|c_0\rangle$ of the quantum video. Therefore, it is necessary to first move them to adjacent positions. This can be achieved through the multi-swap quantum gates shown in Figure 7.7. The circuit in Figure 7.7 completes the process of swapping the secret information

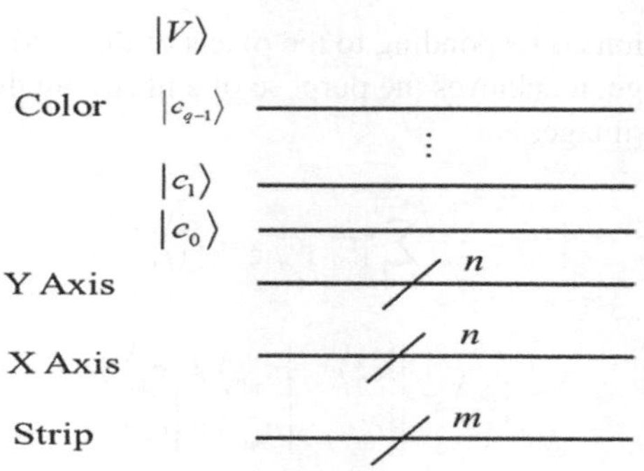

FIGURE 7.5 Quantum circuit of QVNEQR.

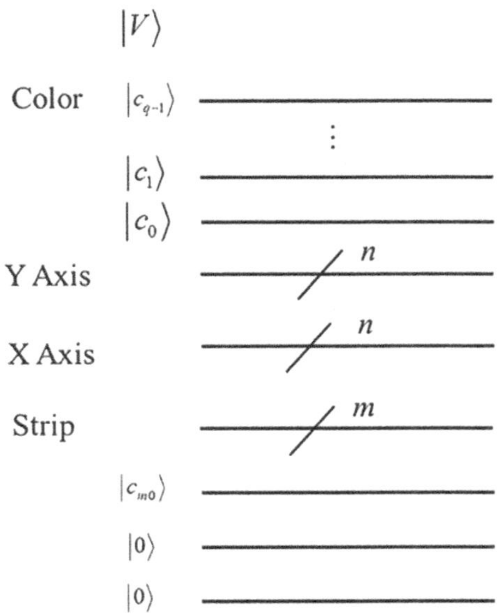

FIGURE 7.6 Quantum circuit for quantum video and quantum qubit secret.

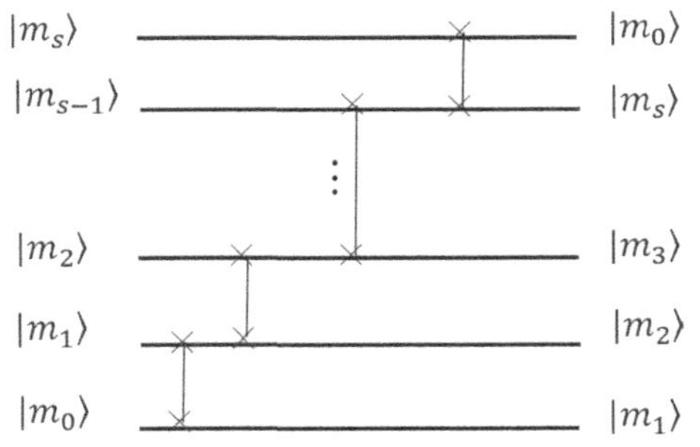

FIGURE 7.7 Quantum multi-swap circuit S.

qubit $|c_{m0}\rangle$ and the two auxiliary quantum qubits $|0\rangle$ to positions adjacent to the lowest effective qubit of the quantum video.

By applying the quantum multi-swap circuit[10–12] S three times, the quantum video state takes the following form:

$$|V_s\rangle = \frac{1}{2^{(\frac{m}{2}+n)}} \sum_{j=0}^{2^m-1} \sum_{i=0}^{2^{2n}-1} \left| c_{q-1}^{j,i} c_{q-2}^{j,i} \ldots c_0^{j,i} \right\rangle |c_{m0}\rangle |0\rangle |0\rangle \otimes |ij\rangle$$

After swapping the positions, to compare the lowest effective qubit $|c_0\rangle$ of QVNEQR with the secret quantum qubit $|c_{m0}\rangle$, the following controlled comparator operation is designed:

$$CC_{k,l} = I^{\otimes q-1} \otimes C \otimes |kl\rangle\langle kl| + I^{\otimes q+3} \otimes \sum_{j=0, j\neq l}^{2^m-1} \sum_{i=0, i\neq k}^{2^{2n}-1} |ij\rangle\langle ij| \quad (7.8)$$

$$l = 0,1,\cdots,2^m-1, \qquad k = 0,1,\cdots,2^{2n}-1$$

Here, the comparator C is the 2-qubit quantum comparator shown in Figure 7.1, which performs a comparison between two quantum qubits. The $CC_{k,l}$ operation enables the comparison between the secret information and the lowest effective qubit of the quantum video QVNEQR.

$$
\begin{aligned}
CC_{k,l}\left(\left|V_s\right\rangle\right) &= I^{\otimes q-1}\otimes C\otimes|kl\rangle\langle kl|\left(\frac{1}{2^{\left(\frac{m}{2}-n\right)}}\sum_{j=0}^{2^m-1}\sum_{i=0}^{2^{2n}-1}\left|c_{q-1}^{j,i}c_{q-2}^{j,i}\cdots c_0^{j,i}\right\rangle\left|c_{m0}\right\rangle|0\rangle|0\rangle\otimes|ij\rangle\right)\\
&+I^{\otimes q+3}\otimes\sum_{j=0,j\neq l}^{2^m-1}\sum_{i=0,j\neq k}^{2^{2n}-1}|ij\rangle\langle ij|\left(\frac{1}{2^{\left(\frac{m}{2}+n\right)}}\sum_{j=0}^{2^m-1}\sum_{i=0}^{2^{2n}-1}\left|c_{q-1}^{j,i}c_{q-2}^{j,i}\cdots c_0^{j,i}\right\rangle\left|c_{m0}\right\rangle|0\rangle|0\rangle\otimes|ij\rangle\right)\\
&=\frac{1}{2^{\left(\frac{m}{2}+n\right)}}\left|c_{q-1}^{j,i}c_{q-2}^{j,i}\cdots c_0^{j,i}\right\rangle\left|c_{m0}\right\rangle\left|c^{l,k}\right\rangle\left|d^{l,k}\right\rangle\otimes|kl\rangle\\
&+\frac{1}{2^{\left(\frac{m}{2}+n\right)}}\sum_{j=0}^{2^m-1}\sum_{i=0}^{2^{2n}-1}\left|c_{q-1}^{j,i}c_{q-2}^{j,i}\cdots c_0^{j,i}\right\rangle\left|c_{m0}\right\rangle|0\rangle|0\rangle\otimes|ij\rangle
\end{aligned}
\tag{7.9}
$$

To compare all quantum qubits, the following transformation is performed:

$$
CC=\prod_{l=0}^{2^m-1}\prod_{k=0}^{2^{2n}-1}CC_{k,l} \tag{7.10}
$$

$$
\begin{aligned}
CC\left(\left|V_s\right\rangle\right) &= \frac{1}{2^{\left(\frac{m}{2}+n\right)}}\sum_{j=0}^{2^m-1}\sum_{i=0}^{2^{2n}-1}\left|c_{q-1}^{j,i}c_{q-2}^{j,i}\ldots c_0^{j,i}\right\rangle\left|c_{m0}\right\rangle\left|c^{j,i}\right\rangle\left|d^{j,i}\right\rangle\otimes|ij\rangle\\
&\triangleq\left|V_{CC}\right\rangle
\end{aligned}
\tag{7.11}
$$

The complete quantum circuit of the quantum video LSQb algorithm is shown in Figure 7.8, which primarily consists of multi-swap gates, controlled comparator gates, LSQb, and inverse multi-swap gates.

7.3.2 Quantum Video Least Significant Bit[13,14] Information Hiding Algorithm

This section primarily focuses on the LSQb algorithm shown in Figure 7.8. The process of the quantum video LSQb algorithm is as follows:

1. If the output result of the quantum comparator is identical to $|c\rangle|d\rangle$, no operation is performed, that is,

$$
U_{k.l}=I^{\otimes q+3}\otimes\sum_{j=0}^{2^m-1}\sum_{i=0}^{2^{2n}-1}|ij\rangle\langle ij| \tag{7.9}
$$

2. If the output result of the comparator differs from $|c\rangle|d\rangle$, the following transformation can be defined to act on the carrier quantum video:

$$
CC_{k,l}=I^{\otimes q-1}\otimes C\otimes|kl\rangle\langle kl|+I^{\otimes q+3}\otimes\sum_{j=0,j\neq l}^{2^m-1}\sum_{i=0,i\neq k}^{2^{2n}-1}|ij\rangle\langle ij| \tag{7.10}
$$

$$l = 0,1,\cdots,2^m-1, \quad k = 0,1,\cdots,2^{2n}-1$$

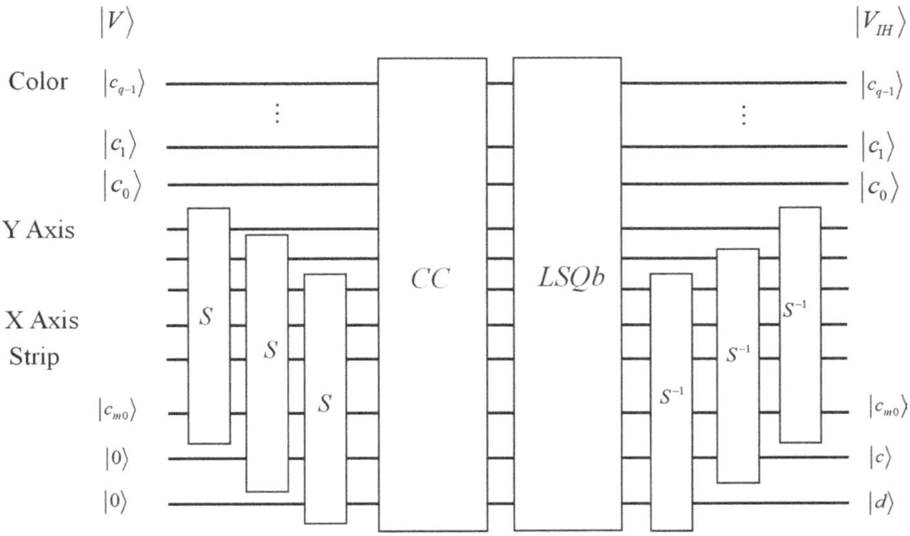

FIGURE 7.8 Quantum video information hiding circuit based on the LSQb algorithm.

Thus, the calculation process is as shown in the following equation:

$$
\begin{aligned}
&U_{k,l}\left(\left|V_{CC}\right\rangle\right) \\
&=\left(I^{\otimes q-1}\otimes U\otimes I^{\otimes 2}\otimes\left|kl\right\rangle\left\langle kl\right|+I^{\otimes q+3}\otimes\sum_{j=0,j\neq l}^{2^m-1}\sum_{i=0,i\neq k}^{2^{2n}-1}\left|ij\right\rangle\left\langle ij\right|\right) \\
&\left(\frac{1}{2^{\left(\frac{m}{2}+n\right)}}\sum_{j=0}^{2^m-1}\sum_{i=0}^{2^{2n}-1}\left|c_{q-1}^{j,i}c_{q-2}^{j,i}\cdots c_0^{j,i}\right\rangle\left|c_{m0}\right\rangle\left|c^{j,i}\right\rangle\left|d^{j,i}\right\rangle\otimes\left|ij\right\rangle\right) \\
&=\frac{1}{2^{\left(\frac{m}{2}+n\right)}}\left(I^{\otimes q-1}\otimes U\otimes I^{\otimes 2}\otimes\left|kl\right\rangle\left\langle kl\right|+I^{\otimes q+3}\otimes\sum_{j=0,j\neq l}^{2^m-1}\sum_{i=0,i\neq k}^{2^{2n}-1}\left|ij\right\rangle\left\langle ij\right|\right) \\
&\left(\sum_{j=0}^{2^m-1}\sum_{i=0}^{2^{2n}-1}\left|c_{q-1}^{j,i}c_{q-2}^{j,i}\cdots c_0^{j,i}\right\rangle\left|c_{m0}\right\rangle\left|c^{j,i}\right\rangle\left|d^{j,i}\right\rangle\otimes\left|ij\right\rangle\right) \\
&=\frac{1}{2^{\left(\frac{m}{2}+n\right)}}\left(\left|c_{q-1}^{k,l}c_{q-2}^{k,l}\cdots c_1^{k,l}\right\rangle U\left(\left|c_0^{k,l}\right\rangle\left|c_{m0}\right\rangle\right)\left|c^{k,l}\right\rangle\left|d^{k,l}\right\rangle\left|kl\right\rangle\right) \\
&+\frac{1}{2^{\left(\frac{m}{2}+n\right)}}\left(I^{\otimes q+3}\otimes\sum_{j=0,j\neq l}^{2^m-1}\sum_{i=0,i\neq k}^{2^{2n}-1}\left|ij\right\rangle\left\langle ij\right|\right)\left(\sum_{j=0}^{2^m-1}\sum_{i=0}^{2^{2n}-1}\left|c_{q-1}^{j,i}c_{q-2}^{j,i}\cdots c_0^{j,i}\right\rangle\left|c_{m0}\right\rangle\left|c^{j,i}\right\rangle\left|d^{j,i}\right\rangle\otimes\left|ij\right\rangle\right) \\
&=\frac{1}{2^{\left(\frac{m}{2}+n\right)}}\left|c_{q-1}^{j,i}c_{q-2}^{j,i}\cdots c_1^{k,l}\right\rangle\left|\tilde{c}_0^{k,l}\right\rangle\left|kl\right\rangle+\frac{1}{2^{\left(\frac{m}{2}+n\right)}}\sum_{j=0}^{2^m-1}\sum_{i=0}^{2^{2n}-1}\left|c_{q-1}^{j,i}c_{q-2}^{j,i}\cdots c_0^{j,i}\right\rangle\left|c_{m0}\right\rangle\left|c^{j,i}\right\rangle\left|d^{j,i}\right\rangle\otimes\left|ij\right\rangle
\end{aligned}
$$

where:

$$\left|\tilde{c}_0^{k,l}\right\rangle = \begin{cases} \left|\bar{c}_0^{k,l}\right\rangle, & \left|c_{m0}\right\rangle \neq \left|c_0^{k,l}\right\rangle \\ \left|c_0^{i}\right\rangle, & \left|c_{m0}\right\rangle = \left|c_0^{k,l}\right\rangle \end{cases}$$

$\left|\bar{c}_0^{k,l}\right\rangle$ represents the complement operation of $\left|c_0^{k,l}\right\rangle$.

$$\left|\bar{c}_0^{k,l}\right\rangle = \begin{cases} |1\rangle, & \left|c_0^{k,l}\right\rangle = |0\rangle \\ |0\rangle, & \left|c_0^{k,l}\right\rangle = |1\rangle \end{cases}$$

Obviously, the unitary transformation $U_{k,l}$ is a CNOT gate, and the secret information needs to be embedded into all positions of the carrier video frames, with each unitary transformation $U_{k,l}$ corresponding to the position (k,l), where $l = 0,1,\ldots,2^m - 1$, and$k = 0,1,\ldots,2^{2n} - 1$. From the above process, it can be concluded that the unitary operation only modifies the last quantum bit of the carrier's color encoding. If the following transformation is applied to the quantum video $\left|V_{CC}\right\rangle$, it can achieve the goal of embedding all secret information into the quantum video:

$$\left|V_{IH}\right\rangle = \frac{1}{2^{(\frac{m}{2}+n)}} \sum_{j=0}^{2^m-1} \sum_{i=0}^{2^{2n}-1} \left|c_{q-1}^{j,i} c_{q-2}^{j,i} \ldots c_1^{j,i}\right\rangle \left|\tilde{c}_0^{j,i}\right\rangle \left|c_{m0}\right\rangle \left|c^{j,i}\right\rangle \left|d^{j,i}\right\rangle \otimes |ij\rangle \tag{7.12}$$

Finally, as shown in Figure 7.8, after performing the CC operation and the LSQb algorithm, the inverse multi-swap transformation is used to restore the relative positions of the qubits.

7.4 QUANTUM VIDEO INFORMATION HIDING ALGORITHM BASED ON MOTION VECTORS

The high-capacity quantum video information hiding algorithm[15–20] can be achieved using the LSQb algorithm described in Section 7.2. However, this method has relatively low security, as anyone can extract and reconstruct the secret information. In classical video steganography, motion vector-based video information hiding leverages the unique characteristics of video media, making it difficult to detect using the existing steganalysis methods. Inspired by the steganographic methods based on motion vectors in classical videos, this section focuses on a quantum video steganography method based on motion vectors.

Using the motion vectors introduced in Section 7.1, the quantum video information hiding algorithm based on motion vectors is described as follows:

1. Assume that the residual difference in the motion vector is $|\Delta B\rangle$.

$$\begin{aligned} |\Delta B\rangle &= \frac{1}{2^n} \sum_{y_i \in \{0,1\}, x_i \in \{0,1\}} |C\rangle \left|t_{b_y}\right\rangle \left|y_2 y_1 y_0\right\rangle \left|t_{b_x}\right\rangle \left|x_2 x_1 x_0\right\rangle \\ &= \frac{1}{2^n} \sum_{y_i \in \{0,1\}, x_i \in \{0,1\}} \left|c_{q-1}^{t_1,t_2} c_{q-2}^{t_1,t_2} \cdots c_0^{t_1,t_2}\right\rangle \left|t_{b_y}\right\rangle \left|y_2 y_1 y_0\right\rangle \left|t_{b_x}\right\rangle \left|x_2 x_1 x_0\right\rangle \end{aligned}$$

Here, t_1 and t_2 represent the indices of the sub-blocks $\left|B_{t_1}^1\right\rangle$ and $\left|B_{t_2}^2\right\rangle$ in the neighboring video frames $\left|I_1\right\rangle$ and $\left|I_2\right\rangle$, respectively.$\left|t_{b_y}\right\rangle$ and $\left|t_{b_x}\right\rangle$ denote the binary-encoded quantum bits of the coordinate axes.

2. Similar to Figure 7.6, three quantum qubit lines are added to represent (a) the quantum qubit secret information and (b) two auxiliary quantum qubits. Then, a three-time quantum multi-swap circuit S is applied to move the secret quantum qubit $\left|c_{m0}\right\rangle$ and the two auxiliary quantum qubits $\left|0\right\rangle$ close to the least significant quantum qubit of the motion vector, resulting in:

$$\left|\Delta B_s\right\rangle = \frac{1}{2^n}\sum_{y_i\in\{0,1\},x_i\in\{0,1\}}\left|c_{q-1}^{t_1,t_2}c_{q-2}^{t_1,t_2}\cdots c_0^{t_1,t_2}\right\rangle\left|c_{m0}\right\rangle\left|0\right\rangle\left|0\right\rangle\otimes$$
$$\left|t_{b_y}\right\rangle\left|y_2y_1y_0\right\rangle\left|t_{b_x}\right\rangle\left|x_2x_1x_0\right\rangle$$

3 Construct the CC transformation:

$$CC_{k,l} = I^{\otimes q-1}\otimes C\otimes\left|t_{b_y}^{k}\right\rangle\left|y_2y_1y_0\right\rangle\left|t_{b_y}^{l}\right\rangle\left|x_2x_1x_0\right\rangle$$
$$+I^{\otimes q+3}\otimes\sum_{y_i\in\{0,1\},x_i\in\{0,1\}}\left|t_{b_y}\right\rangle\left|y_2y_1y_0\right\rangle\left|t_{b_x}\right\rangle\left|x_2x_1x_0\right\rangle$$

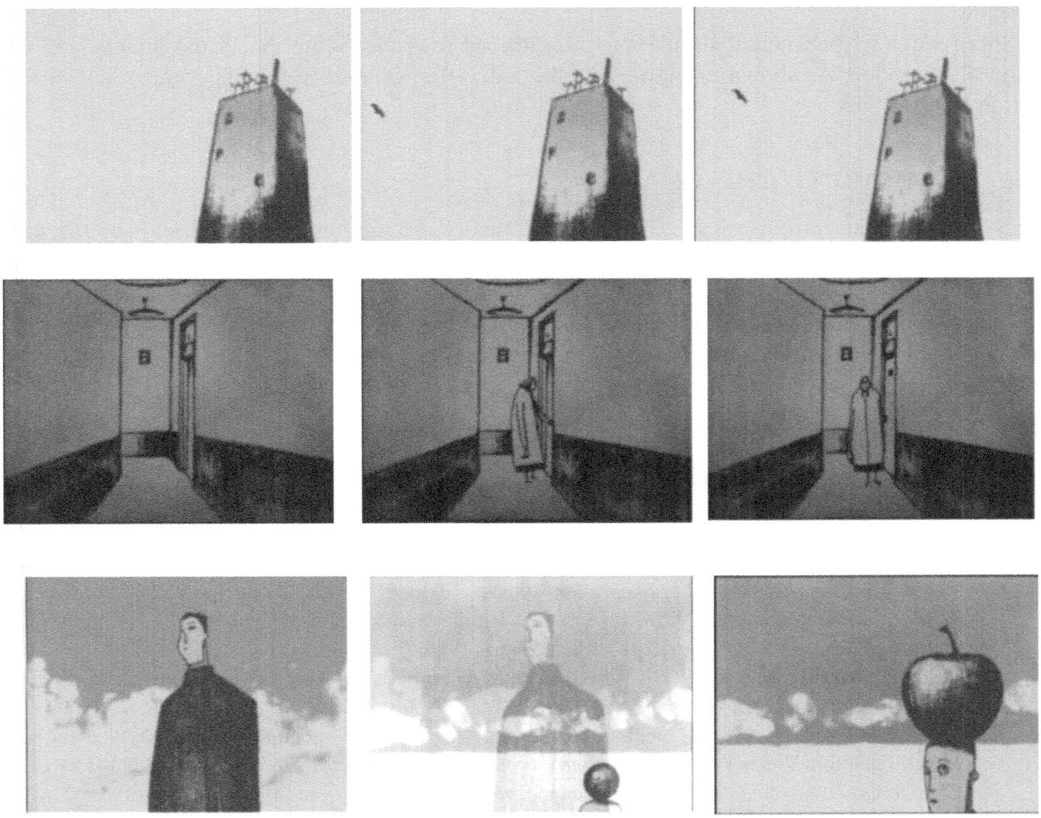

FIGURE 7.9 Selected video frames used for testing.

4. Construct the LSQb transformation:

$$U_{k,l} = I^{\otimes q-1} \otimes U \otimes I^{\otimes 2} \otimes \left| t_{b_y}^{k} \right\rangle \left| y_2 y_1 y_0 \right\rangle \left| t_{b_y}^{l} \right\rangle \left| x_2 x_1 x_0 \right\rangle + I^{\otimes q+3} \otimes \sum_{y_i \in \{0,1\}, x_i \in \{0,1\}} \left| t_{b_y} \right\rangle \left| y_2 y_1 y_0 \right\rangle \left| t_{b_x} \right\rangle \left| x_2 x_1 x_0 \right\rangle$$

5. After performing the CC transformation and the LSQb algorithm, three inverse multi-swap transformations S^{-1} are applied to restore the original encoding.

Some explanations:

Note 1: The specific implementation process of the above five steps is similar to the improved LSQb algorithm introduced in Section 7.2 and will not be elaborated here.

Note 2: The secret quantum bitstream can be accurately extracted using controlled swap gates. During extraction, it is only necessary to extend Figure 7.8 to quantum video or motion vectors, so it will not be repeated here.

7.5 EXPERIMENTAL SIMULATION

In this section, experimental simulations are utilized to demonstrate the visual quality and capacity of the video. As shown in Figure 7.9, the video frames used in the experiment have a resolution of 320×240.

7.6 SUMMARY OF THIS CHAPTER

This chapter first introduces the fundamental theory of motion detection in quantum video frames, which can be utilized to extract motion vectors from quantum videos. Subsequently, the quantum least significant bit (LSB) steganography method is presented. Finally, the quantum LSB steganography method is applied to the motion vectors of quantum videos, achieving a highly secure steganographic method for quantum video motion vectors.

REFERENCES

[1] Yan F, Iliyasu A M, Khan A R, et al. Moving Target Detection in Multi-Channel Quantum Video. *2015 IEEE 9th International Symposium on Intelligent Signal Processing (WISP) Proceedings*. IEEE, 2015: 1–5.

[2] Yan F, Iliyasu A M, Khan A R, et al. Measurements-Based Moving Target Detection in Quantum Video. *International Journal of Theoretical Physics*, 2016, 55: 2162–2173.

[3] Wang S, Song X. Quantum Video Information Hiding Based on Improved LSQb and Motion Vector. *Journal of Internet Technology*, 2017, 18(6): 1361–1368.

[4] Heidari S, Naseri M. A Novel LSB Based Quantum Watermarking. *International Journal of Theoretical Physics*, 2016, 55(10): 1–14.

[5] Qu Z, Chen S, Ji S. A Novel Quantum Video Steganography Protocol with Large Payload Based on MCQI Quantum Video. *International Journal of Theoretical Physics*, 2017, 56(11): 3543–3561.

[6] Chen S, Qu Z. Novel Quantum Video Steganography and Authentication Protocol with Large Payload. *International Journal of Theoretical Physics*, 2018, 57(12): 3689–3701.

[7] Xia H, Li H, Zhang H, et al. Novel Multi-Bit Quantum Comparators and Their Application in Image Binarization. *Quantum Information Processing*, 2019, 18(7): 1–17.

[8] Yewei Y, Chao W, Bei W, et al. An Improved QFT-Based Quantum Comparator and Extended Modular Arithmetic Using One Ancilla Qubit. *New Journal of Physics*, 2023, 25(10): 103011.

[9] Maity H. Design and Implementation of a Two-Qubit Quantum Comparator Circuit (Q-CC). *Journal of Computational Electronics*, 2022, 21(2): 1–5.

[10] Guang Y Y, Chen R H, Bao G X, et al. Measurement-Device-Independent Quantum Key Agreement Based on Entanglement Swapping. *Quantum Information Processing*, 2023, 22(12): 438.

[11] Wang P, Myers R C, Hollenberg L C L, et al. Let the Quantum Creep in: Designing Quantum Neural Network Models by Gradually Swapping out Classical Components. *Quantum Machine Intelligence*, 2025, 7(1): 62–62.

[12] Křižan C, Biznárová J, Chen L, et al. Quantum SWAP Gate Realized with CZ and iSWAP Gates in a Superconducting Architecture. *New Journal of Physics*, 2025, 27(7): 074507.

[13] Jiang N, Zhao N, Wang L. LSB Based Quantum Image Steganography Algorithm. *International Journal of Theoretical Physics*, 2016, 55(1): 107–123.

[14] Sang J Z, Wang S, Li Q. Least Significant Qubit Algorithm for Quantum Images. *Quantum Information Processing*, 2016, 15(11): 1–20.

[15] Ha D, Kim S J. Multi-Player Quantum Data Hiding by Nonlocal Quantum State Ensembles. *Quantum Information Processing*, 2025, 24(5): 132–132.

[16] Kumar V, Singh B, Kumar P. Hidden Quantum State and Signature of Mott Transition in 2D 1T–TaS2. *Physica Status Solidi (RRL) – Rapid Research Letters*, 2025, 19(8): 2500056.

[17] Banik S, Singh D, Roy S, et al. Hidden Bound States and Shannon Information Entropies of Quasi-Exactly Solvable Quantum Oscillators in Curved Space. *The European Physical Journal Plus*, 2025, 140(6): 599–599.

[18] Dalton J B. Underpinning Quantum Theory by a Non-Local Hidden Variable Theory: The Contextuality Loophole. *The European Physical Journal Special Topics*, 2025, (prepublish): 1–9.

[19] Xie Q. Hidden Lie Algebraic Structures and Exact Exceptional Spectrum of the Two-Photon Quantum Rabi Model with Asymmetric Static Bias. *Physica Scripta*, 2025, 100(8): 085124.

[20] Song X, Chen W, Lu J. Research on Financial Stock Market Prediction Based on the Hidden Quantum Markov Model. *Mathematics*, 2025, 13(15): 2505.

Part II

Quantum Communication and Cryptography

8 Principles of Classical and Quantum Information Theory

Bouchra El Alaoui, Abdallah Slaoui, and Rachid Ahl Laamara

8.1 INTRODUCTION

Information is the act of informing, giving knowledge of a fact, or a piece of news about something or someone. This information may be oral, written, or perceived. Today, the classical theory of information proposed by Claude Shannon [1, 2] is applied in the field of communications as well as in other disciplines, such as mathematics, statistical physics, and computer science. Classical information theory relates to encoding, transmission, and quantification of the quantity of information. Shannon's theory is also known in statistics by his famous entropy formula, which is analogous to Boltzmann's entropy formula. The information must be related to the physical properties of the material. In this sense, information is physical. Consequently, information can be encoded in quantum states, and quantum information can be processed. Quantum information theory is based on the fundamental concepts of quantum mechanical theory.

Quantum theory has been an amazing success over the last century. Beyond its explanatory power, it has enabled us to innovate in technology. Lasers, semiconductor devices, solar panels, and magnetic resonance imaging are examples of everyday technologies based on quantum theory and are now classified as "Quantum Technologies." The ultimate exploitation of quantum laws applied to information processing now promises to revolutionize the information and communication technology sector, ushering in the era of quantum technologies [3]. Secure quantum communication, enhanced quantum detection, and prospects of quantum computing were ignited by the realization that quantum characteristics par excellence, such as superposition and entanglement, can be used as resources for encoding and transmitting data in ways that are significantly more efficient or radically new than what is possible with conventional resources alone [4]. It is clear, then, that a deeper understanding of the most authentic features of quantum mechanics holds great promise for application in disruptive technologies.

Remarkably, the question of when a system behaves quantum-like rather than effectively classical still has no universal answer. This is especially true when analyzing composite systems and the nature of the correlations between their subsystems. Until recently, theoretical and experimental attention has been mainly devoted to entanglement between different subsystems [5]: entangled states exhibit a non-classical nature, and some of them may show even stronger deviations from classicality, such as steering [6] and nonlocality [7]. However, even non-entangled states do not lend themselves to a fully classical description in most cases. More general forms of quantum correlations, exemplified by the so-called quantum discord [8, 9], capture fundamental aspects of quantum theory, such as the fact that local measurements on parts of a composite system necessarily

DOI: 10.1201/9781003621423-10

induce a global perturbation of the state. If one recognizes a true "quantity" according to this paradigm, then almost all generally mixed quantum states of two or more subsystems display quantum correlations [10], even in the absence of entanglement.

Let us briefly describe the contents of each part of this chapter. In the first part, we describe the basic concepts of classical information theory that are useful for studying the quantity of information transmitted. We define Shannon's entropy, joint entropy, conditional entropy, and mutual information [11]. This provides us with the basis for the second part, where we study quantum information theory [12], in which information is represented by quantum states and evolves according to the laws of quantum mechanics. We introduce the formalism of the density operator and quantum measurements, then define joint, conditional, and relative von Neumann entropy types. We conclude this chapter with a third section based on several measures of quantum correlations, introducing the notion of entanglement entropy based on the notion of concurrence [13], and focusing on the definition of quantum discord [14], and its geometric version [15], based on the Hilbert–Schmidt norm and also logarithmic negativity [16], which is an entanglement measure for continuous variables.

8.2 CLASSICAL INFORMATION THEORY

Information theory is an application of probabilistic theory to quantify the average information content of a set of messages, particularly in the field of communications or signal transmissions through channels. The communication is first perceived as a transfer of information from the transmitter to the receiver. For reasons of simplification, the general layout of a communication system can be represented as shown in Figure 8.1.

Figure 8.1 shows the transmission of a message from an information source through a noisy channel and received by the receiver. Communication is then based on a chain of five steps as follows:

1. The information source (Source) sent a message.
2. The transmitter (Transmitter) receives and encodes this message in the form of an electrical signal.
3. This electrical signal is then transmitted through a channel that is subject to interference (Noise).
4. The receiver must decode the message received.
5. The message received by the destination (Destination) can be stored (information storage).

Classical information theory is a mathematical and physical representation. It is encoded by classical bits that can take either or two values, 0 or 1. A bit of information is sent with a certain probability. Its transmission may fail due to external noise distorting the signal or interception by an unwanted person. The quantity of information is measured using Shannon's entropy.

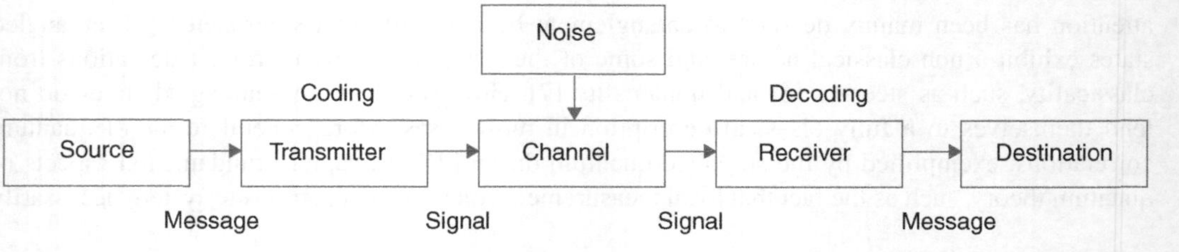

FIGURE 8.1 Shannon's general system of communication.

8.2.1 Definition of Shannon's Entropy

Shannon introduced the entropy of a probability to measure the quantity of information – or uncertainty – in a source of information. His definition was inspired by the concept of entropy originally introduced by Boltzmann and Gibbs to describe statistical mixtures in statistical thermodynamics [17,18], and was adapted to the probabilistic context.

Consider a source of information $\left(\{x_1, p_1\}, \{x_2, p_2\}, \dots, \{x_n, p_n\}\right)$. Let X be a random variable that gives the result x_i with probability p_i, then the function that allows us to measure the quantity of information provided by a random experiment, the so-called Shannon's entropy, is given by:

$$H(X) = -\sum p_{i \in X} \log_2(p_i) \tag{8.1}$$

with the convention $0 \log_2(0) = 0$.

Note that the entropy of a random variable X is a positive quantity and it does not depend on its values (i.e., the elements of the alphabet x_i), but depends solely on the values of the probabilities p_i.

8.2.2 Joint Entropy, Conditional Entropy, and Mutual Information

Joint entropy (respectively, conditional entropy) is a simple extension that measures the uncertainty in the joint distribution of a pair of variables (respectively, the uncertainty in the conditional distribution of a pair of random variables [11]).

The joint entropy is defined by:

$$H(X,Y) = -\sum_{x,y} p(x,y) \log_2(p(x,y)) \tag{8.2}$$

which represents the average quantity of information required to specify the value of two discrete random variables.

This grandeur quantifies the total uncertainty of the pair (X,Y), where $p(x,y)$ is the joint probability. Note that the joint entropy is symmetric $(\text{i.e. } H(X,Y) = H(Y,X))$. Moreover, if X and Y are independent, and through the additive property of Shannon's entropy, the joint entropy can be expressed as a sum of Shannon's entropy of each variable:

$$H(X,Y) = H(X) + H(Y). \tag{8.3}$$

The conditional entropy measures the remaining entropy of the random variable Y when the second random variable X is perfectly known. It represents the entropy of Y conditioned on X. This is denoted as the conditional entropy $H(X/Y)$, which measures the information in X when Y is known, is given by

$$H(X/Y) = -\sum_{x,y} p(x/y) \log_2(p(x/y)), \tag{8.4}$$

where the conditional probability is given by Bayer's law:

$$p(x/y) = \frac{p(x,y)}{p(y)} = \frac{p(y/x).p(x)}{p(y)} \tag{8.5}$$

The conditional and joint entropy types are connected by the following relationship:

$$H(X,Y)=H(X)+H(Y/X)=H(Y)+H(X/Y) \tag{8.6}$$

Indeed,

$$\begin{aligned}
H(X,Y) &= -\sum_{x,y} p(x,y)\,\log_2\big(p(x,y)\big) \\
&= -\sum_{x,y} p(x,y)\,\log_2\big(p(y/x).p(x)\big) \\
&= -\sum_{x,y} p(x,y)\,\log_2\big(p(x)\big)-\sum_{x,y} p(x,y)\,\log_2\big(p(y/x)\big) \\
&= -\sum_{x,y} p(x)\,\log_2\big(p(x)\big)-\sum_{x,y} p(y/x)\,\log_2\big(p(y/x)\big) \\
&= H(X)+H(Y/X)
\end{aligned}$$

Mutual information is the average quantity of information that the received message provides about the transmitted message. The mutual information $I(X,Y)$ measures the deviation from independence between *X* and Y. In other words, it is the quantity of information that one of the two variables in the pair (X,Y) provides about the other variable. The mutual information between the two discrete random variables X and Y, whose joint distribution is defined by $p(x,y)$, is given by the following relationship:

$$I(X,Y)=\sum_{x,y} p(x,y)\,\log_2\left(\frac{p(x,y)}{p(x)p(y)}\right) \tag{8.7}$$

Equation 8.7 can also be expressed in terms of joint entropy and conditional entropy as

$$\begin{aligned}
I(X,Y) &= \sum_{x,y} p(x,y)\log_2\left(\frac{p(x,y)}{p(x)p(y)}\right) \\
&= \sum_{x,y} p(x,y)\log_2\left(\frac{p(x/y)}{p(x)}\right) \\
&= -\sum_{x,y} p(x,y)\log_2(p(x/y))-\sum_{x,y} p(x,y)\log_2\big(p(x)\big) \\
&= -H(x/y)+H(x) \\
&= H(x)+H(y)-H(x,y)
\end{aligned}$$

Hence, the mutual information is written as

$$I(X,Y)=H(x)+H(y)-H(x,y). \tag{8.8}$$

Mutual information satisfies the following properties:

- $I(X, Y) = 0$ if and only if *X* and *Y* are independent random variables.
- It is positive $I(X, Y) \geq 0$.
- It is symmetrical $I(X, Y) = I(Y, X)$.

8.3 QUANTUM INFORMATION THEORY

Quantum information theory [12] is a domain based on classical information theory. The latter gives the mathematical formulation to describe the storage, transmission, and processing of information. In general, quantum information theory is the study of how these tasks can be accomplished, using quantum systems. The latter leads to several new functionalities, not only to improve the execution of certain information processing tasks but also to accomplish impossible or insoluble tasks in the classical domain (Figure 8.2).

8.3.1 Density Operator

In quantum mechanics, a Hilbert space *H* (state space) is associated with every system. It is an n-dimensional vector space with an orthonormal basis $\{|U_i\rangle, i = 0, \ldots, n-1\}$. Any element of the space H is called a ket vector and is denoted by $|\psi\rangle$, and its dimension depends on the considered system. Its dual vector is denoted by $\langle\psi|$. Any state represented by the vector $|\psi\rangle$ in *H*-space can be developed in an orthonormal basis as a linear combination:

$$|\psi\rangle = \sum_{i=0}^{n-1} \upsilon_i |U_i\rangle, \tag{8.9}$$

which is often referred to as superposition by quantum physicists, and the υ_i are complex coefficients.

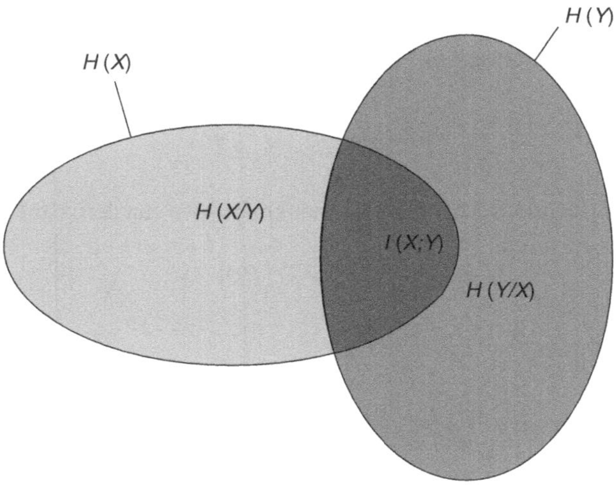

FIGURE 8.2 A traditional representation via a Venn diagram.

The probability of finding the system in state $|U_i\rangle$ is none other than $|\upsilon_i|^2$ with $\sum_{i=0}^{n-1}|\upsilon_i|^2 = 1$. Now, if we have two quantum systems with Hilbert spaces H_1 and H_2, the Hilbert space associated with the combined quantum system is $H = H_1 \otimes H_2$. It is important to note that these two Hilbert spaces could represent entirely different physical systems, for example, the first could be the space of electron spins, while the second could be the space of photon polarizations.

In this section, we discuss the form of the density operator (or density matrix) in pure and mixed quantum states and their corresponding properties. The latter is the tool used to describe all possible quantum states of a given physical system in a single matrix at a given instant. The density operator also allows for simplifying calculations in quantum information theory.

8.3.1.1 Pure State

A general quantum state is represented by a density operator ρ, which is a positive Hermitian operator with unitary trace, namely, an operator that satisfies the following three conditions:

$$\mathrm{Tr}(\rho) = 1, \qquad \rho = \rho^{+} \quad \text{and} \quad \langle U_i|\rho|U_i\rangle \geq 0$$

for all vectors $U_i \in \mathrm{H}$. The state of a quantum system is said to be pure if it has only one non-zero eigenvalue. It is characterized by a maximum of knowledge, and its density matrix corresponds to the projector on the state $|\psi\rangle$. It is written as $\rho = |\psi\rangle\langle\psi|$ whose entities are equal $\rho_{ij} = \langle U_i|\rho|U_j\rangle$. In this case, the knowledge about the preparation of the system is complete, and the preparation generates the desired state with certainty. Using the principles of quantum mechanics, we can obtain the following important properties:

1. If we perform a measurement of the physical quantity A that corresponds to the observable $\hat{A}$, the probability of obtaining the eigenvalue υ_i is

$$P(\upsilon_i) = |\langle U_i|\psi\rangle|^2 = \langle\psi|\hat{P}_i|\psi\rangle, \tag{8.10}$$

where $\hat{P}_i$ represents the projector $|U_i\rangle\langle U_i|$. The mean value of an operator $\hat{A}$ is expressed as a function of the density matrix as:

$$\langle\hat{A}\rangle = \sum_{i,j}\upsilon_i^{*}\upsilon_j A_{ij} = \sum_{i,j}\rho_{ij}\langle U_i|\hat{A}_i|U_j\rangle = \sum_{i}\langle U_i|\rho\hat{A}_i|U_i\rangle = Tr(\hat{A}\rho) \tag{8.11}$$

Therefore, from Equations 8.10 and 8.11_bookmark4 we can derive the following relationship

$$P(\upsilon_i) = Tr(\hat{P}_i\rho)$$

Furthermore, it is easy to check that

$$Tr(\rho) = \sum_{i=0}^{n-1}\langle U_i|\psi\rangle\langle\psi|U_i\rangle = \sum_{i=0}^{n-1}\upsilon_i^{*}\upsilon_i = \sum_{i=0}^{n-1}|\upsilon_i|^2 = 1$$

2. Suppose the measurement result is υ_i and the state of the system after measurement is $|\psi'\rangle = \frac{P_i|\psi}{\sqrt{\psi|P_i|\psi}}$. It is easy to check that the density operator associated with the measure υ_i is given by

$$\rho_{\upsilon_i} = \frac{P_i|\psi\rangle\langle\psi|P_i}{\langle\psi||\psi\rangle} = \frac{P_i\rho P_i}{P(\upsilon_i)}$$

3. The density matrix ρ is a projection operator, so we have $\rho^2 = \rho$ and $Tr(\rho^2) = 1$.
4. If the Hamiltonian describing the system is denoted $\widehat{H}(\mathrm{t})$ then the time evolution of the state $|\psi\rangle$ is

$$|\psi(\mathrm{t})\rangle = \mathrm{U}(\mathrm{t})|\psi(0)\rangle \Leftrightarrow \frac{\partial\|\psi(\mathrm{t})\rangle}{\partial \mathrm{t}} = \mathrm{i}\widehat{H}(\mathrm{t})|\psi(\mathrm{t})\rangle,$$

Where $\mathrm{U}(\mathrm{t})$ and $\widehat{H}(\mathrm{t})$ are related by: $\mathrm{U}(\mathrm{t}) = \exp\left[-\mathrm{i}\int\widehat{H}(\mathrm{t})\ t\mathrm{dt}\right]$. The time evolution of this system, described by the Schrodinger equation, can be written as a function of the density operator as follows:

$$\dot{\rho}(\mathrm{t}) + \mathrm{i}\left[\widehat{H}(t), \rho(t)\right] = 0 \tag{8.12}$$

where $[X,Y]$ represents the commutator of the two operators X and Y.

8.3.1.2 Mixed State

Now, consider the result of the measurement of a mixed quantum state. Suppose we have a mixture of pure quantum states $|\psi_\mathrm{i}\rangle$ with probability p_i, and each $|\psi_\mathrm{i}\rangle$ can be represented by a vector in H-space as Equation 8.9. The density matrix of a mixed state is defined as a statistical mixture of a set of pure states, and after the measurement, the quantum system probably takes one pure state among n possible states. Consequently, the density matrix describing the overall system is given by

$$\rho = \sum_\mathrm{i}\mathrm{p}_\mathrm{i}\rho_\mathrm{i} = \sum_\mathrm{i}\mathrm{p}_\mathrm{i}|\psi_\mathrm{i}\rangle\langle\psi_\mathrm{i}| \tag{8.13}$$

Where p_i is the probability that the system is in the normalized state $|\psi_\mathrm{i}\rangle$ and the sum is over all states accessible to the system. The probability p_i obviously satisfies

$$0 \le \mathrm{p}_\mathrm{i} \le 1, \quad \Sigma\mathrm{p}_\mathrm{i} = 1, \quad \mathrm{p}_\mathrm{i}^2 \le 1$$

For a pure state, only one p_i is equal to unity and all the others are zero. The properties of the pure state are also valid in the case of statistical mixing, except that $\rho^2 = \rho$. Then the density operator is not a projector ($\rho^2 \neq \rho$), and it is also easy to check that $\mathrm{Tr}(\rho^2) < 1$.

8.3.1.3 Reduced Density Operator

Sometimes we are only interested in one of the subsystems of a composite quantum system. This situation is described by the concept of the reduced density operator. Consider a quantum system AB consisting of two subsystems A and B in the bases $\{|\mathrm{k}\rangle, \mathrm{k} = 0,\ldots,\mathrm{d}_\mathrm{A} - 1\} \in \mathrm{H}_\mathrm{A}$ and $\{|\mathrm{m}\rangle, k = 0,\ldots,\mathrm{d}_\mathrm{B} - 1\} \in \mathrm{H}_\mathrm{B}$, respectively. If the total system is described by the density operator ρ_{AB}, then the subsystem A is described by the reduced density operator ρ_A. It is defined by

$$\rho_\mathrm{A} = \mathrm{Tr}_\mathrm{B}(\rho_\mathrm{AB}) = \sum_{\mathrm{m}=0}^{d_A - 1}\langle\mathrm{m}|\rho_\mathrm{AB}|\mathrm{m}\rangle, \tag{8.14}$$

where Tr_B is called the partial trace over the subsystem B. Similarly, for the subsystem B, we have

$$\rho_B = \mathrm{Tr}_A\left(\rho_{AB}\right) = \sum_{k=0}^{d_B-1} k\left|\rho_{AB}\right| k \tag{8.15}$$

The reduced density matrices have the properties of a density matrix. To see this, consider, for example, the case where ρ_{AB} is a density matrix of the pure state

$$\rho_{AB} = \left|\psi_{AB}\right\rangle\left\langle\psi_{AB}\right| \tag{8.16}$$

It is clear that, if $\{|k\rangle_A\}$ and $\{|m\rangle_B\}$ are orthonormal basis vectors of H_A and H_B, respectively, then $\{|k\rangle_A \otimes |m\rangle_B\}$ is an orthonormal basis for $H_A \otimes H_B$ and the state $|\psi_{AB}\rangle$ will be

$$|\psi_{AB} = \sum_{k,m} \alpha_{km} \left|k_A\right\rangle \otimes \left|m_B\right\rangle, \tag{8.17}$$

and the reduced density matrix ρ_A is written as

$$|\rho A\rangle = \sum_{k,k',m} \alpha_{km} \alpha^{*}_{k'm} \quad \left|k_A\right\rangle\left\langle k'_A\right| \tag{8.18}$$

Using the above expression (8.18), it is easy to see that ρ_A satisfies the properties of a density matrix:

1. $\rho_A = \rho_{A+}$
2. $\rho_A \geq 0$, that is, for all $|\phi\rangle \in H_A$, we have $\langle\phi|\rho_A|\phi\rangle \geq 0$,
3. $Tr\rho_A = 1$.

Moreover, using the general relation for the time evolution of the density operator (8.12), we obtain

$$\frac{\partial\rho_A(t)}{\partial t} = -i\left[\hat{H}(t), \rho_A(t)\right], \quad \frac{\partial\rho_B(t)}{\partial t} = -i\left[\hat{H}(t), \rho_B(t)\right] \tag{8.19}$$

8.3.2 Qubit and Bloch Sphere

In information theory, a bit is the basic unit of measurement of information in computing. It also represents the minimum quantity of information transmitted by a message. Quantum information theory uses a quantum entity to encode the elementary unit of information: Two-level quantum systems.

Two-level quantum systems have attracted several researchers because of their potential for manipulating quantum information [19,20] and quantum computation [21,22], and are now called qubits [14]. For example, we can consider the two polarization states of a photon or the two spin states of an electron (i.e., up and down). A qubit is the smallest unit of quantum information. It is also known as a quantum bit. The pure state of a qubit is defined as a linear superposition of calculated base states (conventionally, denoted by $|0\rangle$ and $|1\rangle$); namely, it is represented by the following vector:

$$|\psi\rangle = \alpha|0\rangle + \beta|1\rangle \tag{8.20}$$

where α and β are complex numbers satisfying the normalisation condition $|\alpha|^2 + |\beta|^2 = 1$ and the states $|0\rangle$ and $|1\rangle$ must be orthogonal.

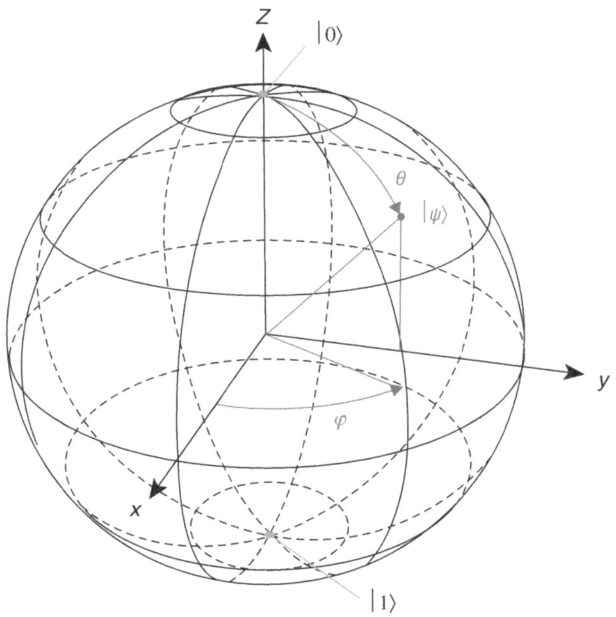

FIGURE 8.3 Geometric representation of a qubit on the Bloch sphere.

The probabilistic nature of the quantum world is reflected in the non-intuitive description of states and transformations associated with quantum systems. In the case of a qubit, the state vector, which consists of a combination of two states $|0\rangle$ and $|1\rangle$, is generally represented by a point on the Bloch sphere (see Figure 8.3). The pure state of a qubit in the Bloch geometric representation is given by the expression as follows:

$$|\psi\rangle = \cos\left(\theta_2\right)|0\rangle + \sin\left(\theta_2\right)e^{i\varphi}|1\rangle \tag{8.21}$$

This equation provides a representation of the state space of a qubit. The Bloch vector for the pure state of a qubit in three-dimensional Euclidean space $\vec{r} = (x, y, z)$ is given by the following coordinates:

$$x = \sin\theta \cos\varphi, \qquad y = \sin\theta \sin\varphi, \qquad z = \cos\theta$$

where the parameters $\theta \in [0, \pi]$ and $\phi \in [0, 2\pi]$ express the relative amplitude and the phase of the base states, respectively.

The density matrix of the pure state system corresponding to $|\psi\rangle$ can be expressed by the Pauli matrices $\vec{\sigma} = \left(\sigma_x, \sigma_y, \sigma_z\right)$ and the Bloch vector $\vec{r}$ such that:

$$\rho = |\psi\rangle\langle\psi|$$

$$= \begin{pmatrix} \cos^2\left(\frac{\theta}{2}\right) & e^{-i\varphi}\sin\left(\frac{\theta}{2}\right)\cos^2\left(\frac{\theta}{2}\right) \\ e^{i\varphi}\sin\left(\frac{\theta}{2}\right)\cos^2\left(\frac{\theta}{2}\right) & \sin^2\left(\frac{\theta}{2}\right) \end{pmatrix}$$

$$= \frac{1}{2}\begin{pmatrix} 1+\cos(\theta) & \sin(\theta)(\cos\varphi - i\sin\varphi) \\ \sin(\theta)(\cos\varphi + i\sin\varphi) & 1-\cos(\theta) \end{pmatrix}$$

$$= \frac{1}{2}\left(I + x\sigma_x + y\sigma_y + z\sigma_z\right)$$

$$= \frac{1}{2}\left(I + \vec{r}\vec{\sigma}\right)$$

8.3.3 Quantum Measurements

Quantum measurement is one of the most difficult and controversial problems in quantum mechanics. Some people think that, in reality, there is no problem at all; others have seen the problem of measurement as one of the great puzzles left unsolved by quantum mechanics. The problem is that quantum mechanics only provides the possibilities of different possible outcomes in an experiment, not a mechanism by which the actual outcome can be determined, which is observed at the end. In general, a quantum measurement is essentially an operation on a qubit, so that the quantum state lies in the superposition of the basis vectors to obtain a classical bit [23,24]. According to quantum theory, any physically measurable quantity associated with an observable (i.e., Hermitian operator) is characterized by a set of operators $\{M_k\}_{k=1}^{N}$ satisfy the constraint $\sum_{k=1}^{N} M_k^{+} M_k = 1$. These operators act on the state space of the system, and the state $|\psi\rangle \in H$ after the measurement becomes

$$|\psi\rangle \rightarrow |\psi_k\rangle \equiv \frac{M_k|\psi\rangle}{\sqrt{p_k}} \tag{8.22}$$

with the probability

$$p_k = \langle\psi|M_k^{+}\, M_k|\psi\rangle = \| M_k | |\psi\rangle \|^2 \geq 0$$

8.3.3.1 Projective Measurement

Projective measurements are a special case of generalized measurements, in which the operators of the measurement $\{M_k\}_{k=1}^{N}$, called projectors, are Hermitian operators. The number of these operators is equal to the dimension of the Hilbert space, and they are reproducible in the sense that if we perform a projective measure once and get the result v_k, repeating the measurement gives the result v_k and does not change the state. As a result of this, $M_k = P_k$ where $P_k P_l = \delta_{kl} P_k$ and $P_k^2 = P_k$ with P_k is a projector operator and δ_{kl} is the Kronecker symbol. By using this, the probability of observation of the result with index k in the measurement of a quantum system in the state $|\psi\rangle$ is

$$p_k = \langle\psi|M_k^{+} M_k|\psi\rangle = \langle\psi|P_k|\psi\rangle \tag{8.23}$$

and the state immediately after the measurement $|\psi\rangle$ is

$$|\psi\rangle \rightarrow |\psi_k\rangle \equiv \frac{M_k|\psi\rangle}{\sqrt{p_k}} \tag{8.24}$$

If we reapply the operator P_k, to the state $|\psi_k\rangle$, there is no change of state. The state after the second application of P_k is

$$|\psi\rangle \to |\psi'_k\rangle \equiv \frac{M_k|\psi\rangle}{\sqrt{p'_k}} \tag{8.25}$$

where $p'_k = \langle\psi|p_k|\psi\rangle$. We can now derive the expression for p'_k as

$$p'_k = \left[\langle\psi|\frac{P_k^+}{\sqrt{P_k}}\right]P_k\left[\frac{P_k}{\sqrt{P_k}}|\psi\rangle\right] = \frac{\langle\psi|P_k|\psi\rangle}{P_k} \tag{8.26}$$

Therefore,

$$\sqrt{p'_k P_k} = \sqrt{\langle\psi|P_k|\psi\rangle} = \sqrt{p_k} \tag{8.27}$$

Finally, we see that the second measure does not affect the state

$$|\psi'_k\rangle = \frac{P_k}{\sqrt{p'_k}} \times \frac{P_k}{\sqrt{p_k}}|\psi\rangle = \frac{P_k}{\sqrt{p_k}}|\psi\rangle = |\psi_k\rangle \tag{8.28}$$

Consequently, we can say that the result observed after a projective measure is deterministic.

8.3.3.2 Positive-Operator Valued Measure

Projective measures are not always possible; they are restrictive; sometimes we destroy a quantum particle in the measurement process, and therefore the repeatability of the projective measurement is violated. For example, a photon may be absorbed by a polarization filter and therefore not available for measurement. When a system is only measured once, the state of the system after the measurement is less important, and it is the probabilities of the measurement results that count.

The positive-operator valued measure (POVM) is a type of generalized measurement that concerns only measurement statistics. POVMs need not be orthogonal or commutative, and allow measurement results associated with non-orthogonal states.

Note that the measurement operators corresponding to non-orthogonal states do not switch and are therefore not simultaneously observable. The number of POVMs may differ from the dimension of the Hilbert space.

Consider the set of measurement operators $\{M_i\}$ that describes a measurement performed on a quantum system in the state $|\psi\rangle$. A POVM has the elements A_i defined by

$$A_i = M_i^+ M_i \quad \text{and} \quad \sum_i A_i = I$$

The set of operators A_i have the following properties:

1. They are defined positively (remember that a positively defined operator has real and positive eigenvalues). This property follows directly from the definition of A_i

2. $p(i) = \langle \psi | A_i | \psi \rangle$ and this is due to the fact that $p(i) = \langle \psi | M_i^+ M_i | \psi \rangle$. So the set of operators $\{A_i\}$ can be used instead of $\{M_i\}$ to estimate the probabilities of different measures.

The elements of a POVM are not measurement operators. Only in the case of projective measurements, all the POVM elements A_i, where

$$A_i \equiv P_i^+ P_i = P_i$$

are the same as measurement operators. POVM measurements prevent us from making identification errors, but sometimes they can provide no information at all. The POVM makes it possible to distinguish between cases where the result of a measurement identifies the state of origin with certainty and those where identification is not possible. POVM measurements can be optimized, meaning that the average information obtained from the measurement is maximized.

Example: Consider a qubit in one of two possible states:

$$|\psi_1\rangle = 0 \Rightarrow |\psi_1\rangle = \begin{pmatrix} 1 \\ 0 \end{pmatrix} \quad \text{or} \quad |\psi_2\rangle = \frac{|0\rangle - |0\rangle}{\sqrt{2}} \Rightarrow |\psi_2\rangle = \frac{1}{2}\begin{pmatrix} 1 \\ -1 \end{pmatrix}.$$

We define a POVM with three operators (elements). The first two are

$$A_1 = \frac{\sqrt{2}}{1+\sqrt{2}} |1\rangle\langle 1| \Rightarrow A_1 = \frac{\sqrt{2}}{1+\sqrt{2}} \begin{pmatrix} 0 \\ 1 \end{pmatrix} \begin{pmatrix} 0 & 1 \end{pmatrix} = \frac{\sqrt{2}}{1+\sqrt{2}} \begin{pmatrix} 0 & 0 \\ 0 & 1 \end{pmatrix}$$

and

$$A_2 = \frac{\sqrt{2}}{1+\sqrt{2}} \frac{(|0\rangle + |1\rangle)(\langle 0| + 1|)}{2} \Rightarrow A_2 = \frac{\sqrt{2}}{2(1+\sqrt{2})} \begin{pmatrix} 1 \\ 1 \end{pmatrix} \begin{pmatrix} 1 & 1 \end{pmatrix} = \frac{\sqrt{2}}{2(1+\sqrt{2})} \begin{pmatrix} 1 & 1 \\ 1 & 1 \end{pmatrix}$$

We define a third operator for perfection

$$A_3 = I - A_1 - A_2 = \begin{pmatrix} 1 & 0 \\ 0 & 1 \end{pmatrix} - \frac{\sqrt{2}}{1+\sqrt{2}} \begin{pmatrix} 0 & 0 \\ 0 & 1 \end{pmatrix} - \frac{\sqrt{2}}{2(1+\sqrt{2})} \begin{pmatrix} 1 & 1 \\ 1 & 1 \end{pmatrix}$$

So,

$$A_3 = \frac{1}{2(1+\sqrt{2})} \begin{pmatrix} 2+\sqrt{2} & -\sqrt{2} \\ -\sqrt{2} & 2\sqrt{2} \end{pmatrix}$$

Let us show that from the three possible measurement results associated with A_1, A_2, and A_3 we can draw the following conclusions about the initial state of the system:

$A_1 \rightarrow$ the original state was $|\psi_2\rangle$

$A_2 \rightarrow$ the original state was $|\psi_1\rangle$

$A_3 \rightarrow$ don't know if the original state was $|\psi_1\rangle$ or $|\psi_2\rangle$

First, let us look at the case where there are results linked to A_1, then, if the original state is $|\psi_1\rangle$, then $p_{A1}(|\psi_1\rangle) = 0$ and $p_{A11}(|\psi_2\rangle) > 0$, that is, the system works only in the state $|\psi_2\rangle$. Indeed,

$$A_1|\psi_1\rangle = \frac{\sqrt{2}}{1+\sqrt{2}}\begin{pmatrix} 0 & 0 \\ 0 & 1 \end{pmatrix}\begin{pmatrix} 1 \\ 0 \end{pmatrix} = \begin{pmatrix} 0 \\ 0 \end{pmatrix}$$

and

$$p_{A_1}(|\psi_1\rangle) = \langle\psi_1|[A_1|\psi_1\rangle] = (1 \quad 0)\begin{pmatrix} 0 \\ 0 \end{pmatrix} = 0$$

Also,

$$A_1(|\psi_2\rangle) = \frac{\sqrt{2}}{1+\sqrt{2}}\begin{pmatrix} 0 & 0 \\ 0 & 1 \end{pmatrix}\frac{1}{\sqrt{2}}\begin{pmatrix} 1 \\ -1 \end{pmatrix} = \frac{1}{1+\sqrt{2}}\begin{pmatrix} 0 \\ -1 \end{pmatrix}$$

and

$$p_{A_1}(|\psi_2\rangle) = \langle\psi_2|[A_1|\psi_2\rangle] = \frac{1}{\sqrt{2}(1+\sqrt{2})}(1 \quad -1)\begin{pmatrix} 0 \\ -1 \end{pmatrix} = \frac{1}{\sqrt{2}(1+\sqrt{2})}$$

If there is a result associated with A_2, in this case, the probability that the initial state is $|\psi_2\rangle$ is equal to zero, $p_{A_2}(|\psi_1\rangle) = 0$ and $p_{A_1}(|\psi_1\rangle) > 0$, so the system could only be in the state $|\psi_1\rangle$. Indeed,

$$A_2(|\psi_2\rangle) = \frac{\sqrt{2}}{1+\sqrt{2}}\begin{pmatrix} 1 & 1 \\ 1 & 1 \end{pmatrix}\frac{1}{\sqrt{2}}\begin{pmatrix} 1 \\ -1 \end{pmatrix} = \frac{1}{2(1+\sqrt{2})}\begin{pmatrix} 0 \\ 0 \end{pmatrix}$$

$$p_{A_1}(|\psi_2\rangle) = \langle\psi_2|[A_2|\psi_2\rangle] = \frac{1}{\sqrt{2}}(1 \quad -1)\frac{1}{2(1+\sqrt{2})}\begin{pmatrix} 0 \\ 0 \end{pmatrix} = 0$$

Also,

$$A_1(|\psi_1\rangle) = \frac{\sqrt{2}}{2(1+\sqrt{2})}\begin{pmatrix} 1 & 1 \\ 1 & 1 \end{pmatrix}\begin{pmatrix} 1 \\ 0 \end{pmatrix} = \frac{\sqrt{2}}{2(1+\sqrt{2})}\begin{pmatrix} 1 \\ 1 \end{pmatrix}$$

$$p_{A_2}(|\psi_1\rangle) = \langle\psi_1|[A_2|\psi_1\rangle] = (1 \quad 0)\frac{\sqrt{2}}{2(1+\sqrt{2})}\begin{pmatrix} 1 \\ 1 \end{pmatrix} = \frac{1}{\sqrt{2}(1+\sqrt{2})}$$

Finally, when there is a result associated with A_3, there is a non-zero probability that the original state is $|\psi_1\rangle$ or $|\psi_2\rangle$. So, $p_{A_3}(|\psi_1\rangle) > 0$ and $p_{A_2}(|\psi_1\rangle) > 0$. In fact,

$$A_3(|\psi_1\rangle) = \frac{1}{2(1+\sqrt{2})}\begin{pmatrix} 2+\sqrt{2} & -\sqrt{2} \\ -\sqrt{2} & 2\sqrt{2} \end{pmatrix}\begin{pmatrix} 1 \\ 0 \end{pmatrix} = \frac{1}{\sqrt{2}(1+\sqrt{2})}\begin{pmatrix} 1+\sqrt{2} \\ -1 \end{pmatrix}$$

$$p_{A_3}(|\psi_1\rangle) = \langle\psi_1|[A_3|\psi_1] = \frac{1}{\sqrt{2}(1+\sqrt{2})}(1 \quad 0)\begin{pmatrix} 1+\sqrt{2} \\ -1 \end{pmatrix} = \frac{1}{\sqrt{2}}$$

In the same way,

$$A_3(|\psi_2\rangle) = \frac{1}{2(1+\sqrt{2})}\begin{pmatrix} 2+\sqrt{2} & -\sqrt{2} \\ -\sqrt{2} & 2\sqrt{2} \end{pmatrix}\frac{1}{\sqrt{2}}\begin{pmatrix} 1 \\ -1 \end{pmatrix} = \frac{1}{\sqrt{2}(1+\sqrt{2})}\begin{pmatrix} 1+\sqrt{2} \\ -1 \end{pmatrix}$$

$$p_{A_3}(|\psi_2\rangle) = \langle\psi_2|[A_3|\psi_2\rangle] = \frac{1}{\sqrt{2}(1+\sqrt{2})}\begin{pmatrix} 1+\sqrt{2} \\ -1 \end{pmatrix} = \frac{1}{\sqrt{2}}$$

Clearly, $0 \le p_{A_3}(|\psi_1\rangle),\ p_{A_3}(|\psi_2\rangle) \ge 1$ and, in this case

$$p_{A_3}(|\psi_1\rangle) = p_{A_3}(|\psi_2\rangle) = \frac{1}{\sqrt{2}}$$

There is never any error in identifying the state of origin when applying the POVM A_1 and A_2. This example shows that when using the POVM element A_3, there is not enough information for reliable identification.

8.3.4 Joint, Conditional, and Relative von Neumann Entropy

Shannon's entropy measures the uncertainty associated with classical probability distribution. Quantum states are described in the same way by replacing the density operator with a probability distribution. This section generalizes Shannon's definition of entropy to quantum states. In 1932 [25] von Neumann defined the entropy of a quantum state ρ by the following formula

$$S(\rho) \equiv -\text{Tr}(\rho\log(\rho)) \tag{8.29}$$

In this formula, logarithms are taken in base two and we define $0log0$ equal to zero. If λ_i are the eigenvalues of ρ, then the von Neumann definition can be rewritten as

$$S(\rho) = -\sum_i \lambda_i \log(\lambda_i) \tag{8.30}$$

Indeed,

$$\begin{aligned} S(\rho) &= -\text{Tr}[\rho\log(\rho)] \\ &= -\text{Tr}\left[\left(\sum_i \lambda_i |\rho_i\rangle\langle\rho_i|\right)\log\left(\sum_i \lambda_i |\rho_i\rangle\langle\rho_i|\right)\right] \\ &= -\text{Tr}\left[\left(\sum_i \lambda_i |\rho_i\rangle\langle\rho_i|\right)\left(\sum_i \log(\lambda_i)|\rho_i\rangle\langle\rho_i|\right)\right] \end{aligned}$$

$$= -\mathrm{Tr}\left[\sum_i \lambda_i \log(\lambda_i)|\rho_i\rangle\langle\rho_i|\right]$$

$$= -\sum_i \lambda_i \log(\lambda_i)$$

$$= H(\lambda_i),$$

where $H(\lambda_i)$ denotes Shannon's entropy.

To calculate von Neumann entropy, the latter formula is generally the most useful. The von Neumann entropy satisfies the following properties:

- Entropy is defined as positive and zero if and only if ρ is a pure state (i.e., $S(\rho) \geq 0$).
- The von Neumann entropy is sub-additive (i.e.,$S(\rho_A \otimes \rho_B) \leq S(\rho_A) + S(\rho_B)$).
- For a pure state ρ_{AB}, we have $S(\rho_A) = S(\rho_B)$.
- Let p_i be a probability distribution and ρ_i a family of states belonging to orthogonal subspaces, then we have

$$S\left(\sum_i p_i \rho_i\right) = H(p_i) + \sum_i p_i S(\rho_i)$$

- The von Neumann entropy is a concave function, that is, if $\lambda_1, ..., \lambda_n \geq 0$ with $\sum_{i=1}^{n} \lambda_i = 1$, then

$$S\left(\sum_{i=1}^{n} \lambda_i \rho i\right) \geq \sum_{i=1}^{n} \lambda_i S(\rho i)$$

- The von Neumann entropy is invariant under a unitary transformation U: $S(\rho) = S(U\rho U^+)$.

Joint, conditional, and relative von Neumann entropy types are similar concepts as defined in classical information theory. Consider two quantum systems, A and B, and the composite system (A, B). Let ρ_A, ρ_B and $\rho_{(A,B)}$ be the density operators of A, B and the composite system (A, B) respectively. The density operators ρ_A and ρ_B are given by the partial traces of the joint density matrix:

$$\rho_A = \mathrm{Tr}_A\left(\rho_{(A,B)}\right) \text{ and } \rho_B = \mathrm{Tr}_B\left(\rho_{(A,B)}\right)$$

The joint von Neumann entropy of the pair of quantum systems A and B is defined as

$$S(A,B) = S\left(\rho_{(A,B)}\right)$$

The conditional von Neumann entropy of system A by the system B is defined as follows

$$S(A/B) = S(A,B) - S(B) = S\left(\rho_{(A,B)}\right) - S(\rho_B) \tag{8.32}$$

The conditional von Neumann entropy $S(A/B)$ is concave.

The relative entropy $S(\rho_1 \| \rho_2)$ for two states described by the density matrices ρ_1 and ρ_2 is defined as

$$S(\rho_1 \| \rho_2) = \mathrm{Tr}\left[\rho_2 \left(\log \rho_2 - \log \rho_1\right)\right]. \tag{8.33}$$

The two density matrices ρ_1 and ρ_2 have the spectral decompositions

$$\rho_1 = \sum_j q_j \left|u_j\right\rangle\left\langle u_j\right| \text{ and } \rho_2 = \sum_i p_i \left|e_i\right\rangle\left\langle e_i\right|$$

where $\{|e_i\}$ and $\{|u_j\}$ are two orthonormal bases of the same physical system. It is easy to see that the quantum relative entropy is positive, $S(\rho_1 \| \rho_2) \geq 0$, more specifically,

$$S(\rho_1 \| \rho_2)\begin{cases} > 0 \; if \rho_1 \neq \rho_2 \\ = 0 \; if \; \rho_1 = \rho_2 \end{cases}$$

Then, the relative entropy can be evaluated in the basis where ρ_2 is diagonal

$$\begin{aligned} Tr\left[\rho_2 \log(\rho_2)\right] - Tr\left[\rho_2 \log(\rho_1)\right] &= \sum_i \left\langle e_i\right|\rho_2 \log(\rho_2)\left|e_i\right\rangle - \sum_i \left\langle u_i\right|\rho_2 \log(\rho_1)\left|u_i\right\rangle \\ &= \sum_i p_i \left(\log(p_i) - \sum_j \left|\left\langle u_i \middle| e_i\right\rangle\right|^2 \log(q_j)\right) \\ &= -\frac{1}{\ln(2)} \sum_i \sum_j p_i \left|\left\langle u_i \middle| e_i\right\rangle\right|^2 \ln\left(\frac{q_j}{p_i}\right) \\ &\geq -\frac{1}{\ln(2)} \sum_i \sum_j p_i \left|\left\langle u_i \middle| e_i\right\rangle\right|^2 \left(\frac{q_j}{p_i} - 1\right) \\ &= 0 \end{aligned}$$

This inequality is strict unless $q_j = p_i, \forall i, j$ satisfying $\left|\left\langle u_i \middle| e_i\right\rangle\right|^2 > 0$. Under this assumption, equality occurs when

$$\sum_i \sum_j \left|\left\langle u_i \middle| e_i\right\rangle\right|^2 \left(p_i - q_j\right) = 0$$

This implies that $\|\rho_2 - \rho_1\| = 0$ or $\rho_2 = \rho_1$.

8.3.5 Quantum Mutual Information

Quantum mutual information is a generalization of the concept of classical mutual information to the quantum case. It is related to the interpretation of the total correlations between two subsystems A and B of a bipartite quantum system AB. It is considered a case of relative entropy between the state ρ_{AB} and the state $\rho_A \otimes \rho_B$. Using Equation 8.33 and the relation

$log_2(\rho_A \otimes \rho_B) = log_2\rho_A \otimes 1_B + 1_A \otimes log_2\rho_B$, we find the quantum extension of mutual information (8.7) as

$$\begin{aligned} I(\rho_{AB}) &= S(\rho_{AB} \| \rho_A \otimes \rho_B) \\ &= Tr[\rho_{AB} log_2\rho_{AB}] - Tr[\rho_{AB} log_2(\rho_A \otimes \rho_B)] \\ &= S(\rho_A) + S(\rho_B) - S(\rho_{AB}) \end{aligned} \tag{8.34}$$

Note that the quantum mutual information is non-negative due to the sub-additivity of the von Neumann entropy, and zero only for a product state $\rho_{AB} = \rho_A \otimes \rho_B$. It can also be written as a function of the quantum conditional entropy by making a direct generalization of Equation 8.8. We find

$$I(\rho_{AB}) = S(\rho_A) - S(\rho_{AB}/\rho_B). \tag{8.35}$$

Despite several similarities between Shannon's entropy and von Neumann's entropy, these entropy types have many distinctly different properties. However, because of the close relationship between these entropies and the operational properties of information, these properties tell us something about the nature of information.

8.4 QUANTIFICATION OF QUANTUM CORRELATION

Quantum information processing depends essentially on various quantum mechanical phenomena [26-28]. Among these, entanglement was considered to be one of the most important elements. Quantum entanglement plays an important role in several areas of quantum information [5], such as quantum teleportation [29-30] and quantum cryptography [31,32], coding [33], and quantum computing, and so on. Quantum entanglement is a form of correlation between two or more particles (that can or cannot be separated). The main aim of the study of entanglement is to find a general way to quantify and characterize the quantity of entanglement of a given quantum state. These are central questions in quantum information theory. Much effort has been devoted to this area of research, and several measures of entanglement have been proposed for bipartite and multipartite quantum systems.

8.4.1 Measuring Entanglement

8.4.1.1 Entanglement Entropy

Entanglement entropy is a measure of entanglement used when the system is prepared in a pure state. It is expressed in terms of von Neumann entropy. This is given by

$$S(\rho) = -Tr\rho log\rho \tag{8.36}$$

Consider a quantum state ρ_{AB} consisting of two subsystems A and B. For each pure state, the entanglement entropy $E(\rho_{AB})$ is defined by

$$E(\rho_{AB}) = -Tr\rho_A \; log\rho_A = -Tr\rho_B \; log\rho_B \tag{8.37}$$

where $\rho_A = Tr_B(\rho_{AB})$.

To illustrate how entanglement entropy is calculated, we consider the Bell state

$$|\Phi^+\rangle_{AB} = \frac{1}{\sqrt{2}}\left(|00\rangle_{AB} + |11\rangle_{AB}\right)$$

The density matrix describing this state is given by

$$\rho_{AB} = |\Phi^+\rangle\langle\Phi^+| \tag{8.38}$$

$$= \frac{1}{2}\left(|0\rangle_A\langle 0| \otimes |0\rangle_B\langle 0| + |0\rangle_A\langle 1| \otimes |0\rangle_B\langle 1| + |1\rangle_A\langle 0| \otimes |1\rangle_B\langle 0| + |1\rangle_A\langle 1| \otimes |1\rangle_B\langle 1|\right)$$

Then the reduced density operator of subsystem A is

$$\rho_A = \frac{1}{2}\left[|0\rangle_A\langle 0| + |0\rangle_A\langle 0|\right] = \begin{pmatrix} \frac{1}{2} & 0 \\ 0 & \frac{1}{2} \end{pmatrix} \tag{8.39}$$

Note that ρ_A is a diagonal matrix. As a result, the diagonal entries represent the eigenvalues. Using Equation 8.37 we can find the entanglement entropy of the system AB as

$$E(\rho_{AB}) = S(\rho_A) = H\left(\frac{1}{2}\right) = -\frac{1}{2}log_2\left(\frac{1}{2}\right) - \frac{1}{2}log_2\left(\frac{1}{2}\right) = log_2 2 = 1$$

From this example, it is clear that the entanglement entropy reaches its maximum value for maximally entangled states and is zero for separable states. We can also easily prove that the entanglement of the subsystem B is always the same as that of subsystem A. Furthermore, for a maximally entangled pure state $|\psi\rangle$, the entanglement entropy has the value

$$\mathrm{E}(|\psi\rangle) = log_2 d \tag{8.40}$$

where d is the dimension of the reduced density matrix.

8.4.1.2 Entanglement of Formation and Concurrence

The formation entanglement (or formation entropy) $E_f(|\psi\rangle)$ of a mixed state is given by

$$E_f(\rho) = min\sum_i p_i E(|\psi_i\rangle) \tag{8.41}$$

where the minimization concerns all possible pure spectral decompositions of the state. This process can be considered a generalization of the previous result obtained in the case of pure states. Indeed, the manipulation of formation entanglement for mixed states is very complicated. An exact expression has been established for systems of 2 qubits. For this type of system, Wootters determined a simple formula for calculating the entropy of formation. It is given by the following formula:

$$E_f(\rho) = E(C(\rho)) \tag{8.42}$$

where $E(C(\rho))$ is defined by:

$$E(C(\rho)) = h\left(\frac{1}{2} + \frac{1-\sqrt{1-C^2}}{2}\right) \tag{8.43}$$

where h is the binary entropy function given by

$$h(x) = -x\,ln(x) - (1-x)\,ln(1-x)$$

and the quantity C is the concurrence [22] of the system. It is given by

$$C(\rho) = max\{0, \lambda_1 - \lambda_2 - \lambda_3 - \lambda_4\} \tag{8.44}$$

where λ_i are the eigenvalues of the following Hermitian matrix

$$\mathrm{R} = \sqrt{\sqrt{\rho}\tilde{\rho}\sqrt{\rho}} \tag{8.45}$$

with the condition that $\lambda_1 \geq \lambda_2 \geq \lambda_3 \geq \lambda_4$. And $\tilde{\rho}$ is the spin-flipped density matrix defined by

$$\tilde{\rho} = (\sigma_y \otimes \sigma_y)\rho^*(\sigma_y \otimes \sigma_y), \qquad \sigma_y = \begin{pmatrix} 0 & -i \\ \mathrm{i} & 0 \end{pmatrix} \tag{8.46}$$

The entropy of formation is a monotonically increasing function of C. When $C = 0$, the state is separable, and when $C = 1$, the state is maximally entangled. For $0 < C \leq 1$, the state is entangled.

8.4.1.3 Logarithmic Negativity

In quantum mechanics, negativity is a measure of quantum entanglement, which is easy to calculate [34]. It is a measure that employs the PPT criterion as a test for separability, also known as the Peres–Horodecki criterion. For an operator A, we define the trace norm as the sum of the singular values of the operator A, denoted by $\|A\|_1 = Tr|A| = Tr\sqrt{A^+A}$. Negativity is defined as

$$N(\rho) = \frac{\|\rho^{\Gamma_A}\|_1 - 1}{2}, \tag{8.47}$$

where ρ^{Γ_A} denotes the partial transpose of ρ with respect to the subsystem A.

Logarithmic negativity is a quantity that is easy to calculate and is defined as

$$E_N(\rho) = log_2\|\rho^{\Gamma_A}\|_1 \tag{8.48}$$

where Γ_A is the partial transpose operator.

It has following relationship with negativity through the formula

$$E_N(\rho) = log_2(2N(\rho) + 1) \tag{8.49}$$

Logarithmic negativity satisfies the following properties:

- It can be zero, even if the state is entangled (if the state is PPT entangled).
- $E_N(\rho \otimes \sigma) = E_N(\rho) + E_N(\sigma)$ (additive over tensor products).
- It does not reduce to pure state entanglement entropy like most other entanglement measures.

8.4.2 Quantum Discord

In general, a composite quantum system can have both classical and quantum correlations that are jointly quantified by the quantum mutual information. Consider a bipartite quantum system AB with a density operator ρ_{AB}. The quantum mutual information can be written as:

$$I(\rho_{AB}) = J(\rho_{AB}) + D(\rho_{AB}) \tag{8.50}$$

where $J(\rho_{AB})$ represents the classical correlations and $D(\rho_{AB})$ the quantum correlations. This quantum part $D(\rho_{AB})$ is called quantum discord [35,36]. This is a type of quantum correlation other than entanglement. Incidentally, separable mixed states (i.e., without entanglement) can have a non-zero quantum discord. Ollivier and Zurek have suggested [8] use of von Neumann measures to quantify the quantum discord. In this context, we assume that the measurement is defined by a set of positive hermetic operators, a positive operator-valued measure (POVM) and is performed on the particle A. The set of POVM elements is denoted by $\mathrm{M}: M_k \geq 0 \sum_{\mathrm{k}} M_k = I$. Thus, a projective measure on the subsystem A will project the system into a statistical set $\left\{p_k^B, \rho_{B_k}\right\}$, such that

$$\rho = \frac{(M_k \otimes I)\rho_{AB}(M_k \otimes I)}{p_k^B} \tag{8.51}$$

where the measurement operation is given by

$$M_k = U\Pi_k U^{+}: k = 0,1 \tag{8.52}$$

where $M_k = |k\,k|(k = 0,1)$ is a rank 1 projector. In Equation 8.52, $U \in SU(2)$ is unitary and the probability p_k^B is given by

$$\mathrm{p}_\mathrm{k}^\mathrm{B} = \mathrm{Tr}\left\{(\mathrm{M}_\mathrm{k} \otimes \mathrm{I})\rho\mathrm{AB}(\mathrm{M}_\mathrm{k} \otimes \mathrm{I})\right\} \tag{8.53}$$

Thus, the quantity of information obtained for particle B is given by

$$S(\rho_B) - \sum_k p_k^B\, \mathrm{S}(\rho_{B_k}) \tag{8.54}$$

Consequently, the classical correlations of a bipartite state can be written as

$$J(\rho_{AB}) = \max_M \left[S(\rho_B) - \sum_k p_k^B\, S(\rho_{B_k})\right]$$

$$= S(\rho_B) - \min_M \sum_{\mathrm{k}} p_k^B\, S(\rho_{B_k})$$

$$= S(\rho_{\mathbf{B}}) - \tilde{S}_{\min}$$

where $\tilde{S}_{\min}$ is the minimum value of the conditional entropy. Quantum discord is the difference between quantum mutual information and classical correlation:

$$D(\rho_{AB}) = I(\rho_{AB}) - J(\rho_{AB}) \tag{8.56}$$

More specifically, quantum discord is worth

$$D(\rho_{AB}) = S(\rho_A) + \tilde{S}_{\min} - S(\rho_{AB}) \tag{8.57}$$

Quantum discord is always positive and zero for classically correlated states, corresponding to correlations of purely quantum origin. In the general case, it is not symmetric under the exchange of the roles of the subsystems A and B. This asymmetry reflects an asymmetry in the distribution of classical and quantum correlations for A and B. It is important to note that almost all quantum states have a non-zero quantum discord. For separable mixed states, the quantum discord can be non-zero. It should be noted that quantum discord is more resilient than entanglement to dissipative environments [37,38]. This has been demonstrated for both Markovian and non-Markovian environments by comparing the dynamics of discord with those of concurrence [39]. Unlike entanglement, which is very fragile in the face of environmental influences, discord is more robust and can even be enhanced by coupling with an environment. This is a particularly interesting resource from an experimental point of view [40,41].

8.4.3 Geometric Measure of Quantum Discord

The evaluation of entropic quantum discord is generally difficult to compute due to minimization, which is not easy. The geometric measure of quantum discord (GMQD) was introduced by Dakic et al. [42]. It quantifies the quantum correlation using the Hilbert–Schmidt distance between the considered state ρ and the nearest null discord state χ [42]. It is given by

$$D_g(\rho) = \min_{\chi\in\Omega_0} \|\rho-\chi\|^2 \tag{8.58}$$

where Ω_0 denotes the set of zero discord states and $\|\rho-\chi\|^2 = Tr(\rho-\chi)^2$ is the Hilbert–Schmidt norm. A state $\chi \in H_A \otimes H_B$ is of zero discord if and only if χ is a classical-quantum state [43]. It is written as

$$\chi = \sum_{k=1}^{m} p_k |\psi_k\rangle\langle\psi_k| \otimes \rho_k \tag{8.59}$$

where p_k is a probability distribution, $\{|\psi_k\rangle\}$ is an arbitrary orthonormal basis in H_A, and ρ_k is a set of arbitrary states (density operators) acting on H_B. Dakic et al. [42] have also obtained an easily computable expression for the geometric measure of quantum discord for a 2-qubit system. Indeed, a bipartite state is written in the Fano–Bloch decomposition as

$$\rho = \frac{1}{4}\left[\mathrm{I}\otimes\mathrm{I} + \sum_{\mathrm{i}}^{3} (\mathrm{x}_\mathrm{i}\sigma_\mathrm{i}\otimes\mathrm{I} + \mathrm{y}_\mathrm{i}\mathrm{I}\otimes\sigma_\mathrm{i}) + \sum_{i,j=1}^{3} \mathrm{R}_{\mathrm{ij}}\sigma_\mathrm{i}\otimes\sigma_\mathrm{j} \right] \tag{8.60}$$

where $x_i = Tr\rho(\sigma_i \otimes I)$, $y_i = Tr\rho(I\otimes\sigma_i)$ are the components of the local Bloch vectors, $\sigma_i\,(i\in\{1,2,3\})$ are the three Pauli matrices, and $R_{ij} = Tr\rho(\sigma_i\otimes\sigma_j)$ are the components of the correlation tensor, for each state ρ, we associate a triple $\{\vec{x},\vec{y},R\}$. For a system of 2 qubits, the state of zero discord has the form

$$\chi = p_1|\psi_1\rangle\langle\psi_1|\otimes\rho_1 + \mathrm{p}_2|\psi_2\rangle\langle\psi_2|\otimes\rho_2 \tag{8.61}$$

where $\{|\psi_1\rangle, |\psi_2\rangle\}$ is an orthonormal basis of a single qubit. The distance between the states ρ and χ is given by:

$$\|\rho-\chi\|^2 = \|\rho\|^2 - 2Tr(\rho\chi) + \|\chi\|^2 \tag{8.62}$$

By minimizing this quantity, the geometric discord can be expressed explicitly by the following formula:

$$D_g(\rho) = \frac{1}{4}\left(\|\mathrm{x}\|^2 + \|\mathrm{R}\|^2 + \mathrm{K}_{\max}\right) \tag{8.63}$$

where $x = (x_1, x_2, x_3)^{\mathrm{T}}$ and $\mathrm{K}_{\max}$ is the largest eigenvalue of the matrix

$$K := xx^T + RR^T$$

K is a 3 × 3 matrix. The eigenvalues of the K matrix are λ_1, λ_2, and λ_3. The geometric measure of quantum discord is given by the following compact formula:

$$D_g(\rho) = \frac{1}{4} min\{\lambda_1 + \lambda_2,\ \lambda_2 + \lambda_3, \lambda_1 + \lambda_3\} \tag{8.64}$$

It follows that to obtain the geometric discord for a 2-qubit state, we only need to diagonalize the K matrix and compare the eigenvalues $\lambda_1, \lambda_2, \lambda_3$.

8.5 CONCLUSION

Quantum information theory was developed to apply Claude Shannon's classical information theory to the quantum world. It is the study of the maximum capacity in quantum systems governed by the laws of quantum mechanics, in particular the principle of superposition and entanglement, to transmit information weakly from sender to receiver while preserving information and correlations. In this chapter, we briefly introduce classical and quantum information theory [44]; the first part describes the basic concepts of classical information theory, which are useful for studying the quantity of information transmitted, and introduces the definitions of Shannon entropy [46], joint entropy, conditional entropy and mutual information [11]. In the second part, we study the quantum theory of information [47,48], where information is represented by quantum states and allowed to evolve according to the laws of quantum mechanics. When we present the formalism of the density operator, quantum measurements, we define joint, conditional, and relative von Neumann entropy and quantum mutual information. We conclude this chapter with a third section based on some measures of quantum correlations, where we introduce the notion of entanglement entropy, based on the notion of concurrency, and focus on the definition of quantum discord and its geometric version based on the Hilbert–Schmidt norm, as well as logarithmic negativity, which is a measure of entanglement for continuous variables.

REFERENCES

[1] C. E. Shannon, A mathematical theory of communication, *Bell Syst. Tech. J.* 27 (1948) 379–423.
[2] E.T. Jaynes, Information theory and statistical mechanics, *Phy. Rev.* 106 (1957) 620.
[3] J. P. Dowling and J. M. Gerard, Quantum technology: The second quantum revolution, *Philos. Trans. R. Soc. Lond. A: Math. Phys. Eng. Sci.* 361(1809) (2003): 1655–1674.

[4] M. A. Nielsen and I. Chuang, Quantum computation and quantum information, Cambridge University Press (2010).

[5] R. Horodecki, P. Horodecki, M. Horodecki, and K. Horodecki, Quantum entanglement, *Rev. Mod. Phys.* 81 (2009) 865.

[6] D. Cavalcanti and P. Skrzypczyk, Quantum steering: A review with focus on semidefinite programming, *Rep. Prog. Phys.* 80(2) (2016) 024001.

[7] N. Brunner, D. Cavalcanti, S. Pironio, V. Scarani, and S. Wehner, Bell nonlocality, *Rev. Mod. Phys.* 86(2) (2014) 419.

[8] H. Ollivier and W. H. Zurek, Quantum discord: A measure of the quantumness of correlations, *Phys. Rev. Lett.* 88(1) (2001) 017901.

[9] L. Henderson and V. Vedral, Classical, quantum and total correlations, *J. Phys. A: Math. Gen.* 34(35) (2001): 6899.

[10] A. Ferraro, L. Aolita, D. Cavalcanti, F. M. Cucchietti, and A. Acín, Almost all quantum states have nonclassical correlations, *Phys. Rev. A.* 81(5) (2010) 052318.

[11] L. Yu, R. B. Griffiths and S. M. Cohen, Efficient implementation of bipartite nonlocal unitary gates using prior entanglement and classical communication, *Phys. Rev. A.* 81 (2010) 062315.

[12] J. M. Renes, Quantum Information Theory: Concepts and Methods, *Berlin, Boston: De Gruyter Oldenbourg*, (2022).

[13] William K. Wootters, Entanglement of formation of an arbitrary state of two qubits, *Phys. Rev. Lett.* 80 (1997) 2245.

[14] B. Schumacher, Quantum coding, *Phys. Rev A.* 51 (1995) 2738.

[15] D. Kurzyk, Introduction to Quantum Entanglement, *Theoretical and Applied Informatics*, 24 (2012) 135-150.

[16] M. B.Plenio, Logarithmic negativity: A full entanglement monotone that is not convex, *Phys. Rev. Lett.* 95 (2005) 090503.

[17] R. Clausius, The nature of the motion which we call heat, *Ann. Phys.* 125(44) (1865) 353.

[18] L. Boltzmann, *Vorlesungen u¨ber Gastheorie*, 2 vols. Leipzig: Barth, 1898 (1896); transl. by S. G. Brush, Lectures on Gas Theory, Cambridge University Press, London (1964).

[19] S. Braunstein, C. Caves, R. Jozsa, N. Linden S. Porescu and R. Schack, Separability of very noisy mixed states and implications for NMR quantum computing, *Phys. Rev. Lett.* 83 (1999) 1054–1057.

[20] S.L. Braunstein, C.M. Caves, R. Jozsa, N. Linden, S. Popescu and R. Schack, Separability of very noisy mixed states and implications for NMR quantum computing, *Phys. Rev. Lett.* 83 (1999) 1054.

[21] G. Brassard, Quantum communication complexity. *Found. Phys.* 33 (2003) 1593–1616.

[22] P. O. Boykin, T. Mor, V. Roychowdhury, F. Vatan and R. Vrijen, Algorithmic cooling and scalable NMR quantum computers, *Proc. Natl. Acad. Sci.* 99 (2002) 3388–3393.

[23] H. M. Wiseman and J. M. Gerard, *Quantum Measurement and Control*, Cambridge University Press (2009).

[24] V. B. Braginsky, B. B. Vladimir and Y. K. Farid, Quantum Measurement, *Cambridge University Press* (1995).

[25] D. Petz, Entropy, von Neumann and the von Neumann Entropy: Dedicated to the memory of Alfred Wehrl, In *John von Neumann and the Foundations of Quantum Physics*, Springer Netherlands (2001) 83-96.

[26] C. H. Bennett, D. P. Divincenzo, C. A. Fuchs, T. Mor, E. Rains, P. W. Shor, J. A. Smolin and W. K. Wootters, Quantum nonlocality without entanglement, *Phys. Rev. A.* 59 (1999) 1070.

[27] M. Horodecki, P. Horodecki, R. Horodecki et al, Local versus nonlocal information in quantum-information theory: Formalism and phenomena, *Phys. Rev. A.* 71 (2005) 062307.

[28] J. Niset and N. J. Cerf, Multipartite nonlocality without entanglement in many dimensions, *Phys. Rev. A.* 74 (2006) 052103.

[29] C. H. Bennett, G. Brassard, C. Crepeau, R. Jozsa, A. Peres and W. K. Wootters, Teleporting an unknown quantum state via dual classical and Einstein–Podolsky–Rosen channels, *Phys. Rev. Lett.* 70 (1993) 1895.

[30] A. Slaoui, N. Ikken, L. B. Drissi and R. A. Laamara, Quantum communication protocols: From theory to implementation in the quantum computer, *In Quantum Computing-Innovations and Applications in Modern Research, IntechOpen*. (2023).

[31] C. Bennett and G. Brassard, Quantum cryptography: Public key distribution and coin tossing, in *Proceedings of IEEE International Conference on Computers Systems and Signal Processing, Bangalore, India* (1984) 175–179.
[32] A. K. Ekert, Quantum cryptography based on Bell's theorem, *Phys. Rev. Lett.* 67 (1991) 661.
[33] C. H. Bennett and S. J. Wiesner, Communication via one- and two-particle operators on Einstein–Podolsky–Rosen states, *Phys. Rev. Lett.* 69 (1992) 2881.
[34] A. Slaoui, M. Daoud, and R. Ahl Laamara, The dynamic behaviors of local quantum uncertainty for three-qubit X states under decoherence channels, *Quantum Inf. Process.* 18 (2019) 250.
[35] M.I. Shaukat, A. Slaoui, H. Terças, et al. Phonon-mediated quantum discord in dark solitons, *Eur. Phys. J. Plus.* 135 (2020) 357.
[36] A. Slaoui, A. Salah, and M. Daoud, Influence of Stark-shift on quantum coherence and non-classical correlations for two two-level atoms interacting with a single-mode cavity field, *Physica A Stat. Mech. Appl.* 558 (2020) 124946.
[37] Y. Y. Xu, W. L. Yang and M. Feng, Dissipative dynamics of quantum discord under quantum chaotic environment, *Eur. Phys. Lett.* 92 (2010) 10005.
[38] Q. L. He and J. B. Xu, Sudden transition and sudden change of quantum discord in dissipative cavity quantum electrodynamics system, *J. Opt. Soc. Am B.* 30 (2013) 251–257.
[39] F. F. Fanchini, T. Werlang, C. A. Brasil, L. G. E. Arruda and A. O. Caldeira, Non-Markovian dynamics of quantum discord, *Phys. Rev. A.* 81 (2010) 052107.
[40] F. Ciccarello and V. Giovannetti, Creating quantum correlations through local nonunitary memoryless channels, *Phys. Rev. A.* 85 (2012) 010102.
[41] F. Ciccarello and V. Giovannetti, Local-channel-induced rise of quantum correlations in continuous-variable systems, *Phys. Rev. A.* 85 (2012) 022108.
[42] B. Dakic, V. Vedral, and C. Brukner, Necessary and sufficient condition for nonzero quantum discord, *Phys. Rev. Lett.* 105 (2010) 190502.
[43] S. Luo, Quantum discord for two-qubit systems, *Phys. Rev. A* 77 (2008) 042303.
[44] C. G. Timpson, Quantum information theory and the foundations of quantum mechanics, OUP Oxford (2013).
[45] O. J. E. Maroney, Information and Entropy in Quantum Theory, *arXiv preprint quant-ph/0411172* (2004).
[46] K. Modi, T. Paterek, W. Son, V. Vedral and M. Williamson, Unified view of quantum and classical correlations, *Phys. Rev. Lett.* 104 (2010) 080501.
[47] A. Slaoui, *Les relations intrinse`que des the´ories des ressources quantiques et leurs ro^les dans la me´trologie quantique*, PhD thesis (2021).
[48] D. C. Marinescu, *Classical and Quantum Information*, 1st ed. San Francisco, CA: Academic (2011).

9 Quantum Teleportation
Concepts and Protocols

Meryem El Kirdi, Hanane El Hadfi, Lalla Btissam Drissi, Rachid Ahl Laamara, and Abdallah Slaoui

9.1 INTRODUCTION

Quantum mechanics, a mathematical and physical theory, seeks to explain phenomena on an incredibly small scale, offering a perspective markedly different from our understanding of the macroscopic world. Commonly referred to as 20th-century physics [1], quantum mechanics has intrigued many physicists and researchers due to its peculiar nature, providing insights into previously unexplained phenomena. Presently, there is growing interest in leveraging the properties of quantum mechanics for coding and information processing systems. This concept gives rise to an emerging field of study and a recent offshoot of physics, known as a discipline that skillfully blends the fundamental tools of classical information theory with the characteristics of quantum mechanics. This integration results in a cohesive framework termed "quantum information theory." Naturally, the need arose to identify distinctions between these two information theories. The primary contrast lies in the nature of the physical medium responsible for encoding information. Classical information theory employs a classical medium, whereas quantum information theory utilizes a quantum medium. Distinct disparities arise in the handling and conveyance of information, guided by the principles of the quantum realm. The overarching objective of probabilistic quantum theory aligns with classical information theory – processing and transmitting information. However, this quantum approach delves into the intricacies of quantum mechanics, aiming to quantify, manipulate, and observe individual elementary entities. The progression of quantum mechanics prompted a crucial demand for enhancing data storage, coding, and transmission through quantum concepts, thereby giving rise to the quantum information storage unit. The qubit, a quantum counterpart to the classical bit, distinguishes itself significantly. Beyond the binary nature of classical bits, a qubit can exist in a superposition of two states, enabling a unit of information to embody combinations of its potential states. This unique capability affords advantages in executing tasks previously unattainable with classical bits. The realm of quantum information not only encompasses superposition but also delves into other fundamental properties of quantum mechanics, notably entanglement and quantum decoherence. Indeed, working on a quantum information application necessitates a thoughtful consideration of these essential quantum phenomena. This emerging domain's significance is evident in its applications within quantum information, with quantum teleportation standing out as a prominent example and a primary focus of this research. Quantum teleportation provides a means to transfer quantum information between distant locations, leveraging entanglement to ensure the transmission of quantum states without energy transfer or consideration of the physical distance between systems. Operating on the principle of quantum non-cloning, these protocols, often complemented by local processes and classical communication channels, hold a central position in quantum information processing. They facilitate the transmission of quantum information within a system without the need for physically moving the system itself. Essentially,

DOI: 10.1201/9781003621423-11

these protocols enable the transfer of quantum information, not physical qubits, between spatially separated communication nodes, authorized solely by classical communication channels. Notably, they highlight the intriguing potential of quantum entanglement – a distinctly quantum phenomenon where distant nodes behave as a unified system – to enable tasks that were previously considered impossible. The pioneering quantum teleportation protocol, introduced by Bennett and team in 1993 [2], proposed a method for quantum information transmission over a certain distance by encoding information in a quantum state using entangled resources (Einstein–Podolsky–Rosen or EPR state). This groundbreaking idea was experimentally verified in 1997 using optical platforms and has since been extended both theoretically and experimentally to cover the teleportation of discrete (DV) and continuous variables (CV) [3, 4]. The standard quantum teleportation process involves two key participants: Alice, who sends the information, and Bob, who receives it. In 1998, Karlsson and Bourennane [5] expanded this protocol by introducing controlled quantum teleportation (CQT), introducing a third collaborator, Charlie, as a supervisor. This extended protocol uses a tripartite entangled state (Greenberger–Horne–Zeilinger (GHZ) state). In CQT, the intervention and authorization of Charlie are crucial for the successful teleportation of coded information. This setup allows secure communication of quantum information among multiple locations, as Bob cannot receive the original target state without Charlie's involvement. Hybrid systems combine continuous variables (taking values in a continuum) and DVs (taking values in a countable set), resembling the dual nature of digital and analog encoding in classic information treatment. Many real-world systems are described in a hybrid manner, prompting research to bridge the gap between CV and DV approaches. Hybrid teleportation, blending CV and DV protocols, has emerged as a key priority in quantum science due to its potential for providing improved solutions compared to individual CV or DV protocols. This approach has been explored both theoretically and experimentally, contributing to advancements in quantum technologies [6]. In practical scenarios, the impact of environmental factors introduces challenges to maintaining maximum entanglement in a quantum channel connecting Alice and Bob. The coupling with the environment causes a reduction in entanglement due to noise, making it difficult to achieve or sustain perfect quantum teleportation. Consequently, the consideration of non-maximally entangled states becomes necessary. In cases where the quantum channel is not maximally entangled, the teleportation fidelity is consistently less than 1. This leads to two options: accepting imperfect quantum teleportation with fidelity less than 1 or opting for perfect quantum teleportation with probabilistic success. The latter choice, known as probabilistic quantum teleportation, ensures reliable quantum teleportation for two out of four possible states [7]. The success probability varies from zero for a non-maximally entangled state to half for a maximally entangled state. The remaining half of successful teleportation occurs when the shared resource and applied measurements are maximally entangled [8].

In the realm of quantum information theory, particularly within the context of quantum teleportation as a process in quantum communication, the handling and measurement of state superposition become imperative. These measurements fundamentally rely on the principle of entanglement, a core concept in quantum teleportation, much like the significance of the no-cloning theorem.

9.2 NON-CLONING THEOREM

In contrast to classical mechanics, where copying a classical bit is straightforward and can be done as many times as needed, quantum mechanics introduces a fundamental constraint stating that it is impossible to create an exact duplicate of an arbitrary unknown quantum state. In classical systems, copying information is straightforward, but in the quantum realm, attempting to replicate a quantum state, especially a qubit, proves unattainable. This theorem underscores the unique and non-replicable nature of quantum information and highlights the impact of measurements on the fragile superposition of quantum states, ultimately preventing the perfect cloning of quantum information. The impossibility of perfectly replicating an unknown and arbitrary quantum state is formally known as the no-cloning theorem. This theorem underscores that a qubit, a fundamental unit in quantum computing, cannot be duplicated. The fragility of a qubit is evident when subjected to measurements; any attempt to precisely determine its state results in its destruction. Post measurement, the qubit collapses into either $|0\rangle$ or $|1\rangle$ with probabilities $|\alpha|^2$ and $|\beta|^2$, leading to the loss of its initial superposition. Given that a qubit exists in a superposition, the act of copying it becomes unattainable, as the original state ceases to exist. This stands in stark contrast to copying classical bits, which is achievable, for instance, through operations like the CNOT gate when the states are known:

$$CNOT|0\rangle|0\rangle \rightarrow |0\rangle|0\rangle, \quad CNOT|1\rangle|0\rangle \rightarrow |1\rangle|1\rangle. \tag{9.1}$$

In the realm of quantum mechanics, it is feasible to copy a classical bit by transferring its information from the first qubit to the second. However, this ease of replication shifts when dealing with quantum states. For any quantum state $|\phi\rangle$, there is no unitary operator U that can be applied to accurately copy the information contained in the state, such as

$$U|\phi,0\rangle = |\phi,\phi\rangle \tag{9.2}$$

This inability to find a U highlights a crucial distinction – while classical bits can be easily duplicated, quantum states resist perfect replication through unitary operations. There are two ways to illustrate this theorem. Firstly, consider the scenario where the goal is to duplicate a qubit in any state, denoted as $|\varphi\rangle$, onto a second qubit initially in an arbitrary state $|\psi\rangle$. In thicase, we assume the existence of a copying device, implemented by a unitary operator U, which operates on the qubits. This operator U is envisioned to satisfy the condition that, for all possible quantum states $|\varphi\rangle$, the duplication process is achieved according to specified criteria

$$U|\phi\rangle|\psi\rangle = |\phi\rangle|\phi\rangle \tag{9.3}$$

Let's take $|\phi\rangle = \frac{1}{\sqrt{2}}(|0\rangle + |1\rangle)$, as an example, by linearity, one found

$$U|\phi\rangle|\psi\rangle = \frac{1}{\sqrt{2}}\left[U|0\rangle|\psi\rangle + U|1\rangle|\psi\rangle\right] = \frac{1}{\sqrt{2}}\left[|0\rangle|0\rangle + |1\rangle|1\rangle\right] \tag{9.4}$$

The EPR state, represented by Equation 9.4, is a maximally entangled state and cannot be expressed in the separable form $|\phi\rangle|\phi\rangle$ as indicated in Equation 9.3. This inconsistency highlights that the initial assumption is incorrect, leading to the conclusion that existence of a cloning machine is not feasible. Alternatively, consider two quantum states represented as $|\phi\rangle$ and $|\psi\rangle$, satisfying the condition U

$$U|\phi\rangle,|0\rangle = |\phi\rangle|\phi\rangle, \qquad U|\psi\rangle,|0\rangle = |\psi\rangle|\psi\rangle \tag{9.5}$$

Here, U is a unitary operator with $U^{+}U = 1$, preserving the scalar product; therefore

$$\begin{aligned}\langle\phi|\psi\rangle &= \langle 0|.\langle\phi|\psi\rangle.|0\rangle = \langle 0|.\langle\phi|U^{+}U|\psi\rangle.|0\rangle \\ &= \langle\phi|.\langle\phi|\psi\rangle.|\psi\rangle = \left(\langle\phi|\psi\rangle\right)^2\end{aligned} \tag{9.6}$$

The inner product $\langle\phi|\psi\rangle$ is either 0 or 1, indicating that states $|\phi\rangle$ and $|\psi\rangle$ are either orthogonal or identical. Consequently, it is possible to replicate only those states that are orthogonal, thus contradicting the initial assumption that $|\phi\rangle$ and $|\psi\rangle$ can be any arbitrary states. In summary, the no-cloning theorem asserts that making an exact duplicate of any unknown quantum state is impossible. Put differently, you cannot replicate a quantum state without changing it, and the process of measuring or copying one state unavoidably affects the other. This principle emerges from the linear and probabilistic aspects inherent in quantum mechanics.

9.3 QUANTUM SUPERPOSITION AND QUANTUM ENTANGLEMENT

Quantum mechanics introduces the concept of quantum superposition, where a particle can simultaneously exist in multiple states until a measurement is made. Upon measurement, the particle collapses into one specific state. This notion challenges classical intuition and is crucial in the context of quantum computing and other quantum technologies. Specifically, quantum superposition implies that the polarization state of a photon can be represented as a combination of two states: $|0\rangle$ and $|1\rangle$. Now, let's examine 2 qubits:

$$|\varphi_A\rangle = a_0|0\rangle_A + a_1|1\rangle_A, \qquad |\varphi_B\rangle = b_0|0\rangle_B + b_1|1\rangle_B, \tag{9.7}$$

with $|a_0|^2 + |a_1|^2 = 1$ and $|b_0|^2 + |b_1|^2 = 1$. The tensor product of these two states is given by

$$|\varphi_A \otimes \varphi_B\rangle = a_0 b_0|00\rangle_{AB} + a_0 b_1|01\rangle_{AB} + a_1 b_0|10\rangle_{AB} + a_1 b_1|11\rangle_{AB} \tag{9.8}$$

Considering any arbitrary vector $|\psi\rangle = \alpha|00\rangle_{AB} + \beta + \gamma|10\rangle_{AB} + \delta|11\rangle_{AB}$, it typically does not follow the tensor product form (9.8), unless $\alpha\delta = \beta\gamma$ is always true, which is not necessarily the case. When the state $|\psi\rangle$ deviates from the form (9.8), we categorize it as entangled or correlated. For two photons, *A* and *B*, sharing identical polarizations, they can exist in states like $|0\rangle_A|1\rangle_B$ or $|1\rangle_A|0\rangle_B$, and they can also be in a superposition of these states simultaneously,

$$|\psi\rangle = \frac{1}{\sqrt{2}}\left[|0\rangle_A + |1\rangle_B + |1\rangle_A|0\rangle_B\right], \tag{9.9}$$

The state exhibits entanglement because the product of α and δ equals the product of β and γ. Photons exhibit entanglement, meaning they cannot be considered as individual entities with distinct quantum states. In an entangled state, the overall properties of the system are well defined, but the properties of each subsystem are not independently determined. For instance, in a two-electron system, we can prepare it so that the total spin state is zero, but the direction of each electron's spin remains unknown. When measuring the spin of one electron, the other immediately assumes the opposite orientation, even if they are separated by considerable distances. This suggests a correlation between measurements, indicating that a measurement on one electron spontaneously determines the state of the other. The behavior of two electrons working together remains mysterious, especially regarding their spin. Measuring one electron's spin vertically does not determine its spin horizontally, highlighting the absence of well-defined rotation axes. These electrons exhibit a correlation beyond classical physics' explanation. When dealing with multiple quantum particles, isolating them is challenging, as interactions with the environment lead to decoherence. Decoherence results in leaked information, causing the initial quantum behavior to shift toward classical. Preserving entanglement, crucial for applications like quantum computers, poses a significant challenge due to decoherence. Bell states, also known as EPR states, exemplify maximally entangled states. Although everything is theoretically entangled, practical observations are hindered by decoherence.

9.4 EXPLAIN THE FUNCTIONING OF THE QUANTUM TELEPORTATION PROTOCOL

The quantum teleportation protocol functions by conveying the quantum characteristics of a particle from one place to another through the utilization of entangled particles and classical communication. This process enables the replication of the initial quantum state at the destination.

9.4.1 Detail the Quantum Teleportation Protocol Step by Step

The quantum teleportation protocol is a sophisticated method for transferring an unknown quantum state between two distant parties located in separate locations, all while avoiding the physical transfer of the quantum information itself. This intricate process leverages non-local quantum correlations facilitated by an entangled pair shared between the two parties through a dedicated quantum channel, coupled with classical communication exchanges conducted via a classical channel. In the standard quantum teleportation protocol, the journey begins with the sender holding the unknown state of the qubit slated for teleportation. This state must be seamlessly transferred to the receiver's qubit. To initiate this cosmic ballet, the sender and receiver establish a prior connection by sharing a maximally entangled state, commonly referred to as the Bell state. Once the groundwork is laid, the protocol unfolds with a structured sequence of steps:

1. Preparation
   - Both the transmitter and receiver possess a qubit earmarked for teleportation.
   - These qubits undergo entanglement with a third qubit, creating the foundational entangled state.
2. Measurement
   - The sender executes a pivotal Bell measurement on the 2 qubits in its possession.
   - This measurement yields one of four possible results, each encoding valuable information into two classical bits (00, 01, 10, 11).
   - Subsequently, the 2 qubits held by the sender are gracefully discarded.
3. Transmission
   - The sender transmits the 2 classical bits to the receiver through the classical communication channel, carrying the encoded quantum essence.
4. Transformation
   - Following the transmission, the receiver's qubit stands at a crossroads, existing in one of four potential states.
   - One state mirrors the original quantum state intended for teleportation, while the other three are closely related.
   - To discern the true identity of the quantum teleportation, the receiver executes a unitary Pauli transformation, rotating the output state.
5. Verification
   - In the concluding act, the transmitter and receiver engage in a meticulous comparison of their qubits.
   - This comparative analysis serves as the cosmic litmus test, ensuring the triumphant success of the quantum state teleportation.

This organized choreography ensures the seamless execution of the quantum teleportation protocol, orchestrating a mesmerizing interplay between entangled quantum states and classical communication.

9.4.2 Exploring Accuracy in Quantum Teleportation: Quantum Fidelity

Quantum teleportation, a captivating process, involves sending the state of a quantum system to another location using entangled particles. The accuracy is theoretically impeccable, because of the bedrock of quantum entanglement and superposition. In practical terms, however, challenges arise – decoherence and noise in quantum channels can slightly compromise fidelity. Persistent efforts by researchers aim to refine the technology, mitigating these challenges and elevating the accuracy of quantum teleportation. Quantum fidelity serves as a crucial metric in the intricate realm of quantum mechanics, acting as the yardstick for gauging the resemblance between two quantum states. This metric not only assesses the degree of similarity but encapsulates the fidelity's essence in elucidating how adeptly one quantum state can transition into another while safeguarding the intricate information meticulously encoded within the quantum system. In the dynamic landscapes of quantum information processing and quantum computing, a soaring fidelity value becomes emblematic of a profound alignment between the states, unveiling a pivotal facet of precision and coherence in the quantum domain. Measuring a state $|\psi\rangle$ in a basis with $|\varphi\rangle$ involves fidelity, which equals the probability of observing $|\varphi\rangle$. It indicates the alignment of states, serving as the likelihood of encountering $|\varphi\rangle$ during measurement. In classical terms, consider two probability distributions, P and Q, associated with a set of events $\{1,\ldots,n\}$

$$P=\{p_1,\ldots,p_n\}, \qquad Q=\{q_1,\ldots,q_n\} \tag{9.10}$$

In this scenario, the measure of fidelity is

$$F(P,Q)=\sum_i \sqrt{p_i q_i}, \qquad F(P,P)=\sum_i p_i = 1 \tag{9.11}$$

The closer the distributions match, the higher the fidelity becomes, approaching unity. In quantum scenarios, the fidelity between two density operators, ρ and σ, is a measure of their similarity:

$$F(\rho,\sigma)=Tr\left[\sqrt{\sqrt{\rho}\sigma\sqrt{\rho}}\right] \tag{9.12}$$

When both ρ and σ exist in their purest forms, $\rho=\rho^2, \sigma=\sigma^2$, $\rho=|\psi\rangle\langle\psi|$, and $\sigma=|\varphi\rangle\langle\varphi|$ in such instances, the fidelity adopts the following expression:

$$\begin{aligned} Tr\left[\sqrt{|\psi\rangle\langle\psi|(|\varphi\rangle\langle\varphi|)|\psi\rangle\langle\psi|}\right] &= Tr\left[\sqrt{(\langle\psi|\varphi\rangle)^2\,|\psi\rangle\langle\psi|}\right] \\ &= |\psi\rangle|\varphi\rangle| Tr\left[\sqrt{|\psi\rangle\langle\psi|}\right] = |\psi\rangle|\varphi\rangle| \end{aligned} \tag{9.13}$$

When dealing with scenarios where only one of the operators attains purity, the expression for fidelity manifests as follows:

$$F(|\psi\rangle\langle\psi|,\sigma)=Tr\left[\sqrt{|\psi\rangle\langle\psi|\sigma|\psi\rangle\langle\psi|}\right]=Tr\left[\sqrt{\langle\psi|\sigma|\psi\rangle|\psi\rangle\langle\psi|}\right]=\sqrt{\langle\psi|\sigma|\psi\rangle} \tag{9.14}$$

If ρ and σ exhibit commutatively, then their mathematical operations can be interchanged without altering the outcome,

$$\rho=\sum_i p_i\,|i\rangle\langle i|, \qquad \sigma=\sum_i q_i\,|i\rangle\langle i| \tag{9.15}$$

We are situated within the conventional definition of fidelity (9.11),

$$F(\rho,\sigma)=Tr\left[\sqrt{\sqrt{\rho}\sigma\sqrt{\rho}}\right]=Tr\left[\sqrt{\sqrt{\sum_i p_i|i\rangle\langle i|}\sum_i q_i|i\rangle\langle i|\sqrt{\sum_i p_i|i\rangle\langle i|}}\right],$$

$$=Tr\left[\sqrt{\sum_i\sqrt{p_i}\,|i\rangle\langle i||i\rangle\langle i|\sum_i q_i\;|i\rangle\langle i|\sum_i\sqrt{p_i}\;|i\rangle\langle i|}\right]=Tr\left[\sqrt{\sum_i p_i q_i\;|i\rangle\langle i|}\right],$$

$$=\sum_i\sqrt{p_i q_i}=F(P,Q)$$

One can demonstrate that fidelity remains unchanged when subjected to a unitary transformation:

$$F\left(U\rho U^+,U\sigma U^+\right)=Tr\left[\sqrt{\sqrt{U\rho U^+}\;U\sigma U^+\sqrt{U\rho U^+}}\right]=Tr\left[\sqrt{U\sqrt{\rho}U^+\,U\sigma U^+\,U\sqrt{\rho}U^+}\right],$$

$$=Tr\left[\sqrt{U\sqrt{\rho}\,\sigma\,\sqrt{\rho}U^+}\right]=Tr\left[U\sqrt{\sqrt{\rho}\;\sigma\sqrt{\rho}}U^+\right]$$

$$=Tr\left[\sqrt{\sqrt{\rho}\;\sigma\;\sqrt{\rho}}\right]=F(\rho\,,\sigma). \tag{9.16}$$

Example:

The expression for the fidelity between $\sigma=(1-p)|0\rangle\langle 0|+p|1\rangle\langle 1|$ and $\rho=|0\rangle\langle 0|$, can be stated as

$$F(\rho,\sigma)=Tr\left[\sqrt{(1-p)\langle 0|0\rangle^2+p\langle 0|1\rangle\langle 0|1\rangle}\right]=\sqrt{1-p}. \tag{9.17}$$

It is obvious that when the shared state is at its maximum entanglement, this phenomenon of quantum teleportation manifests flawlessly, with fidelity of 1.

9.5 SOME RENOWNED CONVENTIONAL QUANTUM TELEPORTATION PROTOCOLS

Bennett and his team played a key role in developing an early quantum teleportation method. In this process, the quantum information of an unknown state was smoothly transmitted from Alice to Bob. They achieved this by using a pre-established entangled quantum channel, enabling the quantum state to disappear at one location and instantly reappear at a distant one through a combination of classical communication and entanglement. It is crucial to note that without entanglement, quantum teleportation would defy fundamental principles of quantum mechanics. This protocol highlights the profound nature of quantum entanglement, highlighting its importance in successfully transferring information between senders (like Alice) and receivers (like Bob). Quantum teleportation has been tested across various physical systems, focusing on its efficiency and success potential across different setups and substantial distances.

9.5.1 Standard Quantum Teleportation Procedure of a Qubit State

The quantum no-cloning theorem establishes that duplicating the exact state of an unknown qubit is impossible. Quantum teleportation is the solution for transmitting such a qubit, a process where Alice aims to send her unidentified qubit to Bob. Despite being unaware of her qubit's state, represented as $|\psi\rangle_C=\alpha|0\rangle_C+\beta|1\rangle_C$, Alice cannot measure α or β directly due to the unknown orientation of

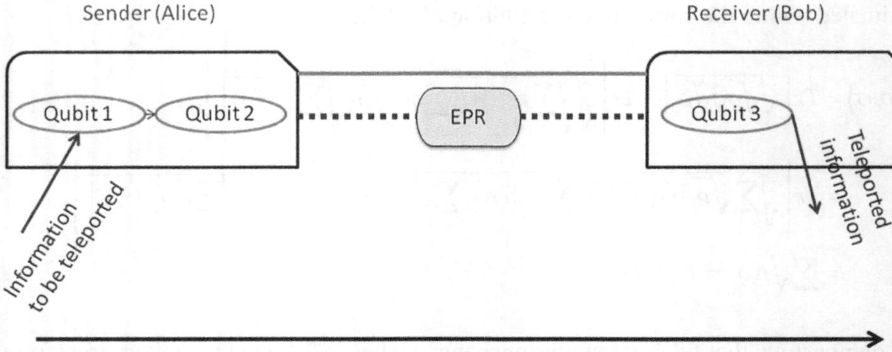

FIGURE 9.1 Illustration depicting the standard quantum teleportation process.

the particle C. Any measurement could alter the state $|\psi\rangle_C$. To achieve their objectives, Alice and Bob share an entangled qubit pair with Alice's unknown qubit. Entangled qubits yield correlated outcomes upon measurement. In a simplified manner, Alice introduces her unknown qubit to a system that interacts with its entangled counterpart for successful transmission. After measuring the unidentified qubit, Alice will share the outcomes with Bob, enabling him to reconstruct the qubit by adjusting his entangled qubit accordingly (see Figure 9.1). If we consider the state of qubit C as $|\psi\rangle_C = \alpha|0\rangle_C + \beta|1\rangle_C$, and qubits A and B are entangled, this process unfolds

$$|\varphi^+\rangle_{AB} = \frac{1}{\sqrt{2}}\left[|00\rangle_{AB} + |11\rangle_{AB}\right], \tag{9.18}$$

Alice will employ the Bell states as the basis to measure the states of qubits A and C. By doing so, the state of the non-entangled qubits A and C can be expressed as a combination of these entangled states forming the Bell state basis,

$$|\varphi^\pm\rangle_{CA} = \frac{1}{\sqrt{2}}\left\{|00\rangle_{CA} \pm |11\rangle_{CA}\right\}, \quad |\psi^\pm\rangle_{CA} = \frac{1}{\sqrt{2}}\left\{|01\rangle_{CA} \pm |10\rangle_{CA}\right\}, \tag{9.19}$$

The initial quantum state of the 3 qubits C, A, and B is represented in the following manner:

$$\begin{aligned}|\psi\rangle_C \otimes |\varphi\rangle_{AB} &= \left\{\alpha|0\rangle_C + \beta|1\rangle_C\right\} \otimes \frac{1}{\sqrt{2}}\left\{|00\rangle_{AB} + |11\rangle_{AB}\right\},\\ &= \frac{1}{\sqrt{2}}\left\{\alpha|00\rangle_{CA}|0\rangle_B + \alpha|01\rangle_{CA}|1\rangle_B + \beta|10\rangle_{CA}|0\rangle_B + \beta|11\rangle_{CA}|1\rangle_B\right\}.\end{aligned} \tag{9.20}$$

Expressed in Equation 9.19 basis and employing the Pauli operators, the state (9.20) can be represented as follows:

$$\begin{aligned}|\psi\rangle_C \otimes |\varphi\rangle_{AB} = \frac{1}{2}\Big[&|\varphi^+\rangle_{CA} \otimes |\psi\rangle_B + |\psi^+\rangle_{CA} \otimes X|\psi\rangle_B\\ &+ |\varphi^-\rangle_{CA} \otimes Z|\psi\rangle_B + |\psi^-\rangle_{CA} \otimes XZ|\psi\rangle_B\Big].\end{aligned} \tag{9.21}$$

Initially, we implement the *CNOT* gate with qubit *A* serving as the target and qubit *C* as the control,

$$CNOT\left|0^A 0^C\right\rangle = \left|0^A\, 0^{A\oplus C}\right\rangle, \qquad CNOT\left|0^A 1^C\right\rangle = \left|0^A 1^{A\oplus C}\right\rangle,$$

$$CNOT\left|1^A 0^C\right\rangle = \left|1^A\, 1^{A\oplus C}\right\rangle, \qquad CNOT\left|1^A 1^C\right\rangle = \left|1^A 0^{A\oplus C}\right\rangle, \tag{9.22}$$

Therefore

$$CNOT|\varphi^+\rangle_{CA} = CNOT\left[\frac{1}{\sqrt{2}}\left\{|00\rangle_{CA} + |11\rangle_{CA}\right\}\right] = \frac{1}{\sqrt{2}}\left[|00\rangle_{CA} + |10\rangle_{CA}\right],$$

$$CNOT|\psi^+\rangle_{CA} = CNOT\left[\frac{1}{\sqrt{2}}\left\{|01\rangle_{CA} + |10\rangle_{CA}\right\}\right] = \frac{1}{\sqrt{2}}\left[|01\rangle_{CA} + |11\rangle_{CA}\right],$$

$$CNOT|\varphi^-\rangle_{CA} = CNOT\left[\frac{1}{\sqrt{2}}\left\{|00\rangle_{CA} - |11\rangle_{CA}\right\}\right] = \frac{1}{\sqrt{2}}\left[|00\rangle_{CA} - |10\rangle_{CA}\right],$$

$$CNOT|\psi^-\rangle_{CA} = CNOT\left[\frac{1}{\sqrt{2}}\left\{|01\rangle_{CA} - |10\rangle_{CA}\right\}\right] = \frac{1}{\sqrt{2}}\left[|01\rangle_{CA} - |11\rangle_{CA}\right], \tag{9.23}$$

The state (9.21) undergoes a transformation

$$= \frac{1}{2}\left[\frac{1}{\sqrt{2}}\left(|00\rangle_{CA} + |10\rangle_{CA}\right)\otimes|\psi\rangle_B + \frac{1}{\sqrt{2}}\left(|01\rangle_{CA} + |11\rangle_{CA}\,|01\rangle_{CA}\otimes X|\psi\rangle_B\right|\right.$$

$$\left. + \frac{1}{\sqrt{2}}\left(|00>_{CA} - |10\rangle_{CA}\right)\otimes Z|\psi\rangle_B + \frac{1}{\sqrt{2}}\left(|01\rangle_{CA} - |11\rangle_{CA}\right)\otimes XZ|\psi\rangle_B\right]. \tag{9.24}$$

Next, as shown in Figure 9.2, Alice utilizes Hadamard gate on qubit *C*,

$$H\left[CNOT\left(|\varphi^+\rangle_{CA}\right)\right] = H\left[\frac{1}{\sqrt{2}}\left(|00\rangle_{CA} + |10\rangle_{CA}\right)\right] = |00\rangle_{CA},$$

$$H\left[CNOT\left(|\psi^+\rangle_{CA}\right)\right] = H\left[\frac{1}{\sqrt{2}}\left(|01\rangle_{CA} + |11\rangle_{CA}\right)\right] = |01\rangle_{CA},$$

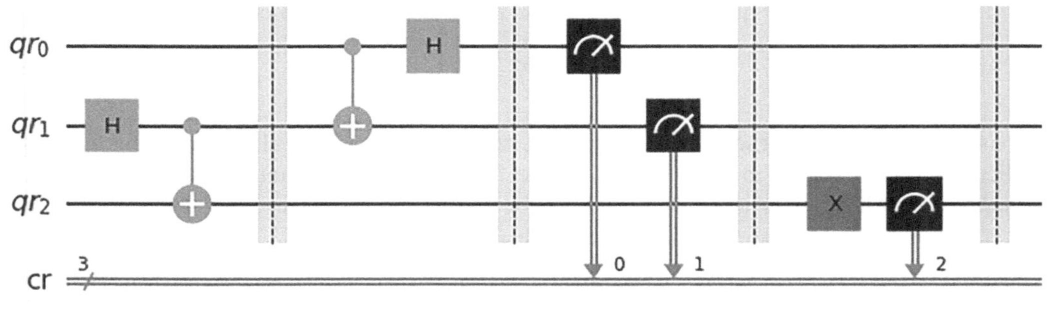

FIGURE 9.2 Quantum circuit illustrating the quantum teleportation process for a qubit.

$$H\left[CNOT\left(|\varphi^-\rangle_{CA}\right)\right]=H\left[\frac{1}{\sqrt{2}}\left(|00\rangle_{CA}-|10\rangle_{CA}\right)\right]=|10\rangle_{CA},$$

$$H\left[CNOT\left(|\psi^-\rangle_{CA}\right)\right]=H\left[\frac{1}{\sqrt{2}}\left(|01\rangle_{CA}-|11\rangle_{CA}\right)\right]=|11\rangle_{CA}, \tag{9.25}$$

Changing the state (9.24) to

$$|\psi\rangle_C\otimes|\varphi\rangle_{AB}=\frac{1}{2}\left[|00\rangle_{CA}\otimes|\psi\rangle_B+|01\rangle_{CA}\otimes X|\psi\rangle_B\right)$$

$$+|10\rangle_{CA}\otimes Z|\psi\rangle_B+|11\rangle_{CA}\otimes XZ|\psi\rangle\big]_B. \tag{9.26}$$

Subsequently, the 2 qubits in Alice's possession take on the forms $|00\rangle_{CA}, |01\rangle_{CA}, |10\rangle_{CA}$ and $|11\rangle_{CA}$. It is important to note that, at this stage, neither measurements nor communication have occurred. Although Bob's qubit may seem influenced by α and β, it is crucial to understand that Bob cannot transform or measure his qubit without knowledge of α and β. The representation of state (9.26) aims to calculate the probability of measurement outcomes and the final state of Bob's qubit following these measurements. In case 1, when Alice observes the state of qubits A and, she obtains the outcome "00" with a certain probability:

$$\left\|\frac{1}{2}\left(\alpha|0\rangle+\beta|1\rangle\right)\right\|^2=\frac{1}{4} \tag{9.27}$$

Based on the obtained outcome, the collective state of qubits C, A, B becomes

$$|00\rangle_{CA}\left\{\alpha|0\rangle_B+\beta|1\rangle_B\right\}. \tag{9.28}$$

Alice sends the binary sequence "00" to Bob. As both bits are "0," Bob decides not to perform any X or Z gate operations on his qubit. Consequently, the state of his qubit remains in the form of $\alpha|0\rangle+\beta|1\rangle$ at the end of the protocol. In the second scenario, Alice performs measurements on the 2 qubits A and C, and obtains the outcome "01." This result is achieved with a certain probability:

$$\left\|\frac{1}{2}\left(\alpha|1\rangle-\beta|0\rangle\right)\right\|^2=\frac{1}{4}. \tag{9.29}$$

Based on this outcome, the quantum state of the 3 qubits C, A, and B becomes

$$|01\rangle_{CA}\left\{\alpha|1\rangle_B+\beta|0\rangle_B\right\}. \tag{9.30}$$

Alice sends the classical bit sequence "01" to Bob. As the first bit is "0" and the second bit is "1," Bob applies the *NOT* or X gate to his qubit. As a result, his qubit transforms into the state $\alpha|0\rangle+\beta|1\rangle$. In scenario three, Alice conducts measurements on the 2 qubits, A and C, revealing the outcome "10.". The occurrence of this outcome is associated with a certain probability:

$$\left\|\frac{1}{2}\left(\alpha|1\rangle-\beta|1\rangle\right)\right\|^2=\frac{1}{4}. \tag{9.31}$$

Following this outcome, the quantum state of the 3 qubits C, A, B undergoes a specific transformation. Put differently, the state of the qubits C, A, B is altered based on the measurement result "10":

$$|10\rangle_{CA}\left\{\alpha|0\rangle_B-\beta|1\rangle_B\right\}. \tag{9.32}$$

Alice transmits the two classic bits "10" to Bob. As the first bit is "1" and the second bit is 0, Bob performs the Z gate on his qubit, and his qubit become $\alpha|0\rangle+\beta|1\rangle$.

In scenario 4, when Alice performs measurements on qubits A and C, she observes the outcome "11" with a certain probability

$$\left\|\frac{1}{2}\left(\alpha|1\rangle-\beta|0\rangle\right)\right\|^2=\frac{1}{4} \tag{9.33}$$

Based on the obtained outcome, the state of qubits C, A, and B is determined by

$$|11\rangle_{CA}\left\{\alpha|1\rangle_B-\beta|0\rangle_B\right\}. \tag{9.34}$$

Alice sends the classical bits "11" to Bob. Upon receiving them, Bob applies operations like X and Z gates on his qubit (see Table 9.1). These operations transform Bob's qubit into a state that precisely matches what Alice initially had. It might sound strange, but teleportation essentially functions as an ideal quantum channel, mimicking the effect of physically sending Alice's qubit to Bob. In this specific case, the fidelity of teleportation is perfect (equals 1) because the shared state is maximally entangled. In real-world scenarios, environmental factors come into play, resulting in a mixed state for the shared quantum state [9–14]. The fidelity of teleportation at this point is determined by a specific formula, as given in Equation 9.12. Consequently, the success of the protocol is closely tied to the degree of entanglement in the state shared between Alice and Bob – higher entanglement leads to a fidelity closer to 1.

9.5.2 CQT Procedure of a Qubit State

The standard protocol facilitates the transfer of information from one party, like Alice, to a distant party, Bob, using entangled bipartite states as a shared resource. However, a limitation arises as the signal weakens and is lost beyond about a hundred kilometers. In traditional fiber optic communication, this issue is addressed by incorporating repeaters along the fiber path. These repeaters receive,

TABLE 9.1
Enumeration of the Set of Individual Operations That Bob Needs to Perform to Restore the Original State, Guided by the Outcome of Alice's Measurement

Alice's Outcome	Bob's Unitary Operator
$\|\phi^+\rangle$	I
$\|\phi^-\rangle$	Z
$\|\psi^+\rangle$	X
$\|\psi^-\rangle$	ZX

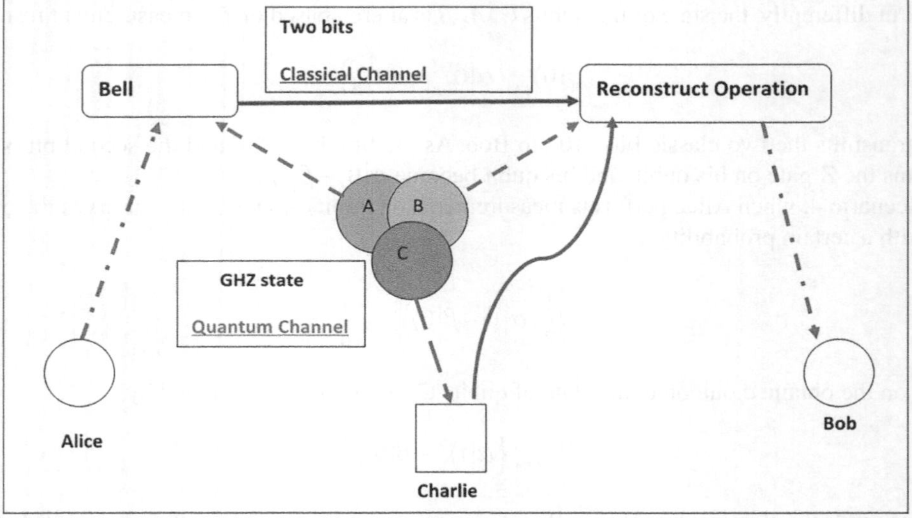

FIGURE 9.3 Illustration of CQT, with Bob serving as the receiver and Charlie taking on the role of the controller.

measure, copy, and amplify the weakened signal for retransmission to the recipient. The process can be repeated by adding more repeaters as needed. The concept of standard quantum teleportation has evolved into CQT, involving a tripartite state shared among three parties. An additional particle is sent to a third party, Charlie, who supervises and controls the teleportation process. This scheme allows the transfer of an unknown state to one or both receivers simultaneously, but only one receiver can reconstruct the quantum state based on the measurement result of the other receiver [15–17]. Importantly, the number of controllers can be increased as required for practical implementation. Suppose that Alice's desired state for teleportation is:

$$|\psi\rangle_1 = \alpha|0\rangle_1 + \beta|1\rangle_1, \tag{9.35}$$

In the above-described protocol, there are two recipients: Bob and Charlie. However, only one of them can ultimately acquire the unknown state based on the measurement outcome of the controller (the other recipient). Alice has the flexibility to select either Bob or Charlie as the recipient. In this illustration, Alice chooses Bob as the actual recipient for the unknown target state, making Charlie the controller in the teleportation process. The process, shown in Figure 9.3, begins with Alice executing a teleportation procedure. To initiate this process, a source of -three- particle entangled states is required, as demonstrated below:

$$|\varphi_G\rangle_{234} = \frac{1}{2}\left(|001\rangle + |010\rangle + |100\rangle + |111\rangle\right)_{234}, \tag{9.36}$$

Particle 2 is specifically designated for Alice, while particles 3 and 4 are specifically assigned to Bob and Charlie, respectively. The state of the overall system, denoted as $|\psi\rangle_1 \otimes |\varphi_G\rangle_{234}$ can be expressed in a different form as outlined below:

$$|\psi\rangle_1 \otimes |\varphi_G\rangle_{234} = \frac{1}{2}[\left(\alpha|001\rangle + \alpha|010\rangle + \beta|101\rangle + \beta|110\rangle\right)_{123} \otimes |0\rangle_4 +$$

$$\left(\alpha|000\rangle + \alpha|011\rangle + \beta|100\rangle + \beta|111\rangle\right)_{123} \otimes |1\rangle_4], \tag{9.37}$$

To enhance clarity in illustrating the transformation, we break down Equation 9.37 into two distinct parts, as outlined below:

$$\begin{aligned}
&\frac{1}{2}\left(\alpha|000\rangle+\beta|100\rangle+\alpha|011\rangle+\beta|111\rangle\right)_{123}\otimes|1\rangle_4=\frac{1}{4}[\left(\alpha|000\rangle+\alpha|110\rangle+\beta|001\rangle+\beta|111\rangle\right)_{123}\\
&+\left(\alpha|000\rangle-\alpha|110\rangle-\beta|001\rangle+\beta|111\rangle\right)_{123}+\left(\alpha|011\rangle+\alpha|101\rangle+\beta|010\rangle+\beta|100\rangle\right)_{123}\\
&+\left(\alpha|011\rangle-\alpha|101\rangle-\beta|010\rangle+\beta|100\rangle\right)_{123}]\otimes|1\rangle_4\\
&=\frac{1}{2\sqrt{2}}[\frac{1}{\sqrt{2}}(|00\rangle+\langle 11|)_{22}\otimes\left(\alpha|0\rangle+\beta|1\rangle\right)_3+\frac{1}{\sqrt{2}}(|00\rangle-|11\rangle)_{22}\otimes(\alpha|0\rangle-\beta|1\rangle)_3+\\
&\frac{1}{\sqrt{2}}(|01\rangle+|10\rangle)_{12}\otimes(\alpha|0\rangle+\beta|1\rangle)_3+\frac{1}{\sqrt{2}}(|01\rangle-|10\rangle)_{12}\otimes(\alpha|0\rangle-\beta|1\rangle)_3]\otimes|1\rangle_4
\end{aligned} \tag{9.38}$$

and

$$\begin{aligned}
&\frac{1}{2}\left(\alpha|000\rangle+\beta|100\rangle+\alpha|010\rangle+\beta|110\rangle\right)_{123}\otimes|0\rangle_4=\frac{1}{4}[\left(\beta|000\rangle+\beta|110\rangle+\alpha|001\rangle+\alpha|111\rangle\right)_{123}\\
&+\left(-\beta|000\rangle+\beta|110\rangle+\alpha|001\rangle+\alpha|111\rangle\right)_{123}+\left(\beta|011\rangle+\beta|101\rangle+\beta|010\rangle+\beta|100\rangle\right)_{123}\\
&+\left(\beta|011\rangle-\beta|101\rangle+\alpha|010\rangle-\alpha|100\rangle\right)_{123}]\otimes|0\rangle_4\\
&=\frac{1}{2\sqrt{2}}[\frac{1}{\sqrt{2}}\left(|00\rangle+|11\rangle\right)_{12}\otimes\left(\beta|0\rangle+\alpha|1\rangle\right)_3+\frac{1}{\sqrt{2}}\left(|00\rangle-|11\rangle\right)_{12}\otimes\left(-\beta|0\rangle+\alpha|1\rangle\right)_3+\\
&\frac{1}{\sqrt{2}}\left(|01\rangle-|10\rangle\right)_{12}\otimes\left(-\beta|0\rangle+\alpha|1\rangle\right)_3+\frac{1}{\sqrt{2}}\left(|01\rangle-|10\rangle\right)_{12}\otimes\left(-\beta|0\rangle+\alpha|1\rangle_3\right)]\otimes|0\rangle_4
\end{aligned} \tag{9.39}$$

These two parts can be broken down using the basis (9.19). The first part (9.38) is rewritten as

$$\begin{aligned}
&\frac{1}{2\sqrt{2}}[|\varphi^+\rangle_{12}\otimes\left(\alpha|0\rangle+\beta|1\rangle\right)_3+|\varphi^-\rangle_{12}\otimes\left(\alpha|0\rangle-\beta|1\rangle\right)_3+\\
&|\psi^+\rangle_{12}\otimes\left(\alpha|0>+\beta|1\rangle\right)_3+|\psi^-\rangle_{12}\otimes\left(\alpha|0>-\beta|1\rangle\right)_3]\otimes|1\rangle_4
\end{aligned} \tag{9.40}$$

Based on the Pauli operators, the first part (9.40) is expressed in the following manner:

$$\frac{1}{2\sqrt{2}}[|\varphi^+\rangle_{12}\otimes|\psi\rangle_3+|\varphi^-\rangle_{12}\otimes Z|\psi\rangle_3+|\psi^+\rangle_{12}\otimes X+|\psi^-\rangle_{12}\otimes ZX|\psi\rangle_3\otimes|1\rangle_4 \tag{9.41}$$

Similarly, Equation 9.39 adopts the ultimate structure provided below:

$$\frac{1}{2\sqrt{2}}\left[|\varphi^+\rangle_{12}\otimes X|\psi\rangle_3+|\varphi^-\rangle_{12}\otimes ZX|\psi\rangle_3+|\psi^+\rangle_{12}\otimes|\psi\rangle_3+|\psi^-\rangle_{12}\otimes Z|\psi\rangle_3\right]\otimes|0\rangle_4 \tag{9.42}$$

The state of the full system (9.37) transforms as follows:

$$\begin{aligned}
&|\psi\rangle_1\otimes|\varphi_G\rangle_{234}=\frac{1}{2\sqrt{2}}[|\varphi^+\rangle_{12}\otimes|\psi\rangle_3+|\varphi^-\rangle_{12}\otimes Z|\psi\rangle_3+|\psi^+\rangle_{12}\otimes X|\psi\rangle_3+\\
&|\psi^-\rangle_{12}\otimes ZX|\psi\rangle_3]\otimes|1\rangle_4+\frac{1}{2\sqrt{2}}[|\varphi^+\rangle_{12}\otimes X|\psi\rangle_3+|\psi^-\rangle_{12}\otimes ZX|\psi\rangle_3+|\psi^+\rangle_{12}\otimes\\
&|\psi\rangle_3+|\psi^-\rangle_{12}\otimes Z|\psi\rangle_3\otimes|0\rangle_4
\end{aligned} \tag{9.43}$$

TABLE 9.2
Categorization of the Various Unitary Transformations

Outcomes of Alice	Outcomes of Charlie	Pauli operation of Bob
$\lvert\varphi^+\rangle_{12}$	$\lvert 1\rangle_4$	I
$\lvert\varphi^-\rangle_{12}$	$\lvert 1\rangle_4$	Z
$\lvert\psi^+\rangle_{12}$	$\lvert 1\rangle_4$	X
$\lvert\psi^-\rangle_{12}$	$\lvert 1\rangle_4$	ZX
$\lvert\varphi^+\rangle_{12}$	$\lvert 0\rangle_4$	X
$\lvert\varphi^-\rangle_{12}$	$\lvert 0\rangle_4$	ZX
$\lvert\psi^+\rangle_{12}$	$\lvert 0\rangle_{4.}$	I
$\lvert\psi^-\rangle_{12}$	$\lvert 0\rangle_{4.}$	Z

Notes: In the scenario where Alice selects Charlie as the recipient, Bob assumes the role of a supervisor in the quantum teleportation protocol, treated similarly to other cases. To precisely restore the initial state based on the eight measurements conducted by Alice and Charlie, unitary transformations are essential. The unitary operator I implies that Bob is not required to perform any additional operations for the recovery process.

Alice utilizes Bell state analyzers to measure particles 1 and 2, and subsequently instructs Charlie to measure particle 4 in the $\{\lvert 0\rangle,\lvert 1\rangle\}$ basis. They share their measurement outcomes with Bob through conventional channels. This ensures that quantum teleportation adheres to causality. If, for instance, Alice obtains the outcome $\lvert\varphi^+\rangle_{12}$ and Charlie measures $\lvert 1\rangle_4$, the resulting state of particle 3 at the receiving end takes the following shape:

$$\lvert\psi\rangle_3 = \alpha\lvert 0\rangle_3 + \beta\lvert 1\rangle_3 \tag{9.44}$$

The state $\lvert\psi\rangle_3$, which Alice initially aims to teleport to Bob, aligns precisely with the outcome depicted in Equations 9.41 and 9.42. Alice and Charlie can yield four and two distinct measurement results, respectively, totaling eight possible outcomes. Yet, for specific results, Bob needs to perform a straightforward unitary transformation for precise reconstruction of the desired quantum state (Table 9.2).

Bob must receive the measurement result from supervisor Charlie through conventional channels for him to reliably obtain the desired quantum state, assuming Charlie is honest during supervision. However, if the control part is uncooperative, there are two scenarios to consider regarding Charlie's behavior. The first scenario involves Charlie not measuring particle 4, while the second scenario occurs when Charlie communicates his measurement result to Bob in a reversed manner. Fidelity, a measure of resemblance between quantum states, is utilized here to determine the probability that the received state matches the desired one. Let's explore the fidelity calculation in the case where Charlie refrains from measuring particle 4 after the Bell state measurement and the state of particles 3 and 4 takes a specific form:

$$\lvert\varphi\rangle_{34} = \alpha\lvert 01\rangle_{34} + \beta\lvert 11\rangle_{34} \tag{9.45}$$

The quantum state that Bob aimed to obtain is represented by Equation 9.44; the reduced density operator of particle 3 is given:

$$\begin{aligned}\rho_3 &= Tr_4\left(|\varphi\rangle_{3434}\langle\varphi|\right)\\ &=|\alpha|^2|0\rangle_{33}\langle 0|+\alpha\beta^*|0\rangle_{33}\langle 1|+\alpha^*\beta|1\rangle_{33}\langle 0|+|\beta|^2|1\rangle_{33}\langle 1|\end{aligned} \tag{9.46}$$

By calculating the fidelity expression, one can derive the following:

$$F_3 = \langle\varphi|_3\,\rho_3\,|\varphi\rangle_3 = \left\{|\alpha|^2+|\beta|^2\right\}^2 \tag{9.47}$$

Fidelity ranges from one half (when both $|\alpha|$ and $|\beta|$ B are1/2) to one (when either $|\alpha|$ is 1 and $|\beta|$ is 0, or vice versa), determined by the coefficients α and β of the quantum state to be teleport. To calculate fidelity, we can express α and β as parameters. This transforms the state (9.44) as follows:

$$|\varphi\rangle_3 = cos(v/2)|0\rangle_3 + e^{i\varphi}sin(v/2)|1\rangle_3 \tag{9.48}$$

The average fidelity, a measure reflecting the extent of faithfulness or accuracy within a specific context, is computed by adding up individual fidelity values and then dividing this sum by the total number of data points or instances being analyzed. These metric functions serve as a numerical gauge, allowing for the evaluation of the overall reliability or proximity of a given set of values. Its significance spans diverse domains, including applications in data analysis, signal processing, and quality assurance, where it serves as a valuable indicator of precision and dependability. The calculated average fidelity is given by the expression below:

$$F_{av} = \frac{1}{4\pi}\int_0^{2\pi} d\varphi \int_0^{\pi} F_3\, sin(v)\, dv = \frac{1}{2} \tag{9.49}$$

When Charlie informs Bob that his measurement outcome is inverted, the quantum state's average limit becomes half, preventing the reconstruction of the receiver particle. Consequently, the collaboration of Alice, Bob, and Charlie is crucial for the protocol's faithful execution, particularly the control aspect. The equality of importance among the three participants is emphasized, as without Charlie's intervention or unbiased measurement, Bob would be unable to reconstruct the initial state. This underscores the necessity of their joint effort in the authentication quantum of users.

9.6 CQT CONNECTING DISCRETE AND CONTINUOUS PHYSICAL SYSTEMS

Quantum teleportation can be categorized into three classes: the first involves a DV entangled resource channel [2, 18], the second employs a CV entangled resource channel [19], and the third pertains to the coupling of DV and CV facilitated by an entangled resource [20, 21]. Hybrid systems emerge through the integration of CVs, which have a range of values, and DVs, which have distinct and countable values. This parallels classical information processing, where digital and analogical encoding methods are employed. Many real-world systems are effectively described using a hybrid framework, necessitating a connection between DVs and CVs. Recent research has focused on developing protocols that address the limitations of individual DV and CV approaches. Consequently, hybrid systems have become a focal point in quantum technology research, aiming to provide improved solutions compared to CV or DV protocols. The combination of CV and DV approaches within the same system has shown both theoretical and experimental advancements, leading to the recognition of hybrid teleportation as a key priority in quantum science [22]. In this section, we delve into the challenge of CQT, specifically dealing with a continuous-variable

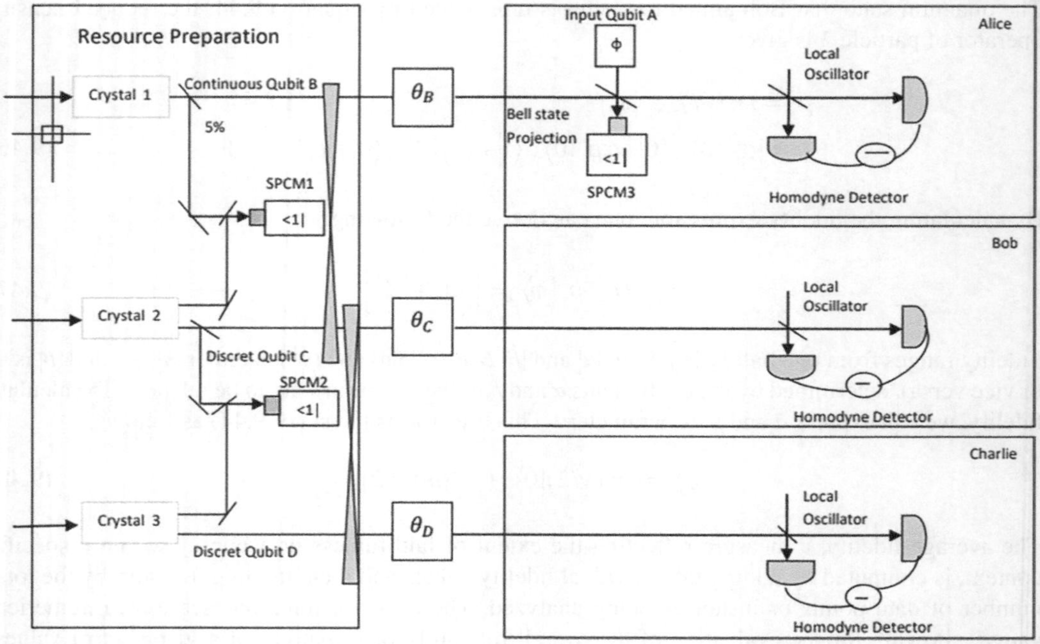

FIGURE 9.4 To experimentally achieve CQT of a continuous mode (9.53), a tripartite hybrid entangled resource state (9.50) is utilized. The preparation of the resource state involves counting clicks from two single-photon counting modules (SPMC), particularly triggered by an SPCM3 click event during the Bell state measurement [23]. After measurements by Alice and Charlie, Bob finalizes the CQT using a homodyne detector in mode C. Additionally, phase measurements are sustained with two other homodyne detectors in modes B and D.

CV qubit characterized by uniform amplitude. This qubit is encoded using a coherent state that reflects a superposition of quantum vacuum and single-photon optical states, highlighting diverse phases. The teleportation process unfolds between two geographically separated entities, Alice and Bob, and is facilitated by the consent of the controller, Charlie. To achieve this objective, we employ an entangled tripartite hybrid state, incorporating interactions between states with DVs and with CVs, as a quantum resource. The coherent part serves as a quantum resource owned by Alice, while the single photon is the possession of both Bob and Charlie. Moreover, Alice has control over the continuous-variable qubit. The interaction between DVs and CVs occurs through a highly transmissive beam splitter. In the experimental implementation of this proposed scheme, linear optical components such as beam splitters, phase shifters, and photon counters are utilized. The successful execution of this protocol depends on the collaborative efforts of all three contributors (see Figure 9.4). To elaborate, we present a method for transmitting a continuous-variable CV qubit state through a tripartite entangled resource state. This state is encoded as a superposition involving a coherent state across two DV modes, representing vacuum and single-photon Fock states. Given our interest in utilizing a hybrid quantum resource for quantum teleportation, our attention shifts to a tripartite resource state described as under:

$$
\begin{aligned}
|R\rangle_{BCD} &= \frac{1}{\sqrt{2}}\left[|\alpha\rangle_B|0\rangle_C\frac{|0\rangle_D+|1\rangle_D}{\sqrt{2}} - |-\alpha\rangle_B|1\rangle_C\frac{|0\rangle_D-|1\rangle_D}{\sqrt{2}}\right] \\
&= \frac{1}{2}\left(\left[|\alpha\rangle_B|0\rangle_C - |-\alpha\rangle_B|1\rangle_C\right]|0\rangle_D + \left[|\alpha\rangle_B|0\rangle_C + |-\alpha\rangle_B|1\rangle_C\right]|1\rangle_D\right)
\end{aligned}
\tag{9.50}
$$

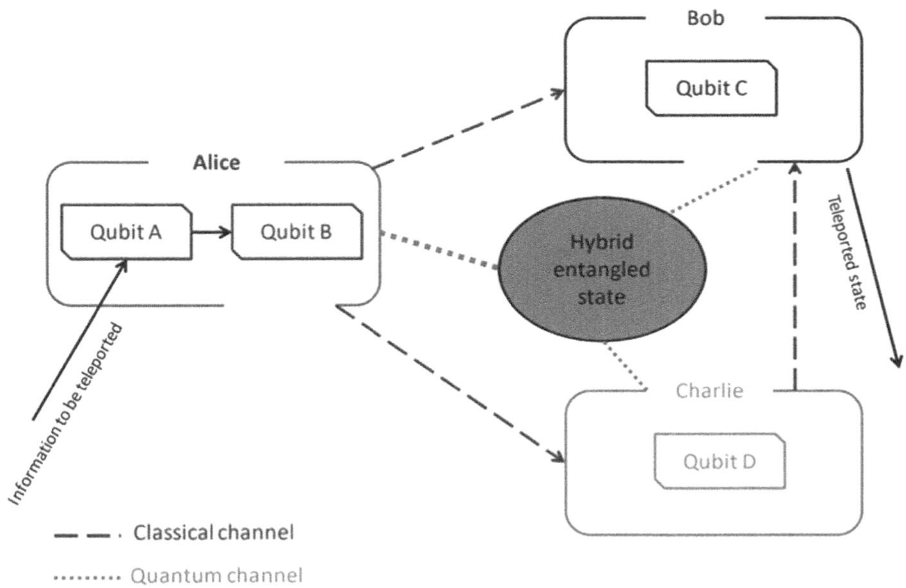

FIGURE 9.5 Description of the conceptual diagram illustrating the discussed protocol.

In Figure 9.4, modes *B, C*, and *D*, correspond to Alice, Bob, and Charlie, respectively. The hybrid entanglement state comprises coherent components with opposing sign amplitudes (assuming $\alpha < 0$). Notably, the single photon occupies both vacuum and single-photon Fock states simultaneously. State (9.50) is characterized as non-maximally entangled due to the non-orthogonality of coherent states. Following the preparation of the intricate quantum channel resource, Alice, Bob, and Charlie are poised to initiate the quantum teleportation protocol, sequentially following the tasks outlined in Figure 9.5.

In the realm of CQT, our trio – Alice, Bob, and Charlie – exists in separate spatial domains yet shares an entanglement facilitated by a quantum channel. This channel serves as a conduit for the transmission of quantum information, complemented by a classical communication channel. The structured sequence of CQT unfolds as follows [22, 24, 8, 7, 25, 26]

1. *State Fusion:* Alice initiates the process by merging the state designated for teleportation with the quantum channel.
2. *Bell Measurement Act:* Following this fusion, Alice adeptly applies a Bell measurement to the two modes under her control.
3. *Photon Counter Performance:* In the subsequent phase, Charlie steps into the spotlight, executing a photon counter (PC) measurement on his mode.
4. *Communication Interplay:* To assist Bob in deciphering the quantum intricacies, Alice and Charlie engage in an exchange, dispatching the outcomes of their measurements to him through traditional communication.
5. *Unitary Transformation Finale:* Armed with the received results, Bob takes center stage, implementing the appropriate unitary transformation – whether it be the flip-bit (X), flip-phase (Z), or flip-bit-phase (Y) – culminating in the acquisition of the coveted initial teleportation state.

When contemplating hybrid entangled states, our immediate association leads us to coherent states. Broadly speaking, coherent states are typically non-orthogonal. Specifically, the overlap between two coherent states, denoted as θ_C and $\theta_B = \theta_C$,, is a key consideration,

$$\left|\langle\alpha|\beta\rangle\right|^2 = e^{-|\alpha-\beta|^2} = x^4 \tag{9.51}$$

With $x = \sqrt{\langle\alpha|\beta\rangle}$. Orthogonal states can be represented through even and odd coherent states:

$$|IMPAIR,\alpha\rangle = N_-\left(|\alpha\rangle - |-\alpha\rangle\right), \qquad |PAIR,\alpha\rangle = N_+\left(|\alpha\rangle + |-\alpha\rangle\right) \tag{9.52}$$

In the realm of Schrödinger's cat states [27], let's consider a scenario where Alice possesses a continuous-variable CV qubit. This qubit is intricately encoded using a combination of coherent states labeled as X and Y. Alice's goal is to transfer or "teleport" this qubit to Bob. Consequently, the initial state can be expressed as follows:

$$|in\rangle_A = X|\alpha> + Y|-\alpha\rangle \tag{9.53}$$

The normalization condition for the state involves the expression $|X|^2 + |Y|^2 + 2x^2Re(X*Y)1$. By employing even and odd coherent states as outlined in Equation 9.52, the input state can be described in the following manner:

$$|in\rangle_A = A_+|PAIR,\alpha\rangle_A + A_-|IMPAIR,-\alpha\rangle_A \tag{9.54}$$

where

$$A_\pm = \frac{X+Y}{2N_\pm}, \qquad N_\pm = \frac{1}{\sqrt{2(1\pm x^2)}} \tag{9.55}$$

Later on, we will substitute the continuous-variable qubit [28], with a coherent state featuring a diverse phase:

$$|in\rangle\_\{A\} = |\alpha e^{i\varphi}\rangle_A = \frac{1}{\sqrt{2}}\left(|0\rangle_A + e^{i\varphi}|1\rangle_A\right) = \frac{1+e^{i\varphi}}{2}|\alpha\rangle_A + \frac{1-e^{i\varphi}}{2}|-\alpha\rangle_A, \tag{9.56}$$

Break down the expression into its individual components, specifically focusing on the term related to a single photon in the Fock basis. Consequently, it can be substituted or represented differently:

$$X = \frac{1+e^{i\varphi}}{2}, \quad Y = \frac{1-e^{i\varphi}}{2} \tag{9.57}$$

When the state $|in\rangle_A$ is represented as $|\pm\alpha\rangle$ with φ taking values in the set $\{0, 2\pi\}$, the teleportation fidelity is maximized. In the coherent state expression $|\pm\alpha\rangle = \frac{1}{\sqrt{2}}(|0\rangle \pm |1\rangle)$ Equation 9.50 transforms accordingly:

$$\begin{aligned}|R\rangle_{BCD} = \frac{1}{2\sqrt{2}}(&|000\rangle_{BCD} + |001\rangle_{BCD} - |010\rangle_{BCD} + |011\rangle_{BCD} \\ &+ |100\rangle_{BCD} + |101\rangle_{BCD} + |110\rangle_{BCD} - |111\rangle_{BCD})\end{aligned} \tag{9.58}$$

9.6.1 Theoretical Study

Now we embark on the implementation of our quantum teleportation protocol, illustrated diagrammatically in Figure 9.4, utilizing a coherent state with varied phase (9.56). This process takes into account the other three phases of the tripartite entangled resource (9.50). In effect, the combined state of the four modes A, B, C. and D. assumes the following form:

$$|\alpha e^{i\varphi}\rangle_A \otimes \left[\hat{U}_{\theta_B\theta_C\theta_D}|R\rangle_{BCD}\right] = |\alpha e^{i\varphi}\rangle_A |\alpha e^{i\theta_B}\rangle_B |0\rangle_C$$

$$\frac{|0\rangle_D + e^{i\theta_D}|1\rangle_D}{2} - |\alpha e^{i\varphi}\rangle_A |-\alpha e^{i\theta_B}\rangle_B e^{i\theta_C}|1\rangle_C \frac{|0\rangle_D - e^{i\theta_D}|1\rangle_D}{2}, \tag{9.59}$$

In this setup, the modes A and B, both falling within the range of Alice, undergo a symmetrical beam splitter, resulting in the transformation of the state (9.59) to

$$\begin{aligned}|\psi\rangle_{ABCD} = &\left|\alpha\frac{e^{i\varphi}+e^{i\theta_B}}{\sqrt{2}}\right\rangle_A \left|\alpha\frac{e^{i\varphi}-e^{i\theta_B}}{\sqrt{2}}\right\rangle_B |0\rangle_C \frac{|0\rangle_D + e^{i\theta_D}|1\rangle_D}{2} \\ &-\left|\alpha\frac{e^{i\varphi}-e^{i\theta_B}}{\sqrt{2}}\right\rangle_A \left|\alpha\frac{e^{i\varphi}+e^{i\theta_B}}{\sqrt{2}}\right\rangle_B e^{i\theta_C}|1\rangle_C \frac{|0\rangle_D - e^{i\theta_D}|1\rangle_D}{2}\end{aligned} \tag{9.60}$$

Next we apply single-photon detection to the state by utilizing a *POVM* positive operatovalued measure) denoted as $\hat{\mathbb{I}}_A \otimes \hat{\mathbb{I}}_B$ in modes A and B,

$$\hat{\mathbb{I}}_A = \Sigma_n[1-(1-\eta)^n]|n\rangle\langle n|, \tag{9.61}$$

Here, $\hat{\mathbb{I}}_B$ represents the identity operator, indicating that the measurement does not alter the state of mode B. Consequently, the bipartite output state of modes C and D, corresponding to the receiver and supervisor, assumes the following configuration:

$$\hat{\rho}_{CD} = Tr_{AB}\left[\hat{\mathbb{I}}_A \otimes \hat{\mathbb{I}}_B |\psi_{ABCD}\rangle\langle\psi_{ABCD}|\right] \tag{9.62}$$

When approximating to the first order of D. the C is significantly reduced to

$$\hat{\mathbb{I}}_A \otimes \hat{\mathbb{I}}_B = |10\rangle_{AB} \tag{9.63}$$

The resulting state after this projection will be

$$\begin{aligned}\langle 10|_{AB}|\psi_{ABCD} = \alpha[&\frac{e^{i\varphi}+e^{i\theta_B}}{\sqrt{2}}|0\rangle_C \frac{|0\rangle_D + e^{i\theta_D}|1\rangle_D}{2} \\ &-\frac{e^{i\varphi}-e^{i\theta_B}}{\sqrt{2}} e^{i\theta_C}|1\rangle_C \frac{|0\rangle_D - e^{i\theta_D}|1\rangle_D}{2}\end{aligned} \tag{9.64}$$

In this protocol, the unique aspect lies in the communication dynamics between Alice and Bob, hinging on the concurrence of the third party, Charlie. At this juncture, Charlie plays a crucial role by implementing a photon countermeasure on mode D, as detailed in Table 9.3. The resulting teleported state assumes its final configuration after undergoing Bob's unitary transformation:

TABLE 9.3
Unitary Transformation Applied by Bob after Receiving Measurement Results from Alice and Charlie

Case	PC result (D)	Unitary operation
1	0	$\hat{U}_1$
2	1	I

$$|out\rangle_C = \frac{1}{2}\left[\left(e^{i\varphi} + e^{i\theta_B}\right)|0\rangle_{\{D\}} + \left(e^{i\varphi} - e^{i\theta_B}\right)e^{i\theta_C}|1\rangle_D\right], \tag{9.65}$$

which defines the ideal expectation. In the domain of quantum mechanics, our foundational understanding hinges on carefully defining ideal expectations. Within this intricate fabric, the fidelity of teleportation serves as a subtle gauge intricately intertwined with the principles of quantum coherence. This fidelity, assessing the faithfulness of teleportation, goes beyond abstraction, encapsulating the merging of quantum states, the graceful dance of entanglement, and the nuanced interplay of information woven throughout the quantum landscape. Determining this fidelity entails a meticulous examination of quantum states at both ends of the teleportation journey, scrutinizing the complexities of superposition and entanglement to reveal the most accurate portrayal of teleportation perfection. Consequently, the fidelity of teleportation operates as a profound code, unraveling the very core of quantum communication's ideal representation in this particular instance. In this case, the fidelity of teleportation is given by

$$F = \frac{1}{4}\left[2 + cos\left(\theta_B - \theta_C\right) - cos\left(2\varphi - \theta_B - \theta_{CB}\right)\right] \tag{9.66}$$

In the classical limit where $F = \frac{1}{2}$ is attained, given by $\cos(\theta_B - \theta_C) - \cos(2\varphi - \theta_B - \theta_{CB}) = 0$, that is, assuming $\phi = \theta_B$, we have $\rho_C = |0\rangle_{CC}\langle 0|$, indicating that Bob receives only half of the state for teleportation. Now, by fixing Bob's receiver mode phase θ_C. and assuming $\theta_B = \theta_C$. we obtain $F = (\cos(2(\phi - \theta_B)))/4$. In Figure 9.6, we illustrate the variation in the fidelity of the first-order CQT of α concerning the phase φ of the teleported coherent state for variousalues of the emitter mode phase θ_B.

9.6.1 Implementation

In our pursuit of implementing the proposed CQT protocol, the goal is to teleport a coherent state characterized by a varied phase (9.56) from the transmitter, Alice, to the receiver, Bob, with the guidance of the controller, Charlie. This process involves utilizing a hybrid entangled tripartite quantum channel along with classical communication. To commence this intricate procedure, we first establish a quantum channel that is pre-shared among the three parties: Alice, Bob, and Charlie. Alice takes the lead in creating the quantum channel, deploying necessary quantum gates and operations, which include delayed measurements. Following the execution of these essential operations, Alice shares the three meticulously prepared qubits with the other involved parties. To ensure the security of the quantum information, Alice promptly destroys her qubits, thereby eliminating the possibility of any further communications. This systematic approach ensures the coherent and secure transmission of quantum information within the framework of the CQT protocol. In the simplest scenario, where the coherent state phase is either 0 or 2π, the quantum channel Equation 9.50 can take the form given by Equation 9.58. Let's initially focus on this configuration

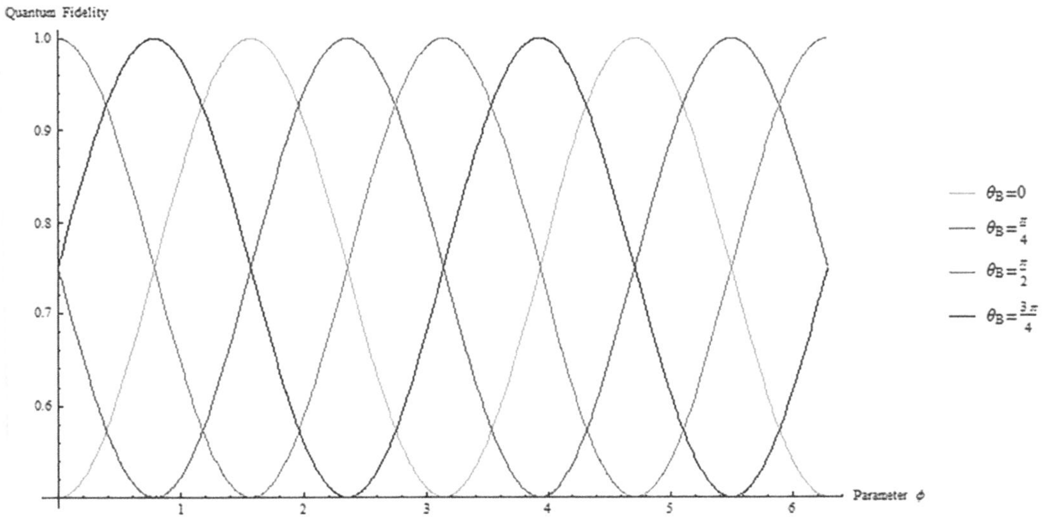

FIGURE 9.6 Fidelity in the first order of α determined after the projection of Alice onto $|01\rangle_{AB}$ (zero- and single-photon detection), with Charlie performing photon counting, at $\theta_B = \theta_C$.

before introducing additional phases. The quantum channel under consideration involves a shared entangled state among 3 qubits, assigned to sender Alice, receiver Bob, and supervisor Charlie. This state is constructed using H gates, CNOT gates, and the implementation of Pauli gates (*X*, *Y*, *Z*) on the "IBM-Qasm" simulator. To illustrate, the 3-qubit entangled state is depicted on the quantum circuit as the product of three states, all initially set to $|0\rangle$.

$$|R_0\rangle_{BCD} = |0\rangle_B \otimes |0\rangle_C \otimes |0\rangle_D \tag{9.67}$$

Qubit *B* belongs to Alice, qubit *C* to Bob, and qubit *D* to Charlie. After the application of gate H on qubit *C*, the initial state $|R_0\rangle_{BCD}$ undergoes transformation to

$$|R_1\rangle_{BCD} = \frac{1}{\sqrt{2}}\left(|000\rangle_{BCD} + |010\rangle_{BCD}\right) \tag{9.68}$$

Applying the Z-gate to qubit *C* results in the transformation of the state $|R_1\rangle_{BCD}$ to a new state:

$$|R_1\rangle_{BCD} = \frac{1}{\sqrt{2}}\left(|000\rangle_{BCD} + |010\rangle_{BCD}\right) \tag{9.69}$$

Executing the CNOT gate with qubit *C* as the control and qubit *B* as the target causes a transformation in the state $|R_2\rangle_{BCD}$

$$|R_3\rangle_{BCD} = \frac{1}{\sqrt{2}}\left(|000\rangle_{BCD} - |110\rangle_{BCD}\right) \tag{9.70}$$

When applying the X-gate to qubit *B*, the state $|R_3\rangle_{BCD}$ takes the below form:

$$|R_4\rangle_{BCD} = \frac{1}{\sqrt{2}}\left(|100\rangle_{BCD} - |010\rangle_{BCD}\right) \tag{9.71}$$

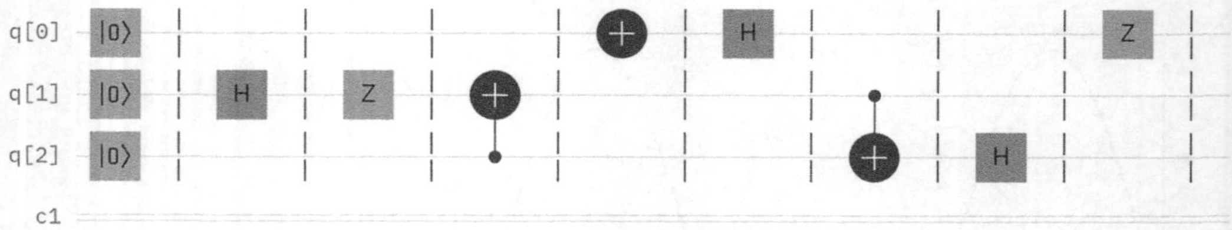

FIGURE 9.7 The quantum circuit depicted illustrates the creation of the tripartite quantum channel, with barriers marking the nine essential steps required for its preparation.

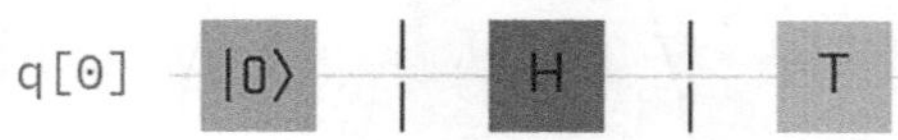

FIGURE 9.8 The quantum circuit demonstrates the creation of the desired state. The coherent state depicted in this diagram is characterized by a diverse phase, specifically $\varphi = \pi/4$

After applying the X-gate to qubit B, if we then apply the Hadamard gate (H-gate) to the same qubit, the resulting state $|R_4\rangle_{BCD}$ becomes

$$|R_5\rangle_{BCD} = \frac{1}{2}\left(|000\rangle_{BCD} - |100\rangle_{BCD} - |010\rangle_{BCD} - |110\rangle_{BCD}\right) \tag{9.72}$$

Executing the CNOT operation on qubits $|\varphi^-\rangle_{12}$ and $|\psi^+\rangle_{12}$ involves qubit $|\psi^-\rangle_{12}$ acting as the control qubit and qubit $|1\rangle_4$ as the target qubit:

$$|R_6\rangle_{BCD} = \frac{1}{2}\left(|000\rangle_{BCD} - |100\rangle_{BCD} - |011\rangle_{BCD} - |111\rangle_{BCD}\right) \tag{9.72}$$

After applying the Hadamard gate (H-gate) to qubit D, the state $|R_6\rangle_{BCD}$ undergoes a transformation:

$$\begin{aligned} |R_7\rangle_{BCD} = &\frac{1}{2\sqrt{2}}\left(|000\rangle_{BCD} + |001\rangle_{BCD} - |100\rangle_{BCD} - |101\rangle_{BCD}\right) \\ &-|010\rangle_{BCD} + |011\rangle_{BCD} - |110\rangle_{BCD} + |111\rangle_{BCD} \end{aligned} \tag{9.73}$$

The state $|R\rangle_{BCD}$ is achieved through the application of a Z gate to qubit B in the final step, as illustrated in Figure 9.7.

In this protocol, we aim to achieve a coherent state characterized by a diverse phase ϕ, as defined in Equation 9.56. To construct this state on a quantum circuit, we begin with an input state represented by a quantum qubit initialized at $|0\rangle$ The process involves applying both the H gate and a phase gate for the reconstruction of the desired coherent state:

$$|in_0\rangle_A = |0\rangle_A \tag{9.75}$$

By applying the H gate to qubit A, the initial state $|in_0\rangle_A$ undergoes a transformation, resulting in a new state:

$$|in_1\rangle_A = \frac{1}{\sqrt{2}}\left(\langle 0|_A + \langle 1|_A\right) \tag{9.76}$$

To reconstruct the input state $|in\rangle_A$ utilize phase gates (S^+, T^+, T, S) as depicted in Figure 9.8.

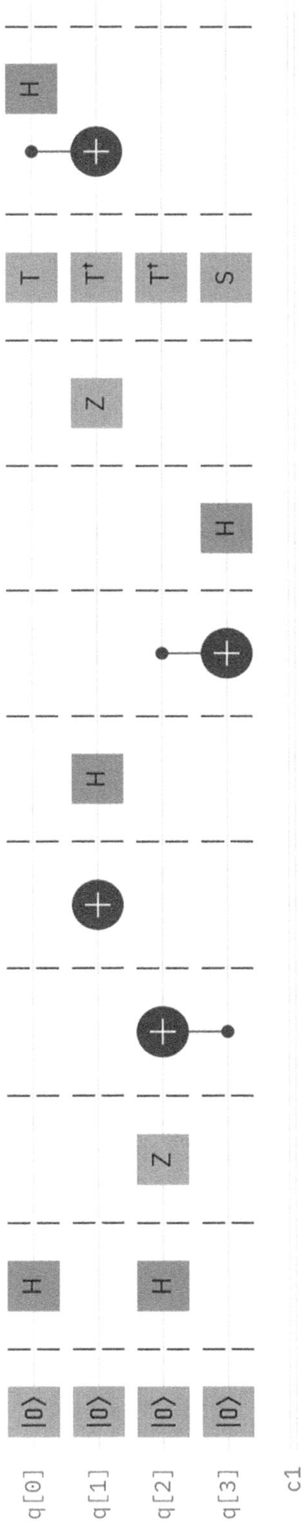

FIGURE 9.9 Quantum circuit illustrating the state combining four subsystems, namely *A* (the input qubit), *B* (Alice's qubit), *C* (Bob's qubit), and *D* (Charlie's qubit). The choice of suitable phases is based on theoretically derived results.

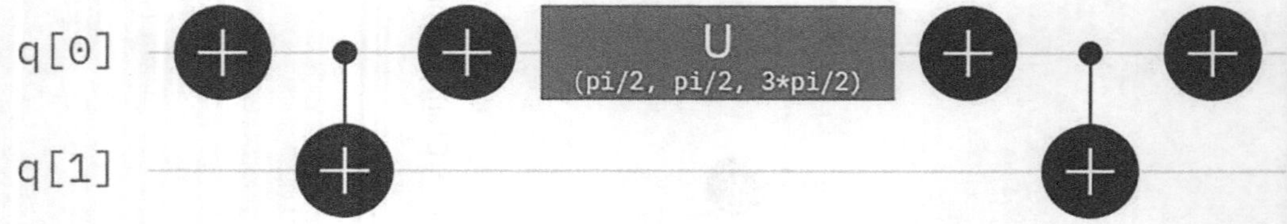

FIGURE 9.10 Illustration of quantum circuit generating coherent state with variable phase $\varphi = \frac{\pi}{4}$.

In the described experiment, the combined state of the target state and a 3-qubit hybrid quantum channel undergo transformations, specifically involving phase gates on all 3 qubits. Subsequently, Alice performs a CNOT operation on qubits I and $\pi/4$, with $\pi/2$ serving as the control qubit and $3\pi/4$ as the target qubit (Figure 9.9). Following this, Alice applies H gate on qubit π, and both $5\pi/4$ and $3\pi/2$ qubits pass through a symmetric beam splitter, implementing CNOT and X gates Figure 9.10. In delayed measurements, CNOT and CZ-gate operations are conducted on various qubit pairs. After these operations, Alice destroys her qubits and communicates measurement results to Bob classically. Charlie also measures his qubit and shares the information with Classic Bob. Upon receiving data from Alice and Charlie, Bob applies appropriate unitary Pauli operators and measures his qubit (Figure 9.11). This entire experiment is carried out on "IBM quantum features," comparing statistical data from teleporting a coherent state with varied phase φ across different IBM simulators ("IBM-Qasm simulator," "IBM-Aer simulator," "IBM-Belem simulator"), as presented in Table 9.4 and Figure 9.12.

9.7 CONCLUDING REMARKS

Quantum teleportation, an intriguing manifestation of quantum non-locality, emerges as a noteworthy application. This process facilitates the transfer of an unknown quantum state between spatially separated entities, bypassing the direct transmission of information. Upon completing the protocol, the initial state of the first entity transforms, aligning the state of the second entity with the originally transmitted information. Bennett's seminal 1993 research served as a catalyst, sparking inquiries, surmounting challenges, and inspiring further exploration to crystallize this groundbreaking idea. As the concept's publication to the present day, quantum teleportation has undergone scrutiny across diverse physical systems. Noteworthy achievements include photon transfer spanning 1400 km between two islands via free space optical links and the teleportation of an atom across a 21-m span. Hybrid techniques have confirmed success, signifying significant progress in the field. While the current distance record using a satellite stands at 1400 km, it is clear that research in this domain is still in its early stages.

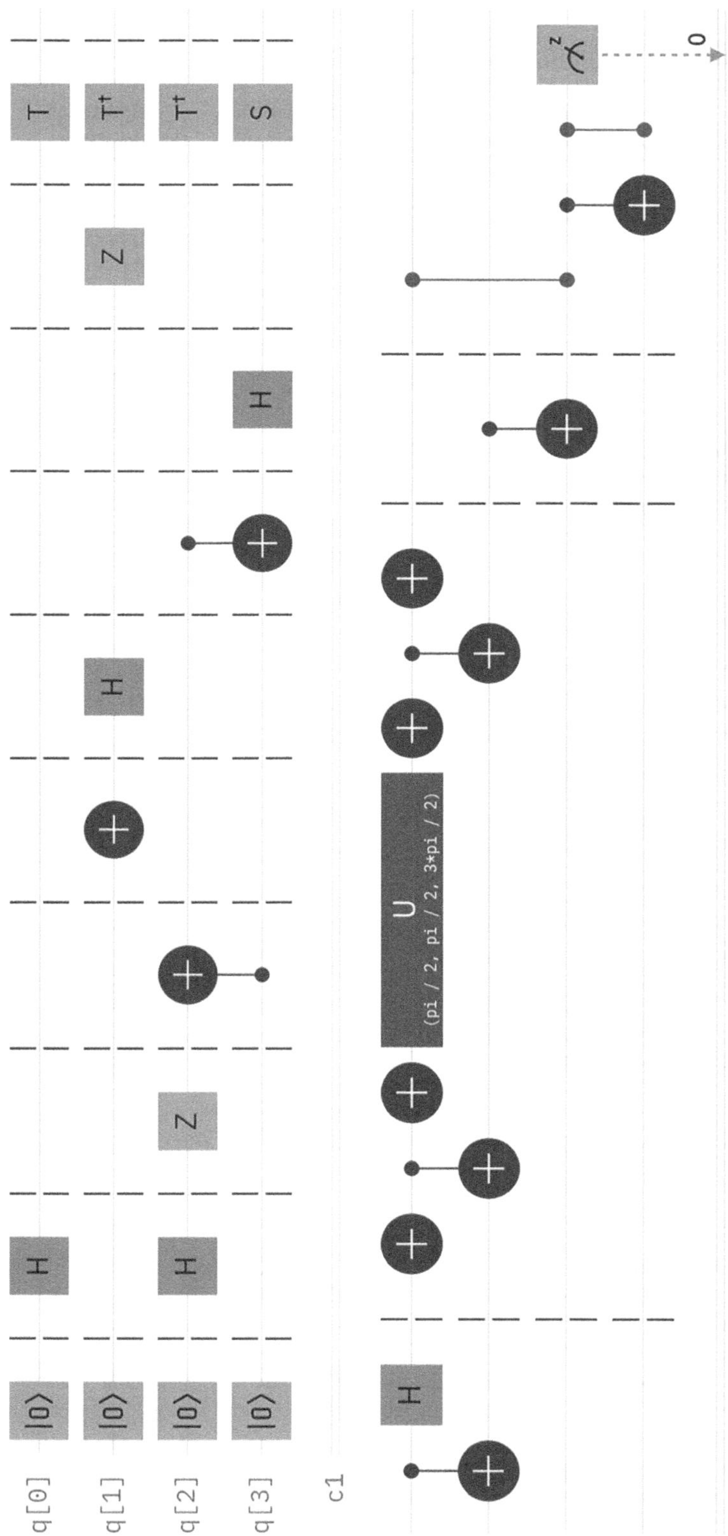

FIGURE 9.11 Quantum circuit demonstrating continuous quantum teleportation of coherent state with variable phase (e.g., $\varphi = \frac{\pi}{4}$ via hybrid tripartite quantum channel.

TABLE 9.4
Contrasting Theoretical Predictions with Experimental Findings

Input's Phase	Input's Probability	Alice's and Bob's Phase	Theor. Output	IBM-Qasm Simulator	IBM-Aer Simulator	IBMQ-Belem Simulator
0	[0.5;0.5]	$-\pi/2$	[0.5;0.5]	[0.502;0.498]	[0.499;0.501]	[0.520;0.480]
$\pi/4$	[0.5;0.5]	$-\pi/4$·	[0.5;0.5]	[0.500;0.500]	[0.498;0.502]	[0.528;0.472]
$\pi/2$	[0.5;0.5]	0	[0.5;0.5]	[0.493;0.507]	[0.499;0.501]	[0.553;0.447]
$3\pi/4$	[0.5;0.5]	$\pi/4$	[0.5;0.5]	[0.499;0.501]	[0.508;0.492]	[0.537;0.463]
π	[0.5;0.5]	$\pi/2$	[0.5;0.5]	[0.490;0.510]	[0.496;0.504]	[0.524;0.476]
$5\pi/4$	[0.5;0.5]	$3\pi/4$	[0.5;0.5]	[0.501;0.499]	[0.498;0.502]	[0.544;0.456]
$3\pi/2$	[0.5;0.5]	π	[0.5;0.5]	[0.495;0.505]	[0.501;0.499]	[0.516;0.484]
$7\pi/4$	[0.5;0.5]	$5\pi/4$	[0.5;0.5]	[0.497;0.503]	[0.496;0.504]	[0.519;0.480]

Notes: Examining probability distributions for coherent quantum transitions in varied phases using the "IBM-Qasm," "IBM-Aer," and "IBMQ-Belem" simulators. Employing 8192 shots enhances result precision and minimizes statistical errors. The shorthand [x;y] denotes qubit output probabilities, where x signifies the likelihood of obtaining $|0\rangle$, and y denotes the likelihood of obtaining $|1\rangle$.

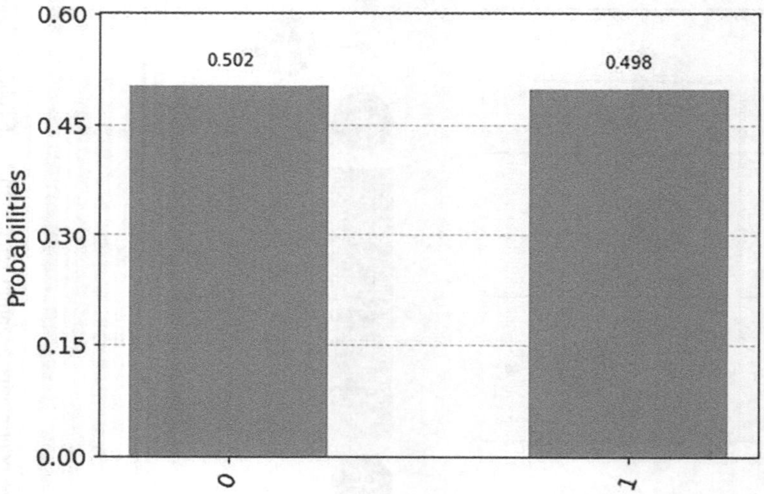

FIGURE 9.12 Histogram: analyzing output probability using IBM-Aer simulator with 1024 shots.

REFERENCES

[1] A. Einstein, B. Podolsky and N. Rosen, Can quantum mechanical description of physical reality be considered complete? *Physical Review*, 47 (1935):777–780.

[2] C. H. Bennett et al., Teleporting an unknown quantum state via dual classical and Einstein–Podolsky–Rosen channels, *Physical Review Letters*, 70 (1993):1895.

[3] D. Boschi, S. Branca, F. De Martini, L. Hardy and S. Popescu, Experimental realization of teleporting an unknown pure quantum state via dual classical and Einstein–Podolsky–Rosen channels, *Physical Review Letters*, 80 (1998):1121.

[4] D. Bouwmeester et al., Experimental quantum teleportation, *Nature*, 390.6660 (1997):575–579.

[5] A. Karlsson and M. Bourennane, Quantum teleportation using three-particle entanglement, *Physical Review A*, 58.6 (1998):4394.

[6] U. L. Andersen et al., Hybrid discrete- and continuous-variable quantum information, *Nature Physics*, 11 (2015):713–719.
[7] M. El Kirdi, A. Slaoui, H. El Hadfi and M. Daoud, Improving the probabilistic quantum teleportation efficiency of arbitrary superposed coherent state using multipartite even and odd j-spin coherent states as resource, *Applied Physics B*, 129.6 (2023):94.
[8] M. El Kirdi, A. Slaoui, H. El Hadfi and M. Daoud, Efficient quantum controlled teleportation of an arbitrary three-qubit state using two GHZ entangled state and one Bell entangled state, *Journal of Russian Laser Research*, 44.2 (2023):1–14.
[9] H. J. Cao and H. S. Song, Teleportation of a single-qubit state via unique W state, *International Journal of Theoretical Physics*, 46.6 (2007):1636–1642.
[10] Y. H. Li et al., Quantum teleportation of three and four-qubit state using multi-qubit cluster states, *International Journal of Theoretical Physics*, 55.3 (2016):1820–1823.
[11] Y. H. Li et al., Quantum teleportation of a four-qubit state by using six-qubit cluster state, *International Journal of Theoretical Physics*, 55.8 (2016):3547–3550.
[12] H. J. Briegel and R. Raussendorf, Persistent entanglement in arrays of interacting particles, *Physical Review Letters*, 86.5 (2001):910–913.
[13] X. S. Ma et al., Quantum teleportation over 143 kilometers using active feed-forward, *Nature*, 489.7415 (2012):269–273.
[14] C. Nölleke et al., Efficient teleportation between remote single-atom quantum memories, *Physical Review Letters*, 110.14 (2013):140403.
[15] S. Takeda et al., Deterministic quantum teleportation of photonic quantum bits by a hybrid technique, *Nature*, 500.7462 (2013):315–318.
[16] J. G. Ren et al., Ground-to-satellite quantum teleportation, *Nature*, 549 (2017):70–73.
[17] S. Popescu et al., Experimental realization of teleporting an unknown pure quantum state via dual classical and Einstein–Podolsky–Rosen channels, *Physical Review Letters*, 80 (1998):1121.
[18] K. Marshall and D. F James, High-fidelity teleportation of continuous-variable quantum states with discrete-variable resources, 31 *JOSA B* (2014):423.
[19] G. Brassard, R. Cleve and A. Tapp, Cost of exactly simulating quantum entanglement with classical communication, *Physical Review Letters*, 83.9 (1999):1874.
[20] J. B. Brask et al., Hybrid long-distance entanglement distribution protocol, *Physical Review Letters*, 105 (2010):160501.
[21] Y. Lim et al., Loss-resilient photonic entanglement swapping using optical hybrid states, *Physical Review A*, 94 (2016):062337.
[22] S. Pirandola and S. L. Braunstein, Physics: Unite to build a quantum internet, *Nature*, 532 (2016):169–171.
[23] A. Feizpour et al., Observation of the nonlinear phase shift due to single post-selected photons, *Nature Physics*, 11.11 (2015):905–909.
[24] M. El Kirdi, A. Slaoui, N. Ikken, M. Daoud and R. Ahl Laamara, Controlled quantum teleportation between discrete and continuous physical systems, *Physica Scripta*, 98 (2023):025101.
[25] H. El Hadfi, M. El Kirdi, L. B. Drissi and R. Ahl Laamara, Large teleportation of two qubit state between non-neighboring nodes based on the utilization of multiple cluster states within quantum networks, *Physica Scripta*, 99.7 (2024):075102.
[26] A. Slaoui, M. El Kirdi, and R. Ahl Laamara et al., Cyclic quantum teleportation of two-qubit entangled states by using six-qubit cluster state and six-qubit entangled state, *Scientific Reports*, 14.1 (2024):15856.
[27] C. C. Gerry and P. L. Knight, Quantum superpositions and Schrödinger cat states in quantum optics, *American Journal of Physics*, 65 (1997):964–974.
[28] E. Ulanov et al., Quantum teleportation between discrete and continuous encoding of an optical qubit, *Physical Review Letters*, 118 (2017):160501.

10 Quantum Teleportation Strategies for Clear and Secure Quantum Networks

Nada Ikken, Abdallah Slaoui, and Rachid Ahl Laamara

10.1 INTRODUCTION

This outstanding quantum entanglement phenomenon has a chance of revolutionary improvements in communication, encryption, and computation. The discovery of methods for generating, transmitting, and reconstructing quantum states, particularly utilizing two-dimensional quantum systems known as qubits, is the focus of Quantum Information and Communication. Entanglement, a fundamental notion in quantum information processing, significantly changes quantum information manipulation and transmission. Bennett et al. [1] introduced the quantum teleportation protocol, which was an important turning point in this field. This protocol, a groundbreaking achievement, changed quantum communication and computing by allowing quantum states to be sent across long distances. Bennett et al.'s [1] initial idea was followed by experimental implementations by Bouwmeester et al. [2] and Boschi et al. [3], proving the protocol's practicality and opening the way for further research.

As the field developed, researchers got into more advanced modifications of the protocol, especially those involving multi-qubit teleportation [4]. Quantum Key Distribution (QKD), another revolutionary information technology, tries to construct secure communication channels using the principles of quantum physics [12]. The BBM92 protocol [5] was a pioneer QKD protocol that allowed two parties to exchange secure cryptographic keys. Contributions to Quantum Error Correction (QEC) codes, such as the Shor code [6] and Steane code [7], were important in protecting quantum computing from mistakes and decoherence. These discoveries set the path for further research into developments such as teleporting multiple qubits utilizing various protocols.

Bidirectional Quantum Teleportation (BQT), which uses quantum teleportation principles, represents an important advance in the field of quantum information processing. Quantum teleportation is a fundamental notion that takes advantage of quantum entanglement that allows for the immediate transfer of quantum states between particles separated by vast distances. The communication of quantum states occurs not only unidirectionally, but also in turn in the setting of BQT. The BQT protocols introduce an additional dimension, allowing for the easy transfer of quantum information between two different locations. This BQT capacity offers promise for interactive quantum communication, distributed quantum computations, and improved quantum networking applications [10].

Quantum teleportation (QT) success has been critical in accelerating the development of distributed quantum computing. This paradigm proposes a networked system in which many quantum computers smoothly interact over long distances. The use of quantum entanglement, a property that allows particles to keep immediate correlations regardless of physical separation, is fundamental to this collaborative method.

QT methods have become a focus of research due to their potential in secure communication and quantum computing. These protocols not only allow for the accurate transfer of quantum states, but they also prepare the way for a wider range of quantum information processing. Recent developments

 DOI: 10.1201/9781003621423-12

have investigated unique issues such as multipartite entanglement and higher-dimensional quantum systems. Researchers are working hard to improve the efficacy as well as durability of quantum teleportation by addressing issues including noise, decoherence, and scale. Their continued efforts are focused on improving current techniques and developing new ways to overcome present limitations. These outstanding quantum protocols illustrate the diverse aspects of quantum information processing, from protecting communication channels to error correction and quantum computation.

MultiHop quantum teleportation is an exciting notion that expands the concepts of quantum teleportation to include many intermediary hops, allowing quantum information to be sent over larger distances or across advanced quantum systems. This method frequently makes use of entangled states, such as the W state, to help in the teleportation process. The W state is a sort of entangled quantum state in which three or more quantum particles are involved. The W state is a resource for extending quantum information all over the network in the context of MultiHop teleportation [8]. The entanglement shared by the particles in the W state allows quantum states to be transmitted from one location to another via intermediary nodes. The main idea behind MultiHop teleportation is to break down the teleportation process into stages, each of which includes the transmission of quantum information between related nodes. The entanglement present in the W state ensures the preservation of quantum correlations during these intermediate phases, eventually leading to the quantum state's successful teleportation to its final destination [9].

10.2 QUANTUM MECHANICS BACKGROUND

Let's examine the fundamental ideas and postulates of quantum physics that are required to appreciate the complex nature of creating a quantum network. A qubit is a discrete two-level state capable of adopting two distinctive (orthogonal) states, namely the zero or ground state and the one or excited state, which are commonly represented as 0 and 1, respectively. For example, if we use a qubit to indicate the state of a photon, we may specifically represent its horizontal and vertical polarizations by employing the two base states.

10.2.1 Entanglement

The concept of quantum entanglement, a type of correlation with no classical analog, is the most significant difference between classical and quantum physics. Simply explained, entanglement is an individual representation of superposition involving many qubits in which the whole quantum state defies explanation in terms of the individual quantum states of each qubit, exceeding the classical concept of independence.

To understand the entanglement notion, consider Figure 10.1, which illustrates the famous EPR pair – a pair of maximally entangled qubits. The qubits are in a superposition in this state, with an equal balance of zero and one. When the outcomes are measured separately, they form a random mix of zeros and ones, each having an equal probability of happening. However, when the results of these measures are compared, a stunning connection is found. It becomes obvious that if qubit A records a zero, qubit B consistently replicates the result, and vice versa when qubit A records a

FIGURE 10.1 This figure represents quantum entanglement: an EPR state with perfectly linked qubits. Observe the entangled qubits' correlated behavior, in which one's outcome instantly affects the other's, demonstrating the interesting non-locality of quantum entanglement.

one. This amazing event, which is strongly placed in quantum physics principles, highlights the unmatched interconnectedness between these entangled systems.

- General form of an entangled state for 2 qubits:

$$|\Psi\rangle = \left(1/\sqrt{2}\right)\left(|0\rangle A \otimes |1\rangle B - |1\rangle A \otimes |0\rangle B\right) \tag{10.1}$$

- Entanglement measurement outcome correlations:
  In an entangled state, the correlation between the measurements of qubit A and qubit B is often defined as:

$$P(A = 0, B = 0) = P(A = 1,\ B = 1) = 1/\sqrt{2}$$

$$P(A = 0,\ B = 1) = P(A = 1,\ B = 0) = 0$$

If qubit A is measured to be in state $|0\rangle$, it follows that qubit B will also be in state $|0\rangle$. Similarly, if qubit A is measured to be in state $|1\rangle$, qubit B will also be in state $|1\rangle$. After detecting quantum entanglement, Einstein, Podolsky, and Rosen (EPR) were inspired to begin addressing the EPR problem. This odd phenomenon implies that measuring one entangled qubit changes the state of the other, independent of their physical distance. This appears to violate relativity and propose an instantaneous exchange of information at first glance. This contradiction, however, derives from a misunderstanding of the nature of entanglement. It is critical to recognize that entanglement does not require classical information transmission between particles. Instead, it reveals an unusual connection in which the measurement of one particle instantly dictates the state of the other. This correlation is formed at the time of entanglement. Consider two dice that are somehow connected to further understand this notion. When one dice shows a particular number, the other always shows a complimentary number. In basic terms, the result of one dice determines the end result of the other. However, it should be noted that this does not imply that information is being sent at a velocity faster than the speed of light. The association between the dice was set from the beginning and has no immediate impact on the outcome of each particular dice. Similarly, quantum entanglement functions inside the complex framework of quantum physics, and any concept of faster-than-light communication becomes untenable when we dive more into the non-local.

10.2.2 No-Cloning Theorem

The no-cloning theorem is based on fundamental quantum mechanics concepts and claims that an unknown qubit cannot be reproduced. This is because nature has set constraints on quantum system changes, requiring them to be unitary. This constraint is critical because it leads to the no-cloning theorem [13], which has important consequences in the field of communication engineering. The capacity to duplicate information is, in fact, a key requirement in classical communication operations. This assumption, however, is invalid in quantum physics due to the no-cloning theorem. Because of nature's constraint on unitary transformations, secure copying of information is not possible, underlining the fundamental contrasts between classical and quantum communication.

We can describe it as follows:

$$U|\Psi\rangle \otimes |0\rangle = U|\Psi\rangle \otimes |0\rangle = |\Psi\rangle \otimes |\Psi\rangle \tag{10.2}$$

The unitary operator U performs a cloning method in this instance, with the state being an unknown quantum state and 0 being an additional state that is initially set to a standard reference state

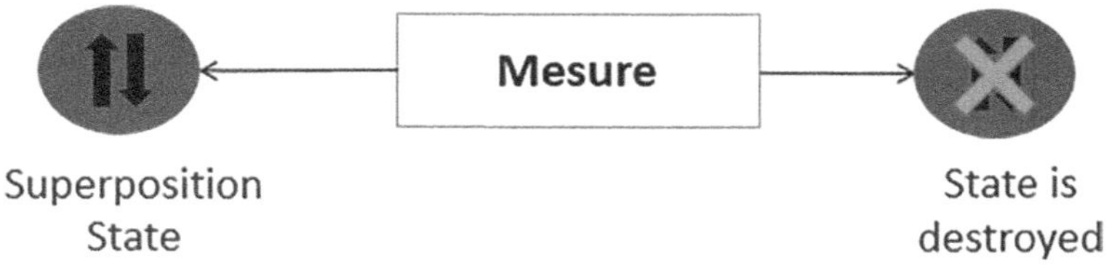

FIGURE 10.2 The no-cloning theorem, an important subject in quantum physics, is shown here. Simply put, this theorem maintains that it is impossible to exactly copy an unknown quantum state as every attempt will result in a disturbance. This is a harsh reminder of the fundamental constraint of reproducing quantum information.

(Figure 10.2). The no-cloning theorem says that no unitary operator U can perfectly clone unknown quantum states. In other words, attempting to completely recreate a quantum state leads in undesirable interference or change, making an exact clone impossible to achieve. This theorem shows a fundamental limitation in our capacity to copy quantum information properly.

10.2.3 Quantum Measurements

When measuring a system with multiple potential outcomes, such as a qubit in its dual states, one of the fundamental principles of quantum mechanics claims that the act of observation causes the original quantum state to permanently collapse into one of the measured states [14]. As a result, the measured state can only be zero or one, resulting in a fundamentally unpredictable conclusion. Each result is determined by the balance of zeros and ones in the initial superposed quantum state.

The formalism of quantum mechanics, in which the state of a quantum system is represented by a wave function, is one of the most frequent methods to explain quantum measurements. The use of operators and eigenstates defines the measuring method. These fundamental equations offer a convincing explanation for the physics of quantum measurements:

- The projection of the quantum state Ψ on one of the eigenstates a_i yields to the result of a quantum measurement of the measurement operator A. The projection postulate expresses this mathematically:

$$P(a_i) = \left|\langle a_i | \Psi \rangle\right|^2 \tag{10.3}$$

  This equation shows us the probability that we will get the measurement result a_i while the system is in the state Ψ with certainty and engagement.
- The estimated or typical value of an observable "A" in this condition, represented as, can be defined as follows:

$$\langle A \rangle = \langle \Psi | A | \Psi \rangle \tag{10.4}$$

  Here the value is the mean outcome that would be obtained if the A observable was measured many times on identical systems established in the $|\Psi\rangle$ state.
- Heisenberg's uncertainty principle establishes the basic limits of precision when measuring two observables at the same time. In basic terms, it says:

$$\Delta A \cdot \Delta B \geq \hbar/2 \tag{10.5}$$

Consider the uncertainties in measuring the observables A and B, denoted as ΔA and ΔB, respectively. Let us not forget about the reduced Planck constant $\hbar$.

In conclusion, the unexpected and confusing properties of quantum measurements, such as the quick collapse of a quantum state upon observation, indicate the importance of measurement within the quantum domain. The delicate balance of the original overlaid quantum state is reflected in the shift of the state into a definite consequence of either zero or one. The wave function is at the core of quantum mechanics, functioning as a basic tool for understanding the complexity of quantum measurements. This approach provides an improved understanding of the complex measurement process by utilizing operators and eigenstates. The essential equations that govern quantum measurements, such as the probabilistic projection postulate (Equation 10.1) and the definition of expectation value (Equation 10.2), give a comprehensive and robust basis for evaluating this phenomenon.

10.3 QUANTUM COMPUTING

The methods employed to manipulate quantum states can manifest in diverse physical phenomena. For instance, when qubits are encoded in particles with quantum mechanical spin, manipulating logic involves varying spin through changes in an applied magnetic field at different orientations. Alternatively, if a qubit is encoded in the internal excitation state of an ion, gate operations can be realized by adjusting the duration of exposure to a laser beam or by modifying the wavelength of the laser light. To implement a quantum gate physically, it must correspond to the quantum mechanical evolution of an isolated quantum system. The transformation it induces is dictated by Schrödinger's equation:

$$i\text{h}\frac{\partial|\Psi(t)\rangle}{\partial t} = H|\Psi(t)\rangle \tag{10.6}$$

$$U(t) = \exp\left(\frac{-i\,H\,t}{\text{h}}\right) \tag{10.7}$$

$$|\Psi(t)\rangle = U(t)|\Psi(0)\rangle \tag{10.8}$$

where H represents the Hamiltonian specifying the physical fields and forces at play. The quantum mechanical evolution governed by this equation is unitary as long as no measurements are made, and no unintended interactions occur with the environment. In such cases, starting from an initial state $|\Psi(0)\rangle$, the isolated quantum system's evolution is described by a unitary transformation. This implies that a quantum logic gate, acting on an isolated quantum computer, will unitarily transform the state until an observation is made, ensuring the logical reversibility of quantum gates.

Early visionaries in the field of quantum computing, including Richard Feynman and David Deutsch [11], identified striking similarities between classical reversible gates and their quantum counterparts. Their insight was grounded in the recognition that matrices associated with reversible classical gates, specifically permutation matrices, also qualified as unitary matrices. This observation allowed for their interpretation as operators orchestrating the evolution of an initial quantum state (representing input) into a final quantum state (representing output) in alignment with Schrödinger's equation. Consequently, classical reversible gates like NOT, controlled-NOT (CNOT), and SWAP emerged as the most analogous counterparts to quantum logic gates in the classical realm. However, it is crucial to note that the range of classical reversible computing gates is confined to unitary gates with matrix representations as permutation matrices. Conversely, deterministic quantum computing permits the inclusion of any gate featuring a unitary matrix, irrespective of whether it is a permutation matrix.

10.3.1 Quantum Gates

10.3.1.1 Pauli Matrices

The Pauli matrices, often known as the Pauli spin matrices, are complex matrices introduced by Pauli in his study of spin within the framework of quantum mechanics. These matrices have the dual characteristics of being Hermitian and unitary in the setting of single qubits, represented as (1, X, Y, Z). Matrix algebra is used to express the transformational dynamics. The Pauli matrices are defined in the context of bipartite entanglement as follows:

$$\mathrm{I}=\begin{pmatrix}1 & 0\\ 0 & 1\end{pmatrix},\mathrm{X}=\begin{pmatrix}0 & 1\\ 1 & 0\end{pmatrix},\mathrm{Y}=\begin{pmatrix}0 & -i\\ i & 0\end{pmatrix},\mathrm{Z}=\begin{pmatrix}1 & 0\\ 0 & -1\end{pmatrix} \tag{10.9}$$

Various Hamiltonians seen in practical applications include $H = Z^{(1)}Z^{(2)}$, representing the Ising interaction, and $H = X^{(1)}X^{(2)} + Y^{(1)}Y^{(2)}$ for the XY interaction. Another example is $H = 2X^{(1)}X^{(2)} + Y^{(1)}Y^{(2)}$, where the superscript labels specify the qubit upon which the operator acts (Figures 10.3 and 10.4).

Pauli matrices X and Z are often referred to as bit-flip and phase-flip matrices, respectively. The Y matrix represents the combination of both bit and phase flip. The Identity matrix I has no effect on the qubit's state. Any communication protocol requires at least two parties, namely the transmitter and the receiver. The proposed protocol has been established in such a way that any party participating in the protocol can function as either a sender or a receiver at any moment. As a result, it is assumed that the participating entities have the necessary skills. This would need connectivity to the quantum channel, the capacity to measure the qubit on the basis, and eventually the ability to produce and entangle the qubits in the right state.

10.3.1.2 Hadamard Gate (H)

The Hadamard gate, referred to as H, is one of the most useful single-qubit gates, if not the most beneficial. The Hadamard gate's matrix representation is as follows:

$$\mathrm{H}=\frac{1}{\sqrt{2}}\begin{pmatrix}1 & 1\\ 1 & -1\end{pmatrix} \tag{10.10}$$

The importance of the Hadamard gate in quantum computing cannot be highlighted. When a qubit starts in a definite state, either $|0\rangle$ or $|1\rangle$, the Hadamard gate turns it into a superposition of: $\mathrm{H}|0\rangle = (|0\rangle + |1\rangle)$ and $\mathrm{H}|1\rangle = \frac{1}{\sqrt{2}}(|0\rangle - |1\rangle)$.

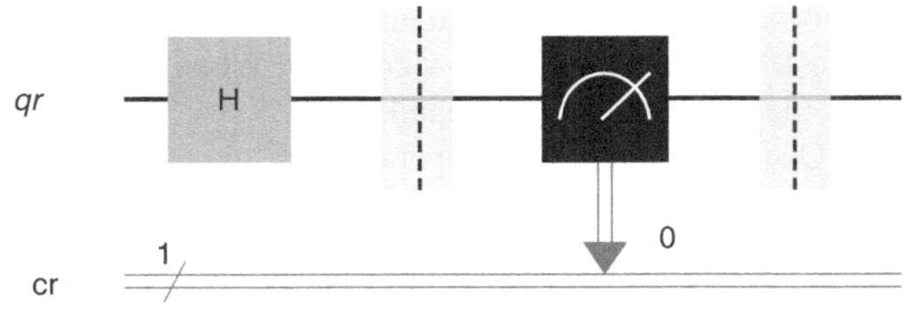

FIGURE 10.3 The application of a Hadamard gate in IBM.

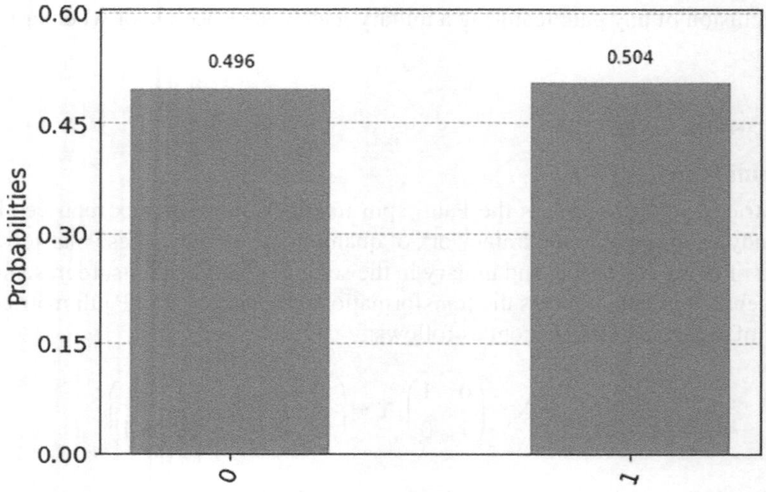

FIGURE 10.4 Probability histogram in Qasm simulator.

We will use the IBM simulator to illustrate the application of a Hadamard gate on a $|0\rangle$s tate qubit while verifying the results. The histogram will demonstrate the outcomes of performing the circuit 100 times. Each run is independent, with the qubit being reset to the $|0\rangle$ state, passing through the gate, and then measured. This method is performed 1024 times, with each histogram point illustrating the frequency/probability of measuring $|0\rangle$ or $|1\rangle$. Because of statistical inaccuracy, the Hadamard gate effectively generates a superposition state of both $|0\rangle$ and $|1\rangle$ on a single qubit. Increased collection of data brings the outcome closer to a 50/50 split. This tendency is analogous to flipping a coin and calculating heads or tails, where a higher number of flips approaches a 50/50 chance of observing heads or tails. It is important to remember that measurement causes the superposition to collapse. Because only one classical state can be observed, all other quantum information is destroyed. This event, known as the quantum computing no-cloning theorem, illustrates the difficulty of copying a qubit's state after measurement (no cloning). When a superposition state is measured, it simply turns into one of the basic states and cannot, therefore, be replicated (no cloning). The complexities of how and why measurement collapse occurs are unclear.

10.3.1.3 NOT Gate (X)

The Pauli X matrix is identical to the traditional (reversible) NOT gate:

$$X = \begin{pmatrix} 0 & 1 \\ 1 & 0 \end{pmatrix} \tag{10.11}$$

The NOT gate in classical computing changes the value of a single input in the inverse direction. It converts a 0 bit to a 1 bit or a 1 bit to a 0 bit. A quantum X gate works in a similar way, causing a qubit in the important state $|0\rangle$ to change to $|1\rangle$ and vice versa. When a qubit exists in a superposition of all possible base states, the superposition is flipped.

You may use the IBM Q simulator to test the operation of this process. All qubits on the IBM Q machine (or any other quantum simulator) start in the $|0\rangle$ state. Simply drag the X gate into any qubit to perform this simple gate. Add the measurement operation at the end for result analysis, creating what is known as a quantum circuit, the quantum analog to classical circuits. A circuit defines how a qubit evolves during computing, depending on the gates that act on it (Figures 10.5–10.8).

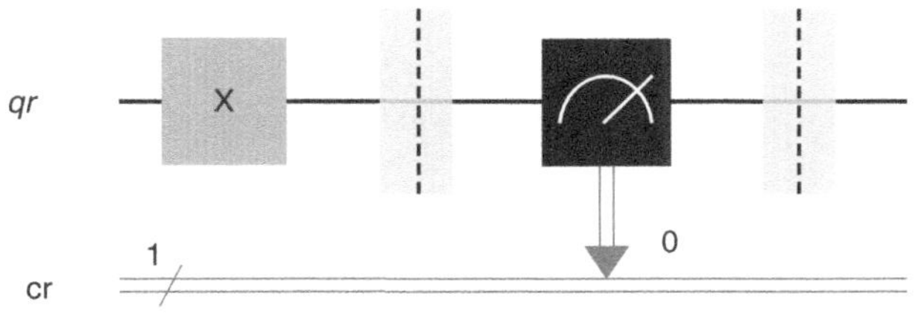

FIGURE 10.5 The application of the NOT gate in IBM.

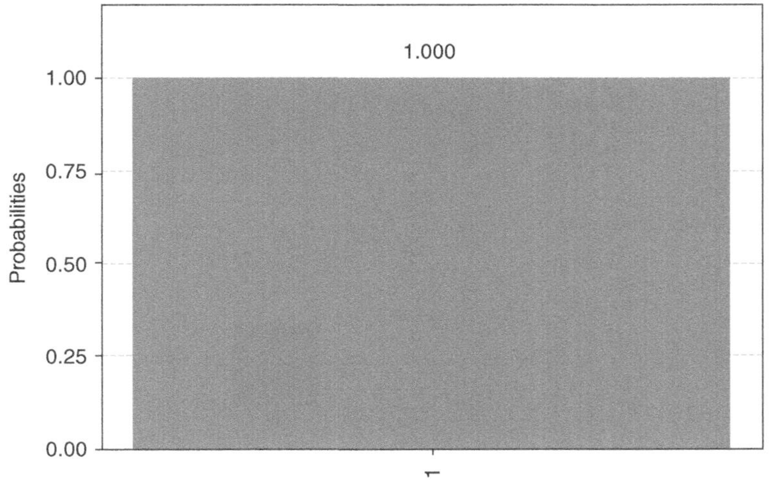

FIGURE 10.6 Probability histogram in Qasm simulator.

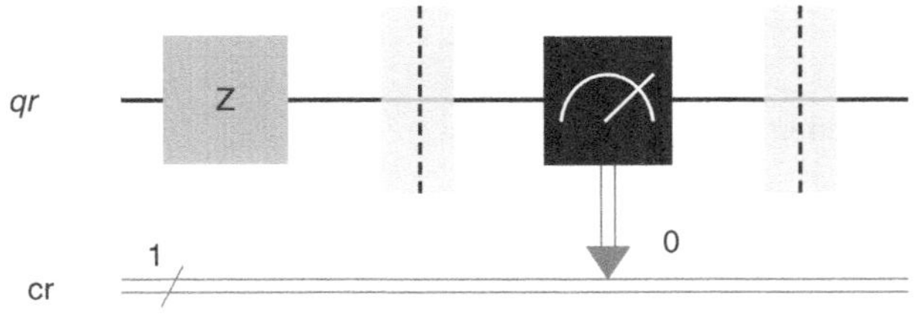

FIGURE 10.7 The application of the Z gate in IBM simulator.

Hardware errors can change the quantum state of a qubit. These errors may be caused by difficulties in properly controlling electromagnetic field interference, temperature changes, or loss of energy. Accidental and incorrect change of a qubit's state causes errors, which are frequently referred to as noise or decoherence. Because of the effect of these hardware-induced errors on quantum computers, where measurements depend on the state of the qubit, identifying between accurate and incorrect measurements becomes difficult.

10.3.1.4 Z Gate (Z)

The Z-gate matrix is represented as follows:

$$X = \begin{pmatrix} 1 & 0 \\ 0 & -1 \end{pmatrix} \tag{10.12}$$

FIGURE 10.8 Probability histogram in Qasm.

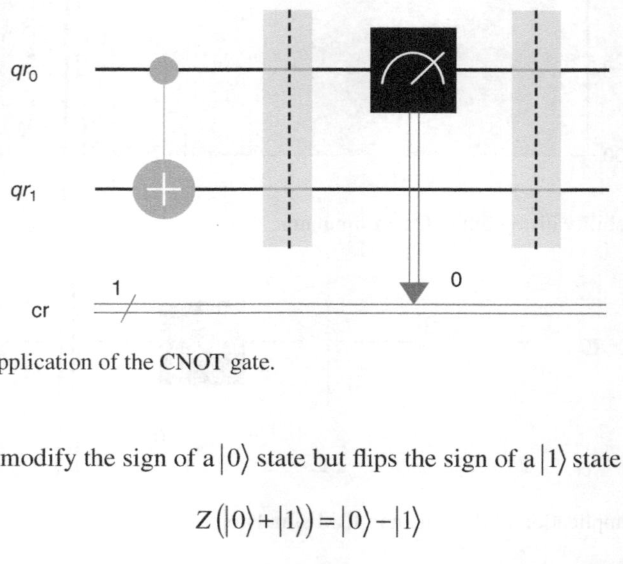

FIGURE 10.9 The application of the CNOT gate.

The Z gate does not modify the sign of a $|0\rangle$ state but flips the sign of a $|1\rangle$ state to $-|1\rangle$.

$$Z(|0\rangle+|1\rangle)=|0\rangle-|1\rangle \tag{10.13}$$

In simplest terms, applying the Z gate to a superposition state $|0\rangle+|1\rangle$ changes it to $|0\rangle-|1\rangle$. This highlights the importance of the Z gate in changing the phase of the qubit state, especially by adding a phase inversion for the $|1\rangle$ component. The matrix representation represents this transformation in an easy-to-understand way, with the diagonal components representing the preservation of $|0\rangle$ and the change of sign for $|1\rangle$. The capacity of the Z gate to cause phase changes is a critical feature of quantum computing operations.

10.3.1.5 CNOT Gate (CNOT or C-X)

The CNOT gate is a 2-qubit multi-qubit gate made up of a control qubit (the first qubit) and a target qubit (the second qubit). When the control qubit is in the state $|1\rangle$, the CNOT gate reverses the target qubit's state from $|0\rangle$ to $|1\rangle$ or vice versa (Figures 10.9 and 10.10).

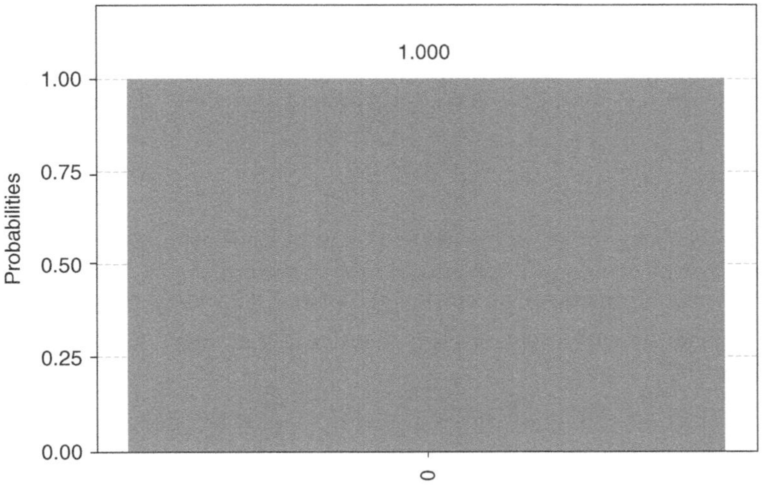

FIGURE 10.10 Probability histogram in Qasm simulator.

The following matrix represents the CNOT gate's operating behavior:

$$\text{CNOT} = \begin{pmatrix} 1 & 0 & 0 & 0 \\ 0 & 1 & 0 & 0 \\ 0 & 0 & 0 & 1 \\ 0 & 0 & 1 & 0 \end{pmatrix} \tag{10.14}$$

When the unitary operator U represents the Pauli-X operator, the CNOT gate is successfully a controlled-X operator designated as CX, with the first qubit serving as the control and the second qubit serving as the target. CNOT performs a conditional X gate operation on the second qubit, dependent on the state of the first qubit. Particular cases of CNOT action on a 2-qubit register: CNOT leaves the state $|00\rangle$ unmodified, changes $|01\rangle$ to $|01\rangle$, flips $|10\rangle$ to $|11\rangle$, and swaps $|11\rangle$ to $|10\rangle$. This gate performs a controlled operation that modifies the target qubit depending on the state of the control qubit. This CNOT gate is fundamental in quantum processing and is used in a number of quantum algorithms and circuits.

10.3.2 Bloch Sphere Representation

After a study of particular cases of specialized quantum logic gates, such as the NOT and H gates, our focus changes to understanding the most general type of quantum gate applicable to a single qubit. To address this issue, we provide a group of quantum gates that may rotate around the Bloch sphere's three mutually perpendicular axes. A pure state appears as a point on the surface of the Bloch sphere in the context of a single qubit. A single qubit gate acting on this state changes it into another point on the Bloch sphere. Notably, gates allowing rotations around the (x, y, z) axes are of particular interest because every given 1 qubit quantum gate may be decomposed into series of such rotation gates. To define a reference frame for arbitrary single qubit pure states, we define three perpendicular to one another axes x, y, and z or similarly, three polar coordinates: a radius (unity for all points on the Bloch sphere) and two angles: (latitude, measured counter-clockwise as we rotate around the z-axis) and (longitude, measured clockwise as we rotate around the z-axis). As a result, each point on the Bloch sphere may be defined using either (x, y, z) coordinates or (r, θ, φ)

coordinates. It is crucial to note, however, that the general qubit state must also include an overall phase component, which we must ignore for the time being.

$$x = r\ sin(\theta)\ cos(\phi),$$
$$y = r\ sin(\theta)\ sin(\phi), \qquad (10.15)$$
$$z\ = r\ cos(\theta)$$

We propose that the gates shown in Figure 10.11 can be built using the Pauli (x, y, z) matrices. The fourth Pauli matrix, chosen as I, can also be used to perform a global overall phase shift. More specifically, we present the unitary matrices $R_x(\theta), R_y(\theta), R_z(\theta)$. These matrices are produced from Hamiltonians that correspond to the four Pauli matrices X, Y, Z, and I, in that order. After applying the R_x gate with $\theta = \pi/_2$.

$$R_X(\theta) = \begin{pmatrix} \cos(\theta/2) & -\sin(\theta/2) \\ -\sin(\theta/2) & \cos(\theta/2) \end{pmatrix} \qquad (10.16)$$

The $R_y(\theta)$ gate is a rotation gate that rotates the Bloch sphere by an angle around the y-axis. This gate is an important element in quantum circuits, allowing qubit states to be controlled. We will look at the influence of the $R_y()$ gate on a qubit's state and visualize its effect on the Bloch sphere using Qiskit. We may examine how the qubit state changes along the y-axis by adjusting the rotation angle $\theta = \pi/_2$, giving important understanding into the behavior of quantum gates:

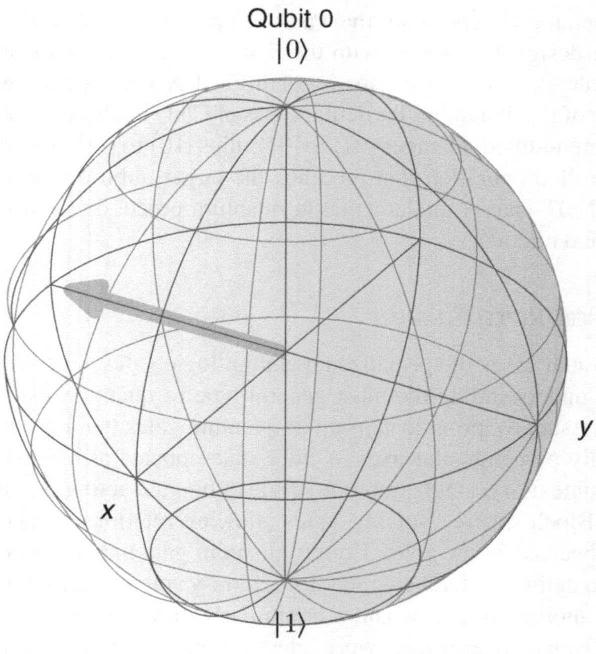

FIGURE 10.11 The R_x, R_y, R_z gates are a single-qubit rotation around the x-, y-, z-axis at an angle of $\theta = \pi/2$ (radians).

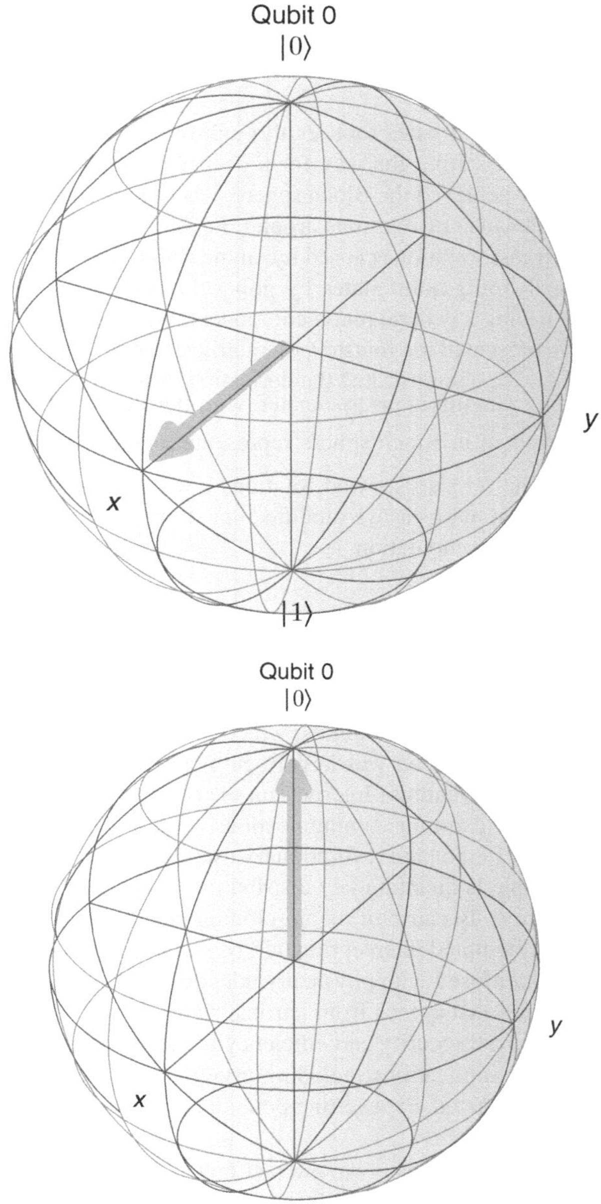

FIGURE 10.11 (Continued)

$$R_X(\theta) = \begin{pmatrix} \cos(\theta/2) & -\sin(\theta/2) \\ \sin(\theta/2) & \cos(\theta/2) \end{pmatrix} \tag{10.17}$$

The R_z gate is important, allowing rotations around the z-axis of the Bloch sphere. The Bloch sphere, a visual representation of a qubit's state, reveals the complex transformations generated by quantum gates. Qiskit explore the R_z gate, revealing the visual rotation of quantum states on the Bloch sphere as we apply rotations and see the dynamic evolution of qubits.

$$R_y(\theta) = \begin{pmatrix} e^{-i(\theta/2)} & 0 \\ 0 & e^{i(\theta/2)} \end{pmatrix} \quad (10.18)$$

Finally, our examination of quantum gates took us through the complexities of specialized logic gates such as NOT and H gates, setting the way for a deeper understanding of the most general single-qubit quantum gates. We achieved the Bloch sphere, where pure states appear as points on its surface. We uncovered the relevance of rotations around the Bloch sphere's three orthogonal axes x, y, and z by observing the transformations caused by single-qubit gates.

We created the basis for defining qubit states by going through the Bloch sphere coordinates (x, y, z) or (r, θ, ϕ).Despite having a general phase component, we set it aside for the time being for our current investigation. We presented the fourth Pauli matrix (I) for global phase shifts, building on the Pauli matrices (X, Y, Z, I). We demonstrated the derivation of the unitary matrices $R_x(\theta)$, $R_y(\theta)$' and $R_z(\theta)$ from corresponding Hamiltonians associated with the Pauli matrices. Rotations around the x and y axes were highlighted in Bloch sphere representations of R_x and R_y gates, explaining their effect on qubit states. The R_z gate showed rotations about the z-axis on the Bloch sphere, showing the dynamic evolution of qubits as we modified quantum states with Qiskit, improving our understanding about the quantum computation.

10.4 QUANTUM COMMUNICATION

Quantum communication, based on quantum mechanics principles, provides an innovative means of transferring information. Despite classical communication, which relies on bits, quantum communication makes advantage of the power of qubits, which may exist in multiple states at the same time. Previously, quantum communication was largely employed for specific applications such as QKD and superdense coding. The Quantum Internet, on the other hand, gives up new possibilities by allowing the transmission of quantum communications (qubits). As quantum communication develops, the opportunities for different and revolutionary uses expand far beyond just relaying classical bits. A well-known example is quantum key distribution, which employs quantum principles to provide strong encryption keys. Because of the no-cloning theorem and quantum entanglement, this technique guarantees any attempted interception will be recognized. There is also quantum teleportation, which is frequently confused with physically teleporting objects [15]. Instead, it involves sharing the quantum information that comes from particle states. As researchers continue to make progress, the potential for improved security and efficiency in quantum communication arises. This offers a wide range of possible applications including cryptography, data transport, and network security. As more research is conducted, the exciting potential of practical quantum communication systems is becoming a reality approach.

The Quantum Internet transforms the way we interact by using distant nodes' unique capability to exchange quantum states. However, quantum physics principles impose limits on the duplication or secure measurement of qubits. Although a qubit may be encoded onto a photon and immediately sent to another node over a fiber link, problems occur when the photon becomes reduced or disrupted by noise. The postulate of quantum measurement and the no-cloning theorem make it impossible to recover the lost data by measurement or duplication, leading to a permanent loss of quantum information. Consequently, the immediate transfer of qubits by photons is no longer possible.

10.4.1 QUANTUM TELEPORTATION

The idea of quantum teleportation could be linked back to quantum entanglement, a notion initially examined in 1935 by EPR [16] and Schrödinger [17]. These scientists showed the surprising properties of quantum systems. Based on EPR, quantum theory allows for the development of interesting correlations between two separate components of a quantum system. When these correlations are available, it is possible to predict the outcome of a measurement on a particular aspect of the system by examining the other, demonstrating the mutual dependence of entangled quantum phenomena. According to the EPR argument, if quantum theory is complete and matches to locality, then the predicted quantity must have a definite value even before it is measured. The argument showed up from the notion that quantum theory falls short of wholeness from a classical position by not allowing established values prior to measurement. Schrödinger's explanation of this argument accurately embodies modern quantum mechanics thinking, suggesting that the wave function includes all relevant information about a quantum system. Entangled quantum states, according to Schrödinger [18, 19], have a characteristic in which the entire system has a definite state while the individual components do not. The fundamental concept of pure-state entanglement is effectively captured by this idea. In answer to the EPR problem, Bell discovered correlations that contradict classical models while staying consistent in any model that respects both locality and completeness [20]. The wave function φ, which fully defines every quantum system by identifying properties such as spatial location, energy, angular and linear momentum, and spin, makes this possible.

10.4.2 Quantum Teleportation via Greenberger–Horne–Zeilinger State

The Greenberger–Horne–Zeilinger (GHZ) state is an interesting and very non-classical quantum state represented by the interaction of several qubits in a system. Specifically, the GHZ state can be represented as:

$$|GHZ\rangle = \frac{|0\rangle^{\otimes n} + |1\rangle^{\otimes n}}{\sqrt{2}} \tag{10.19}$$

when n qubits are available. This state is distinguished from classical states by outstanding non-local correlations between each of its qubits. While the GHZ state has certain similarities with the Bell state often utilized in quantum teleportation, it differs itself by involving all three parties in the teleportation process [21]. The simplest form of the GHZ state requires just three qubits, showing the intricate entanglement between multiple particles:

$$|\mathrm{GHZ}\rangle = \frac{1}{\sqrt{2}}\left(|000\rangle + |111\rangle\right) \tag{10.20}$$

The 3-qubit GHZ state has complex levels of entanglement, making it an important resource for studying quantum processes involving several players [22]. These states are important in quantum information processing protocols such as quantum communication and computation, where the use of non-local correlations allows tasks that are beyond the capability of classical computers. Entangled states are important in the rapidly increasing field of quantum information science as a key building element.

Consider Alice's intention to teleport an unknown qubit to Charlie with the support of Bob. Alice is holding the unknown state, where $|\alpha|^2 + |\beta|$.

$$|\Psi\rangle_a = \alpha|0\rangle_a + \beta|1\rangle_a \tag{10.21}$$

Alice combines the unknown state Ψ_a with the entangled state (GHZ) and measures the Bell states of her two particles. The whole state is described before Alice's Bell measurement as:

$$|\Psi\rangle_T = |\Psi\rangle_a \otimes |GHZ\rangle_{ABC} \tag{10.22}$$

Which can be written in term of Bell basis:

$$|\Psi\rangle_T = [|\Psi_+\rangle_{aA}(\alpha|00\rangle+\beta|11\rangle)+|\Psi_-\rangle_{aA}(\alpha|00\rangle-\beta|11\rangle) + |\Phi_+\rangle_{aA}(\beta|00\rangle-\alpha|11\rangle)+|\Phi_-\rangle_{aA}(\beta|00\rangle-\alpha|11\rangle)]^2 \tag{10.23}$$

With Bell measurements are:

$$|\Psi_\pm\rangle_{aA} = \frac{1}{\sqrt{2}}\left(\alpha|00\rangle_{aA}+\beta|11\rangle_{aA}\right) \tag{10.24}$$

$$|\phi_\pm\rangle_{aA} = \frac{1}{\sqrt{2}}\left(\alpha|01\rangle_{aA}+\beta|10\rangle_{aA}\right) \tag{10.25}$$

Following Alice's Bell measurement, the above equation separates the whole state into four states. The interesting part is that the quantum information of the unknown state is now shared between Bob and Charlie. The information is then recovered at Charlie's location because of Bob's measurement and Charlie's unitary operation. Bob offers two classical bits of information beside one from Alice by measuring his GHZ particle in the x-direction and announcing the result. With this information, Charlie may use a unitary operation on his qubit to recover the initial unknown state as follows:

$$|\Psi^+\rangle_a A|x^+\rangle^B \rightarrow I,\quad |\Psi^+\rangle_a A|x^-\rangle^B \rightarrow \widehat{\sigma z} \tag{10.26}$$

$$|\Phi^+\rangle_a A|x^+\rangle^B \rightarrow \widehat{\sigma x},\quad |\Phi^+\rangle_a A|x^-\rangle^B \rightarrow \widehat{\sigma x}\widehat{\sigma z}$$

$$|\Psi^-\rangle_a A|x^+\rangle^B \rightarrow \widehat{\sigma z},\quad |\Psi^-\rangle_a A|x^-\rangle^B \rightarrow I$$

$$|\Phi^-\rangle_a A|x^+\rangle^B \rightarrow \widehat{\sigma x}\widehat{\sigma z},\quad |\Phi^-\rangle_a A|x^-\rangle^B \rightarrow \widehat{\sigma x}$$

In conclusion, the application of the GHZ state in quantum teleportation gives fascinating exploration into the highly non-classical. These states, represented by the GHZ state, are distinguished from classical states by their non-local correlations between qubits. Despite the frequently utilized Bell state in the standard teleportation, the GHZ state includes all three players in the teleportation process: Alice, Bob, and Charlie. In the presented situation, Alice has the initial control of the unknown state Ψ_a, which she desires to communicate to Charlie with the help of Bob.

Before Alice's Bell measurement, the combined state "A" is given as a tensor product of the unknown state and the entangled GHZ state. The total state is described in terms of the Bell basis, showing the entanglement structure, which is compatible with Alice's Bell measurement. In particular, Alice's Bell Measurement causes the whole state to be divided into four different states, with the quantum information of the unknown state now shared by Bob and Charlie. The subsequent recovery of this information at Charlie's location through a combination of Bob's measurement and Charlie's unitary operation is interesting. Bob includes two classical bits of information, along with one from Alice, by measuring his GHZ particle and sending the results. Charlie's unitary operation then permits the initial unknown state to be recovered. This procedure demonstrates the collaborative aspect of quantum teleportation via GHZ states, in which entanglement plays an important part in transferring quantum information between several participants. The complicated entanglement patterns seen in the GHZ state highlight its importance as an important resource in the field of quantum information processing, with applications in quantum communication and computation.

10.4.3 Bidirectional Quantum Teleportation through the Multipartite Glauber Coherent State

BQT protocol allows for the transmission of quantum states between two parties in both directions. It requires the parties to have their own trigger qubits, which store the unknown states they want to share. The protocol also requires a shared entangled state and a classical communication channel. To initiate the BQT protocol, certain gates are applied to the trigger qubits to create the desired states. Then, both parties perform an entanglement measurement on their trigger qubits and the shared entangled state. The measurement results are communicated through the classical channel. Based on the measurement results, both parties perform correction operations on their trigger qubits, effectively transferring the quantum states from one party to the other. The success rate of QT is probabilistic but can approach 100% with enough attempts. In the case of the proposed BQT protocol for even and odd coherent states, the multipartite Glauber coherent state is used as the quantum resource. By controlling the overlap and number of probes, the teleportation efficiency of the teleported states in both directions can be maximized. Overall, the BQT protocol enables the bidirectional transmission and reconstruction of quantum states by utilizing entanglement, measurement, and classical communication between the parties. The equation of a coherent state, denoted as α, is a fundamental concept in quantum mechanics, particularly in the context of quantum optics and quantum information. Theoretically, Glauber proved that these states can be defined as eigenvectors of the bosonic annihilation operator and can be obtained by applying a displacement operator $D(\alpha) = \exp^{(\pm a\dagger - \pm^* a)}$ to the vacuum state of the harmonic oscillator as $|\alpha\rangle = D(\alpha)|0\rangle = \exp^{\frac{-|\alpha|}{2}} \sum_n \frac{\alpha^n}{\sqrt{n!}}|n\rangle$ where α is the complex dimensionless amplitude of the coherent state and n stands for the Fock state. When the concept is applied to a quantized monochromatic electromagnetic wave, n represents the number of elementary quanta of energy, known as photons.

The introduction of an orthonormal basis, comprising even and odd coherent states [24, 25], is defined as:

$$|Even\rangle \equiv |\alpha_e\rangle = N_+(|\alpha\rangle + |-\alpha\rangle), \quad |Odd\rangle \equiv |\alpha_o\rangle = N_-(|\alpha\rangle - |-\alpha\rangle) \tag{10.27}$$

with normalization factors $N_\pm$ are given by

$$N_\pm = [2(1 \pm p)]^{\frac{-1}{2}}$$

These even and odd coherent states serve as the foundation for encrypting logical qubits in the two-dimensional Hilbert space, mapping $|\mathbf{0}\rangle_L \to |\alpha_e\rangle$ and $|\mathbf{1}\rangle_L \to |\alpha_o\rangle$. The states $|\alpha\rangle$ and $|-\alpha\rangle$ are then expressed as a superposition of logical qubits:

$$|\alpha\rangle \equiv \beta|\mathbf{0}\rangle_L + \gamma|\mathbf{1}\rangle_L, \qquad |-\alpha\rangle \equiv \mathbf{a}|\mathbf{0}\rangle_L - \mathbf{b}|\mathbf{1}\rangle_L \tag{10.28}$$

where

$$\beta = \sqrt{\frac{1+p}{2}} \qquad \gamma = \sqrt{\frac{1-p}{2}} \tag{10.29}$$

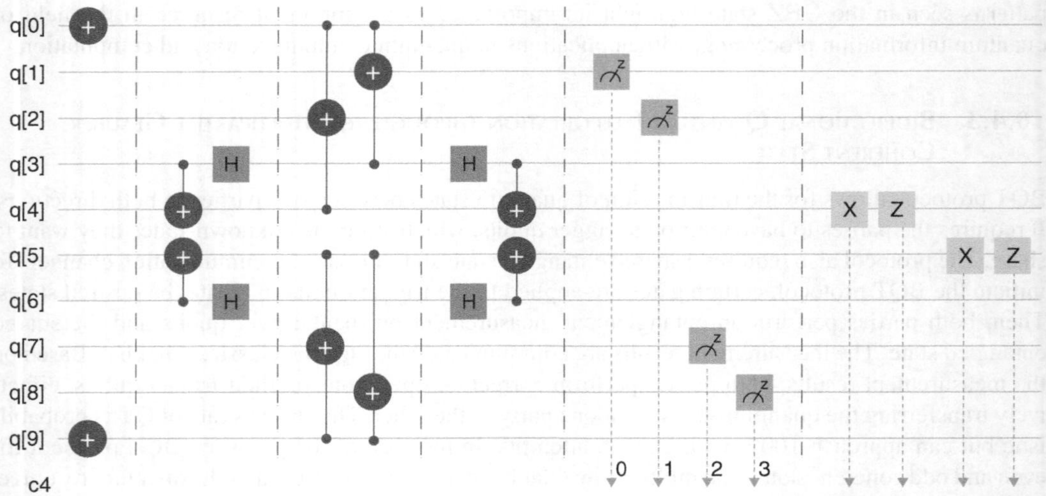

FIGURE 10.12 Qiskit circuit for implementing BQT with independent trigger qubits. X-gates set the initial and final qubits to |1⟩, followed by CNOT gates for entanglement. Controlled-controlled-NOT (CCNOT) gates and additional Hadamard and phase shift gates generate entanglement with tunable parameters m and p for high-fidelity accuracy [26].

Our initial choice of a bipartite split of the multipartite coherent state serves as the system's entangled state, which we later reduce to a bipartite state for application as an entangled state:

$$|\psi\rangle = N|\alpha\rangle_r \otimes |\alpha\rangle_{n-r} + e^{im\pi}|-\alpha\rangle_r \otimes |-\alpha\rangle_{n-r} \tag{10.30}$$

Here is the circuit we created. The implementation of BQT with two independent triggers qubits using Qiskit. The X-gates are applied to set the first and last qubits to the state 1. Then, we perform CNOT gates between two pairs of qubits; these CNOT gates create entanglement between the qubits. The application of controlled-controlled-NOT (CCNOT) gates between three qubits are used to teleport the state of one qubit to another qubit, and the entangled state of the entire system is generated by putting Hadamard gates (H), controlled-Z gates, and phase shift gates, the values of two variables, m and overlap p, are adjusted to achieve a high-fidelity accuracy.

The Qiskit simulations showed that the protocol was successful in transmitting information in both directions (Figure 10.12). The simulation probabilities fit high-fidelity values very well, suggesting the efficacy of the teleportation procedure. With relatively few mistakes, the output states were found to be very similar to the original quantum states. The approach exhibited remarkable fidelity and the possibility of BQT. We show that the probability derived using the Belem, Qasm, and Aer simulators very well fit high-fidelity values. Furthermore, the accuracy of these probabilities is significantly affected by certain p and m values. The great agreement between the probability and high-fidelity numbers shows that the teleportation procedure was successful. With a few mistakes, the output states are found to be very near to the establishing quantum states. You always go back to the original article to check more details [26].

10.5 QUANTUM NETWORK PROTOCOLS: MULTIHOP

Communication in quantum wireless MultiHop networks plays an important role in the context of global quantum networks. Our proposed technique uses QT between distant nodes which do not have initial entanglement to allow effective quantum communication in such networks. We use entanglement swapping methods based on many types of Werner states to achieve the necessary entanglement between intermediary nodes. After that, all measurement results and Werner state needs are individually communicated to the target node. The final recovery of the initial quantum state is accomplished by equivalent local operations, representing a significant achievement in the field of quantum wireless MultiHop networks. The absence of a direct quantum channel between source and destination nodes requires a hop-by-hop transmission through intermediary nodes in quantum communication scenarios. The Werner state, a frequently used quantum communication resource, was chosen as the quantum channel for this procedure. This method includes teleporting classical information to a next-door node, which is understood to be a single quantum qubit at the source node [27].

When dealing with mixed states, the representation gets more intricate, often needing the application of a density matrix. We construct a precise equation for the recovered qubit in the case of One-Hop teleportation. This recovered qubit has multiple uses, serving as both the result of the current teleportation step and the input for the upcoming hop teleportation. This gradual teleportation across intermediary nodes improves understanding of quantum communication dynamics and sets foundations for more advanced quantum communication protocols.

10.5.1 MultiHop Quantum Teleportation

Figure 10.13 demonstrates a unique quantum MultiHop system. This system design includes distributed quantum nodes, each of which may store quantum particles individually. Unlike classical systems, which use classical bits to transfer information, our proposed quantum system takes advantage of the unique qualities of qubit.

The qubit, the fundamental building block of quantum information, is described as: $|\Psi\rangle = \alpha|0\rangle + \beta|1\rangle$, where α and β are complex probability amplitudes. The condition $|\alpha|^2 + |\beta|^2 = 1$ assures that the qubit maintains its normalization. This quantum representation allows information encoding and transmission in a manner unique from classical networks.

Basically, our quantum MultiHop system brings in a paradigm change by using quantum mechanics principles to enable the safe and efficient transition of quantum information between nodes. The closest nodes are the paired quantum nodes connected by a line, and the quantum wireless channel only functions between the nodes situated next to each other. The shared quantum channel between neighboring nodes, specifically using a Werner state, which is a convex mixture

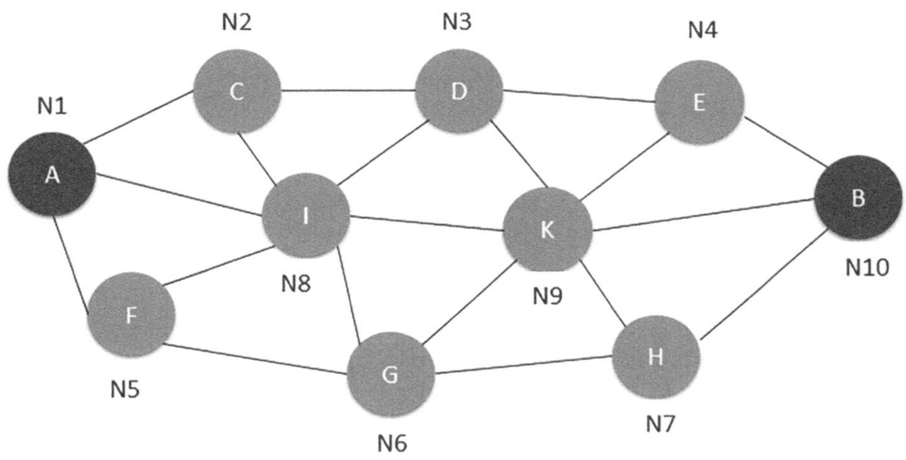

FIGURE 10.13 Quantum wireless MultiHop.

of a maximally mixed separable state and a maximally entangled pure state, is important in our discussion. There are several versions of the Werner state; here, we introduce our entangled version.

State (W state):

$$|W\rangle = (|100\rangle + |010\rangle + |001\rangle) \tag{10.31}$$

We take ρ_W as the quantum channel [28]:

$$\rho_W = \varepsilon(|\Psi^+\rangle\langle\Psi^+|) + (1/4 - \varepsilon)I \tag{10.32}$$

Here $0 < \varepsilon < 1$ and $|\Psi^+\rangle = (1/\sqrt{2})(|01\rangle + |10\rangle)$ is a maximally entangled pure state. The Werner states show different entanglement properties depending on the singular weight ε. The Werner states may be entangled $(\varepsilon > 1)$ or separable $(\varepsilon \leq 1)$. The Werner state is pure only when $\varepsilon = 1$. There is no direct quantum channel combining the sender and the receiver during information transfer between nodes, as described in [13]. Multiple routes connect the source and destination nodes, and nodes can transfer information through quantum states if and only if both a hop-by-hop quantum channel and a wireless channel exist at the same time. The quantum route is defined by the quantum channel, while the wireless route is defined by the wireless channel. It is important to note that the nodes engaged in the quantum route may differ from those involved in the wireless route.

10.5.2 Single-Hop Quantum Teleportation

Alice, Bob, and Charlie share the W state. Alice combines an unknown state with her W qubit and performs Bell measurements on these qubits. Before Alice's measurement, the total state $|\Psi_T\rangle = |\Psi\rangle_A \otimes |W\rangle_{abc}$ can be expressed in terms of the Bell basis as follows:

$$\begin{aligned}\Psi_T = \frac{1}{\sqrt{3}}|\Psi\rangle_{+Aa}&\{|0\rangle_b(\beta|0\rangle_c + \alpha|1\rangle_c) + |1\rangle_b(\alpha|0\rangle_c)\} - |\Psi_-\rangle_{Aa}\{|0\rangle_b(\beta|0\rangle_c - \alpha|1\rangle_c)\\ &- |1\rangle_b(\alpha|0\rangle_c)\} + |\phi_+\rangle_{Aa}\{|0\rangle_b(\alpha|0\rangle_c + \beta|1\rangle_c) + |1\rangle_b(\beta|0\rangle_c)\}\\ &+ |\phi_-\rangle_{Aa}\{|0\rangle_b(\alpha|0\rangle_c - \beta|1\rangle_c) - |1\rangle_b(\beta|0\rangle_c)\}]\end{aligned} \tag{10.33}$$

The total state is reduced into one of the four states as shown in Equation 10.30_bookmark2 after Alice's measurement. We describe our ideas in the form of a quantum circuit that uses Qiskit to accomplish One-Hop teleportation.

The quantum circuit shown in Figure 10.14_bookmark3 illustrates the One-Hop teleportation using the Werner state. Qiskit, a quantum computing framework, is used to implement the circuit. Let's briefly go over the key components of the circuit:

- *Entanglement State:* The shared W state among Alice, Bob, and Charlie is initially prepared.
- *Alice's Operations:* Alice combines the unknown state Ψ_A with her part of the W state and performs Bell measurements, collapsing the total state into one of the four terms mentioned in Equation 10.30.
- *Quantum Teleportation Circuit:* The circuit involves a series of quantum gates, including CNOT, Hadamard (H), controlled-Z (CZ), X, and Z gates. These gates manipulate the qubits to transfer the unknown state from Alice to Bob.
- *Measurements:* Quantum measurements are performed on Alice's qubits, and the measurement results are transmitted classically to Bob. These results guide the application of subsequent quantum gates.

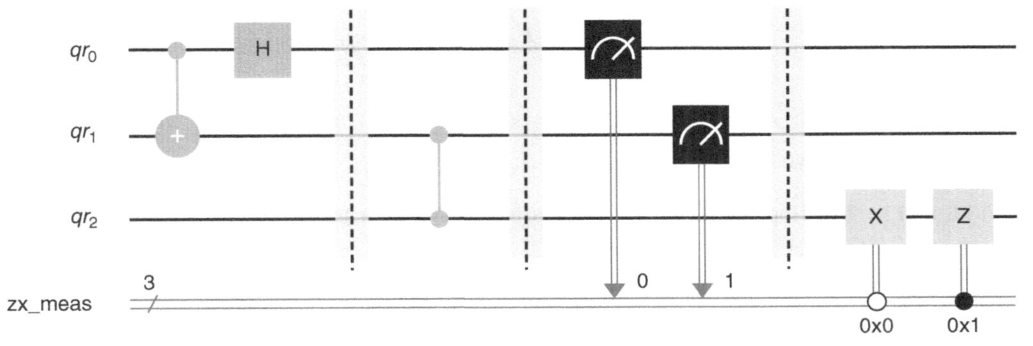

FIGURE 10.14 Qiskit quantum circuit showing One-Hop teleportation using the Werner state.

- *Final State:* Bob applies the necessary gates (X and Z) based on the received classical information to recover the teleported state.

The initial density matrix $\rho_{in} = |\Psi\rangle\langle\Psi|$ that describes the first qubit is:

$$\rho_{in} = \begin{pmatrix} |\alpha|^2 & \alpha\beta^* \\ \alpha^*\beta & |\beta|^2 \end{pmatrix} \tag{10.34}$$

The second and third qubits are described by Equation 10.29, which describes the quantum channel W. Concerning One-Hop quantum teleportation fidelity and efficiency, fidelity is a measure of how well the transported state matches the original state. The fidelity of QT procedures is determined by factors such as the quality of the entangled states, measurement accuracy, and the efficiency of the quantum gates.

The fidelity between the original state ρ_{in} and the teleported state ρ_{out} can be calculated using the following formula:

$$F = Tr\left(\sqrt{\sqrt{\rho_{in}}\,\rho_{out}\sqrt{\rho_{in}}}\right)^2 \tag{10.35}$$

Here, Tr denotes the trace of the matrix and ρ_{out} is the density matrix of the teleported state. The density matrix ρ_{out} is the density matrix of the state that Bob obtains after applying the necessary gates to his qubit. To calculate the fidelity, you need to find ρ_{out} and substitute it into the fidelity formula. High fidelity implies that the teleported state closely matches the original state. The fidelity may be affected by various factors such as decoherence, imperfections in gates, and measurement errors.

10.5.3 Two-Hop Quantum Teleportation

Two-Hop quantum teleportation improves on the principle by adding an intermediary node (Jack) between the transmitter and the receiver. Alice, Charlie, and Bob are now sharing entangled particles, resulting in a multipartite entangled state. Alice entangles her particle with the unknown state, takes measurements, and sends it back to Jack. Given this information, Jack performs additional steps and delivers a better set of information to Bob. Finally, Bob applies quantum gates to his particle, resulting in an accurate reconstruction of the original quantum state.

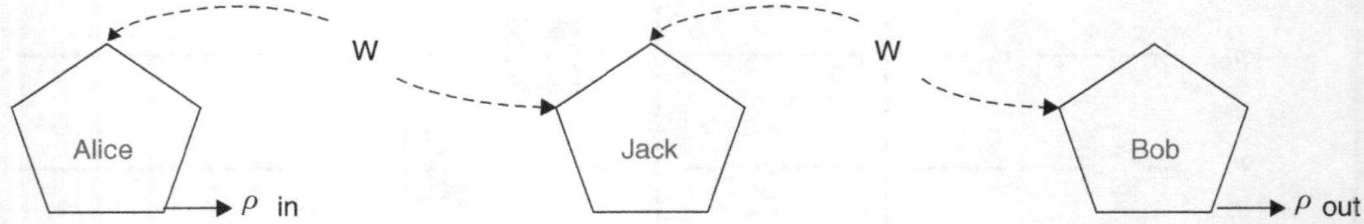

FIGURE 10.15 Two-Hop quantum teleportation using the Werner state.

Alice uses the Werner state to transfer the quantum state ρ_{in} into Jack (Figure 10.15). Following that, Jack uses the W state as the entangled state to teleport the quantum state to Bob, resulting in the recovery of the state ρ_{out} by Bob.

The advantages of Two-Hop quantum teleportation are as follows: Two-hop teleportation provides an additional degree of protection. Even if one of the nodes fails, the information alone is enough to reassemble the quantum state. The protocol provides for mistake correction and increased fidelity in the teleportation process with the addition of an intermediary node. The Two-Hop strategy, on the other hand, brings additional difficulty in terms of resource needs and the necessity for strong quantum channels. A Two-Hop quantum teleportation circuit's illustration shows complicated entanglement procedures and many quantum gates.

The given quantum circuit is created for a Two-Hop quantum teleportation protocol (Figure 10.16). It involves the teleportation of quantum states from qubit 0 to qubit 4 in the first hop and from qubit 2 to qubit 4 in the second hop, using a total of 5 qubits and 4 classical bits. Based on measurement results, the circuit uses CNOT gates, Hadamard gates, CZ gates, and conditional operations. Different qubits are measured, and following gates are conditionally applied according to the classical bit results. The general design of the circuit represents a two-step teleportation procedure, allowing quantum information to be transmitted between nodes.

The initial state can be expressed as:

$$\left(\alpha|0\rangle \mathrm{A}+\beta|1\rangle \mathrm{A}\right)\otimes\left(1/\sqrt{3}\right)\left(|001\rangle \mathrm{abc}+|010\rangle \mathrm{abc}+|100\rangle \mathrm{abc}\right) \tag{10.36}$$

Which evolves into:

$$\begin{aligned}&1/6\ [|\Phi^{+}\rangle \mathrm{Aa}\left(\alpha|01\rangle \mathrm{bc}+\alpha|10\rangle \mathrm{bc}+\beta|00\rangle \mathrm{bc}\right)\\&+|\Phi^{-}\rangle \mathrm{Aa}\ \left(\alpha|01\rangle \mathrm{bc}+\alpha|10\rangle \mathrm{bc}-\beta|00\rangle \mathrm{bc}\right)\\&+|\Psi^{+}\rangle \mathrm{Aa}\ \left(\alpha|00\rangle \mathrm{bc}+\beta|01\rangle \mathrm{bc}+\beta|10\rangle \mathrm{bc}\right)\\&+|\Psi^{-}\rangle \mathrm{Aa}\ \left(\alpha|00\rangle \mathrm{bc}-\beta|01\rangle \mathrm{bc}-\beta|10\rangle \mathrm{bc}\right)]\end{aligned} \tag{10.37}$$

Table 10.1 shows the measurement results of Alice's and Bob's Pauli gate operation used to recover the state.

The results of these operations, which include various entangled states such as $|\Phi^{+}\rangle, |\Phi^{-}\rangle, |\Psi^{+}\rangle$, and $|\Psi^{-}\rangle$, define the specific gates used by Bob to reconstruct the transmitted quantum information. This provides a thorough exploration of the quantum circuit's design and functionality in the context of Two-Hop quantum teleportation.

10.5.4 N-Hop (MultiHop)

Now we consider the case of N-Hop quantum teleportation. As shown in Figure 10.17, this scheme has extended the Two-Hop scheme into N-Hop, where there are $(n-2)$ intermediate nodes between Alice and Bob. The entanglement exists between two consecutive nodes, between Alice and her

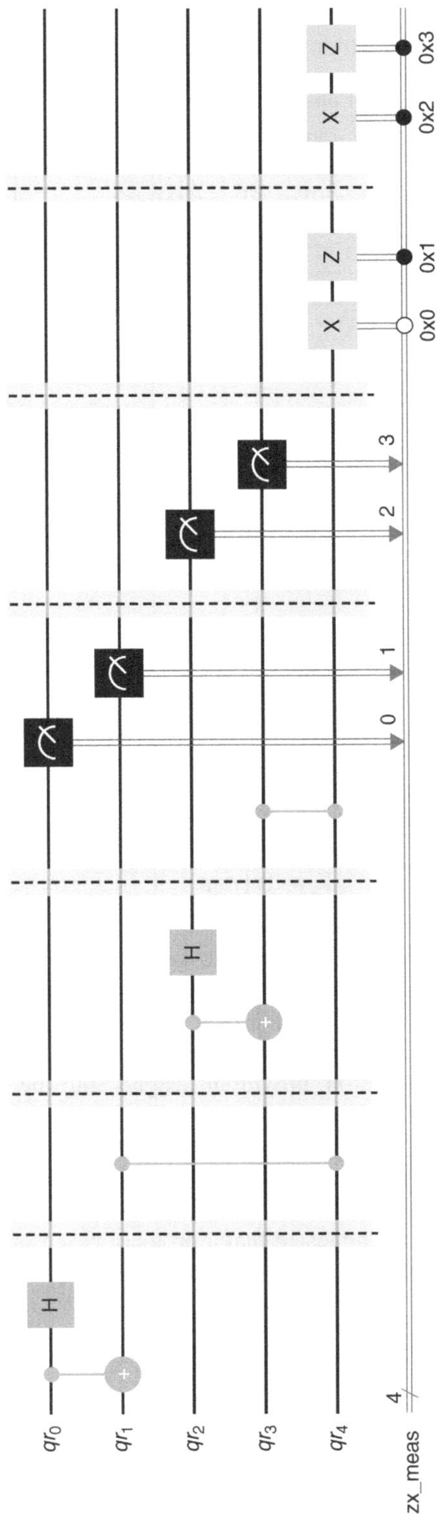

FIGURE 10.16 IBM quantum circuit of Two-Hop quantum teleportation using the Werner state.

TABLE 10.1
Measurement Outcomes and Recovery Operations

Alice's Bell Measurement on Aa	Collapsed State of Qubits bc	Bob's Recovery Operation
$\|\Psi^+\rangle$	$\alpha\|00\rangle + \beta(\|01\rangle + \|10\rangle)$	Identity I_1
$\|\Psi^-\rangle$	$\alpha\|00\rangle - \beta(\|01\rangle + \|10\rangle)$	Logical Z_1
$\|\Phi^+\rangle$	$\alpha(\|01\rangle + \|10\rangle) + \beta\|00\rangle$	Logical X_1
$\|\Phi^-\rangle$	$\alpha(\|01\rangle + \|10\rangle) - \beta\|00\rangle$	Logical X_1 Z_1

Note: This table shows the Bell measurement outcomes obtained by Alice on qubits A and a, and the corresponding logical Pauli operations applied by Bob on the encoded qubit in the b c subsystem to recover the state $|\psi\rangle = \alpha|0\rangle + \beta|1\rangle$.

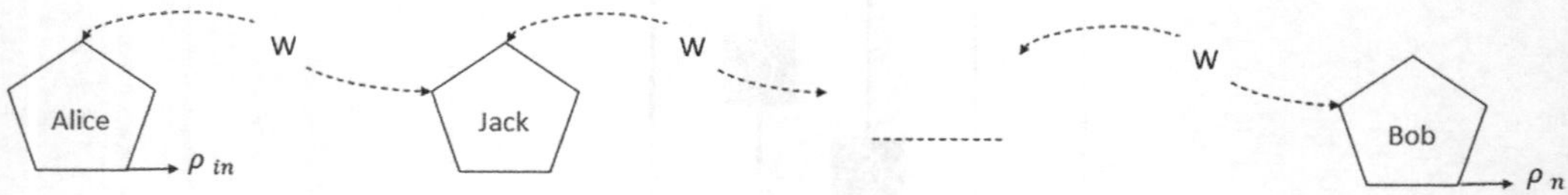

FIGURE 10.17 N-Hop quantum teleportation.

neighbor node and the intended receiver Jack and his adjacent node, where each of these (n – 1) pairs shares a W entangled state. The main difference between N-Hop and Two-Hop quantum teleportation is the number of intermediate nodes. In the Two-Hop scheme, there is only one intermediate node (Jack) between Alice and Bob. However, in the N-Hop scheme, there can be multiple intermediate nodes between them.

According to what is mentioned above, the fidelity of N-Hop is:

$$F_n = \mathrm{Tr}\left(\rho_{in}\rho_{n-1}\right) \tag{10.38}$$

We should know that with the increase in the hops, the fidelity decreases, and it remains close to 1. Figure 10.18 illustrates the N-Hop quantum teleportation circuit. The circuit demonstrates the gradual teleportation of quantum information over a series of nodes, highlighting the process's dependent nature. The fidelity calculation is an important matric for evaluating the efficiency and security of N-Hop quantum teleportation, providing insights into the limitations associated with extended network designs.

The fidelity results discussed for the N-Hop scenario indicate that a quantum teleportation MultiHop system based on the Werner state behaves similarly to a classical system. The probability of successful transmission decreases as both distance and the number of hops increase. In combination with the One-Hop circuit, it is suggested that a protocol based on the minimal hops concept should be employed to achieve optimal transmission efficiency from the source node to the destination node.

10.6 CONCLUSION

The addition of a link node in the N-Hop system improves protection and error correction. Even if a node fails, the information available is sufficient to rebuild the quantum state, offering extra security. This expanded design, however, offers issues in terms of resource needs and the requirement for strong quantum channels. The circuit's operation is based on measurement results from

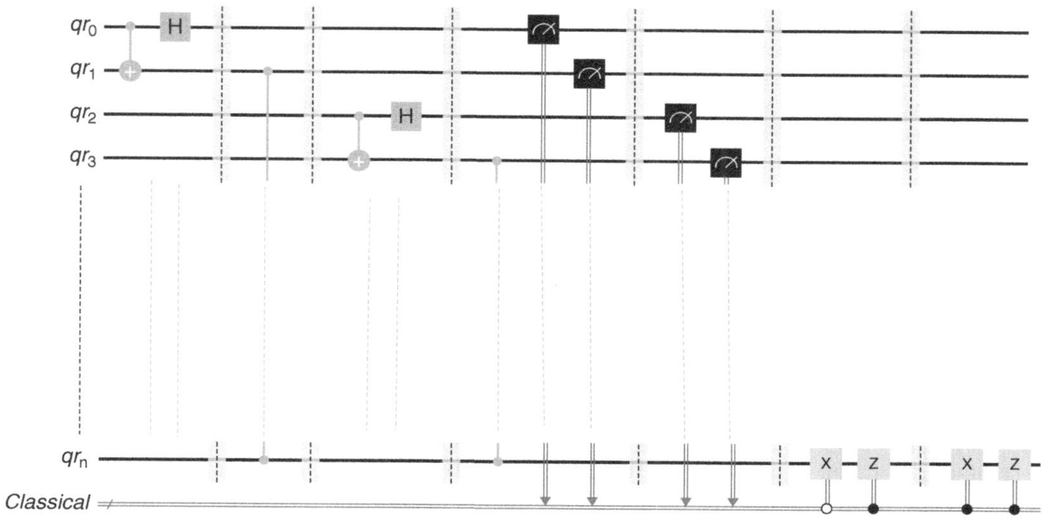

FIGURE 10.18 N-Hop quantum teleportation circuit.

Alice's qubits, which are classically shared to Bob and direct subsequent quantum gate operations. Bob's gates for rebuilding communicated quantum information are determined by the measurement outcomes. The N-Hop quantum teleportation circuit requires a greater number of quantum channels and resources than the Two-Hop method. The fidelity of teleportation decreases as the number of hops increases, emphasizing the difficulties of creating an extended system. The N-Hop quantum teleportation expands the Two-Hop technique by integrating extra intermediary nodes, which enhances security and error correction. However, this enhancement increases complexity and resource requirements. The dependence on measurement outcomes from Alice's qubits is at the center of the circuit's operation. These results are typically expressed to Bob and used as a guide for coming quantum gate operations. Bob's use of certain quantum gates for the reconstruction of communicated quantum information is strongly related to the measurement results. This dynamic interaction shows the essential function that classical communication plays in driving quantum events within the teleportation circuit.

Because of its wide design, the N-Hop quantum teleportation circuit requires more quantum channels and resources than the more simplified Two-Hop method. The quality of the teleportation process decreases as the number of hops increases, demonstrating the fundamental problems connected with handling a coherent and efficient extended system. Finally, N-Hop quantum teleportation is a considerable development of the Two-Hop approach, including extra intermediate nodes that significantly improve security and error correction capabilities. This improvement, however, comes at the expense of more complexity and higher resource needs, illustrating the importance of carefully considering choices in the effort of developing quantum teleportation capabilities.

REFERENCES

[1] Bennett, C. H., Brassard, G., Crépeau, C., Jozsa, R., Peres, A., Wootters, W. K. (1993). Teleporting an unknown quantum state via dual classical and Einstein–Podolsky–Rosen channels. *Physical Review Letters*, 70(13), 1895.

[2] Bouwmeester, D., Pan, J. W., Mattle, K., Eibl, M., Weinfurter, H., Zeilinger, A. (1997). Experimental quantum teleportation. *Nature*, 390(6660), 575–579.

[3] Boschi, D., Branca, S., De Martini, F., Hardy, L., Popescu, S. (1998). Experimental realization of teleporting an unknown pure quantum state via dual classical and Einstein–Podolsky–Rosen channels. *Physical Review Letters*, 80(6), 1121.

[4] Ikram, M., Zhu, S. Y., Zubairy, M. S. (2000). Quantum teleportation of an entangled state. *Physical Review A*, 62(2), 022307.

[5] Bennett, C. H., Brassard, G., Crépeau, C., Maurer, U. M. (1992). Generalized privacy amplification. *IEEE Transactions on Information Theory*, 41(6), 1915–1923. doi:10.1109/18.165464

[6] Shor, P. W. (1995). Scheme for reducing decoherence in quantum computer memory. *Physical Review A*, 52(4), R2493–R2496. doi:10.1103/PhysRevA.52.R2493

[7] Steane, A. M. (1996). Multiple particle interference and quantum error correction. *Proceedings of the Royal Society A: Mathematical, Physical and Engineering Sciences*, 452(1954), 2551–2577. doi:10.1098/rspa.1996.0114

[8] Lee, S. W., Park, J., Jeong, H. (2006). MultiHop quantum teleportation using hybrid entanglement of W and GHZ states. *Journal of Physics A: Mathematical and General*, 39(2), 523–529.

[9] Yeo, Y., Park, J., Lee, S. W. (2008). MultiHop quantum teleportation using graph states. *Journal of Physics A: Mathematical and Theoretical*, 41(45), 455308.

[10] Gou, Y. T., Shi, H. L., Wang, X. H., Liu, S. Y. (2017). Probabilistic resumable bidirectional quantum teleportation. *Quantum Information Processing, 16(11), 278.*

[11] Shor, P. W. (1994, November). Algorithms for quantum computation: discrete logarithms and factoring. In *Proceedings 35th Annual Symposium on Foundations of Computer Science* (pp. 124–134). IEEE.

[12] Lo, H. K., Zhao, Y. (2008). Quantum cryptography. *arXiv preprint arXiv:0803.2507.*

[13] Nielsen, M. A., Chuang, I. L. (2011). *Quantum Computation and Quantum Information*, 10th ed. Cambridge University Press.

[14] Amelino-Camelia, G. (2000). Quantum theory's last challenge. *Nature*, 408(6813), 661–664.

[15] Gisin, N., Thew, R. (2007). Quantum communication. *Nature Photonics*, 1(3), 165–171.

[16] Einstein, A., Podolsky, B., Rosen, N. (1935). Can quantum-mechanical description of physical reality be considered complete? *Physical Review*, 47(10), 777.

[17] Schro¨dinger, E. (1935, October). Discussion of probability relations between separated systems. In *Mathematical Proceedings of the Cambridge Philosophical Society* (Vol. 31, No. 4, pp. 555–563). Cambridge University Press.

[18] Schrödinger, E. (1980). 1.11 The Present Situation in Quantum Mechanics: A Translation of Schrödinger's" Cat Paradox' Paper. *Proceedings of the American Philosophical Society*, *124*(5), 323-38.

[19] Bashar, M. A., Chowdhury, M. A., Islam, R., Rahman, M. S., Das, S. K. (2009, March). A review and prospects of quantum teleportation. In *2009 International Conference on Computer and Automation Engineering* (pp. 213–217). IEEE.

[20] Bell, J. S. (1964). On the Einstein–Podolsky–Rosen paradox. *Physics Physique Fizika*, *1*(3), 195.

[21] Yang, K., Huang, L., Yang, W., Song, F. (2009). Quantum teleportation via GHZ-like state. *International Journal of Theoretical Physics*, 48, 516–521.

[22] Karlsson, A., Bourennane, M. (1998). Quantum teleportation using three-particle entanglement. *Physical Review A*, 58(6), 4394.

[23] Williams, C. P. (2010). *Explorations in Quantum Computing*. Springer Science and Business Media.

[24] Dodonov, V. V., Malkin, I. A., Man'Ko, V. I. (1974). Even and odd coherent states and excitations of a singular oscillator. *Physica*, 72, 597–615.

[25] Ansari, N. A., Di Fiore, L., Man'ko, M. A., Man'ko, V. I., Solimeno, S., Zaccaria, F. (1994). Quantum limits in interferometric gravitational-wave antennas in the presence of even and odd coherent states. *Physical Review A*, 49, 2151.

[26] Ikken, N., Slaoui, A., Ahl Laamara, R., Drissi, L. B. (2023). Bidirectional quantum teleportation of even and odd coherent states through the multipartite Glauber coherent state: Theory and implementation. *Quantum Information Processing*, 22(10), 391.

[27] Lee, J., Kim, M. S. (2000). Entanglement teleportation via Werner states. *Physical Review Letters*, 84(18), 4236.

[28] Prakash, H., Verma, V. (2013). Quantum teleportation of single qubit mixed information state with Werner-like state as resource. *arXiv preprint arXiv:1305.4259.*

11 Quantum State Dynamics and Information Transmission in Open Systems

Safae Gaidi, Yassine Dakir, Abdallah Slaoui, Mohammed El Falaki, and Rachid Ahl Laamara

11.1 INTRODUCTION

Quantum mechanics marks a profound departure from classical physics by introducing intrinsically probabilistic behaviors and non-classical correlations that defy intuitive understanding. At the core of this theoretical framework are quantum states, which are mathematically described by density operators acting on complex Hilbert spaces. These operators encapsulate both the statistical mixtures and coherent superpositions of quantum systems, serving as the foundation of modern quantum theory.

To understand, manipulate, and control quantum systems, whether in the context of computation, communication, or metrology, it is essential to study how these quantum states evolve over time or in response to external operations. The mathematical description of such evolution is provided by linear maps that preserve key physical properties: trace (ensuring total probability is conserved), positivity (ensuring valid probabilities), and Hermiticity (maintaining observable quantities). Among these, completely positive trace-preserving (CPTP) maps represent the most general physical transformations permissible within quantum mechanics. These maps are crucial for modeling realistic processes, especially when systems are not isolated but interact with an external environment.

This leads naturally to the study of open quantum systems, where the system of interest is coupled to its surrounding environment. Unlike idealized closed systems, open systems are susceptible to decoherence and dissipation, effects that are critical in both theoretical considerations and practical applications. In particular, when the environment retains a memory of past interactions, the system's evolution may become non-Markovian, meaning its future dynamics depend not only on its present state but also on its history. This memory effect challenges the standard assumptions of Markovian dynamics and has significant implications for information processing and error correction in quantum technologies.

The characterization and quantification of non-Markovianity have thus emerged as essential themes in quantum information theory, offering insights into both fundamental physics and practical implementations. Understanding these effects allows researchers to harness, rather than merely mitigate, the influence of the environment, potentially leading to more resilient quantum protocols.

Additionally, quantum operations such as teleportation, quantum key distribution, and error correction rely heavily on non-classical resources like entanglement and the formalism of CPTP maps. These protocols exemplify the powerful capabilities of quantum mechanics in enabling tasks that are classically infeasible. For instance, teleportation uses entangled states and local operations combined with classical communication to transmit unknown quantum states between distant parties, a process impossible in any classical setting.

DOI: 10.1201/9781003621423-13

This chapter introduces the foundational principles underlying quantum dynamics, with a focus on the role of CPTP maps, the implications of open system dynamics, and the emergence of non-Markovian effects. Special attention is given to how these concepts are applied in quantum information science, offering both a theoretical perspective and a pathway toward real-world quantum technologies.

11.2 QUANTUM STATES, LINEAR MAPS, AND OPEN QUANTUM SYSTEMS

11.2.1 Quantum States

Let H be a finite-dimensional Hilbert space, and $L(H)$ be the space of linear operators acting on H. To avoid technical complications, we restrict ourselves to finite-dimensional quantum systems, although many results naturally extend to the infinite-dimensional case.

A quantum state is described by a density operator $\rho \in L(H)$ [1], which satisfies the following conditions:

- $\rho \geq 0$: ρ is positive semi-definite (i.e., $\langle x|\rho|x\rangle \geq 0$ for all $|x\rangle \in H$)
- $\mathrm{Tr}(\rho) = 1$: the trace condition ensures normalization.

Given a fixed orthonormal basis $\{|k\rangle\}_{k=1}^{d}$, the diagonal entries $\langle k|\rho|k\rangle$ form a classical probability distribution. Thus, the density matrix can be interpreted as encoding a family of classical distributions dependent on the measurement basis. This justifies the interpretation of density matrices as "quantum probabilities" [2].

Density matrices provide a unified framework for describing both pure and mixed states [3]:

- A pure state is represented by $\rho = |\psi\rangle\langle\psi|$, which corresponds to a state with maximum knowledge about the system.
- A mixed state corresponds to an ensemble $\{p_i, |\psi_i\rangle\}$, and is expressed as:

$$\rho = \sum_i p_i |\psi_i\rangle\langle\psi_i|. \tag{11.1}$$

Two distinct ensembles $\{p_i, |\psi_i\rangle\}$ and $\{q_i, |\phi_i\rangle\}$ represent the same physical state if they yield the same density matrix

$$\sum_i p_i |\psi_i\rangle\langle\psi_i| = \sum_i q_i |\phi_i\rangle\langle\phi_i|. \tag{11.2}$$

This property illustrates the equivalence of different descriptions of quantum states, known as *ensemble representations*. Such representations are especially useful when dealing with thermal states, noisy environments, and statistical mixtures in practical quantum systems.

Density operators are crucial in quantum theory, particularly in describing entanglement [3]. For a bipartite pure state $|\psi_{AB}\rangle \in H_A \otimes H_B$, the reduced density operators are given by

$$\rho_A = \mathrm{Tr}_B\left(|\psi_{AB}\rangle\langle\psi_{AB}|\right), \quad \rho_B = \mathrm{Tr}_A\left(|\psi_{AB}\rangle\langle\psi_{AB}|\right), \tag{11.3}$$

which are generally mixed even if $|\psi_{AB}\rangle$ is pure. The non-factorization of these reduced states reflects an entanglement between subsystems A and B, a cornerstone of quantum information theory [1].

11.2.2 Linear Maps in Quantum Dynamics

A central question in quantum mechanics is: How do quantum states evolve over time? That is, how is the transformation $\rho \mapsto \rho_t$ described?

To ensure consistency with the no-signaling principle, all ensembles that result in the same initial state must evolve to ensembles that yield the same final state. This requirement implies that the evolution map Φ must be:

- Linear.
- Convexity-preserving: for any $p_1 + p_2 = 1$,

$$\Phi\left(p_1\rho_1 + p_2\rho_2\right) = p_1\Phi\left(\rho_1\right) + p_2\Phi\left(\rho_2\right). \tag{11.4}$$

This linearity condition ensures that quantum evolution respects the principles of quantum superposition and coherence. The convexity-preserving property reflects the fact that probabilistic mixtures of states evolve according to the same probabilities. Hence, quantum evolutions are described by linear maps $\Phi : L(H) \to L(H')$ that must preserve:

- Hermiticity: $\Phi\left(X^\dagger\right) = \Phi(X)^\dagger$.
- Trace: $\mathrm{Tr}\left(\Phi(X)\right) = \mathrm{Tr}(X)$.
- Positivity: $X \geq 0 \Rightarrow \Phi(X) \geq 0$.

Such maps are called positive trace-preserving (PTP) maps. However, this class is insufficient to describe general quantum dynamics – particularly in composite systems [2].

Suppose $\Phi_1 : L(H_1) \to L(H_{1'})$ and $\Phi_2 : L(H_2) \to L(H_{2'})$ are two PTP maps. Then their tensor product

$$\Phi_1 \otimes \Phi_2 : L\left(H_1 \otimes H_2\right) \to L\left(H_{1'} \otimes H_{2'}\right) \tag{11.5}$$

may fail to be positive when applied to entangled states. This issue highlights the necessity of extending PTP maps to completely positive (CP) maps to ensure physical validity in composite quantum systems [4].

11.2.3 CP Maps

To handle entangled states properly, we require a stronger condition known as complete positivity. A map Φ is CP if $\mathrm{id}_k \otimes \Phi$ is positive for all $k \in \mathbb{N}$ [5].

Complete positivity ensures that the evolution of quantum states is valid even when the system is entangled with an auxiliary subsystem. This is crucial for describing phenomena like quantum teleportation, entanglement distillation, and quantum error correction, where entangled states are manipulated [1].

CPTP maps have the following key properties:

- *Contractivity:* CPTP maps exhibit contractivity, which means that they reduce the distinguishability between quantum states under certain mtrics. Specifically, for any two density operators ρ and σ, the trace distance between their images under a CPTP map Φ is always less than or equal to the trace distance between ρ and σ [2]:

$$T(\rho,\sigma) \geq T\left(\Phi(\rho),\Phi(\sigma)\right), \tag{11.6}$$

where the trace distance $T(\rho,\sigma)=\frac{1}{2}\mathrm{Tr}\left|\rho-\sigma\right|$. Additionally, the quantum relative entropy $D(\rho\,\|\,\sigma)$ also obeys a similar inequality:

$$D(\rho\,\|\,\sigma)\geq D\big(\Phi(\rho)\,\|\,\Phi(\sigma)\big), \tag{11.7}$$

which is a consequence of the Uhlmann inequality. Furthermore, CPTP maps do not decrease the fidelity between states:

$$F(\rho,\sigma)\leq F\big(\Phi(\rho),\Phi(\sigma)\big), \tag{11.8}$$

where $F(\rho,\sigma)=\mathrm{Tr}\sqrt{\sqrt{\rho}\sigma\sqrt{\rho}}$.

- *Invariant States:* Every CPTP map Φ admits at least one invariant state π, which satisfies:

$$\Phi(\pi)=\pi. \tag{11.9}$$

This follows from Schauder's fixed-point theorem. In this case, Φ is strictly contractive, meaning that for any ρ and σ,

$$T(\rho,\sigma)>T\big(\Phi(\rho),\Phi(\sigma)\big). \tag{11.10}$$

The map admits a unique stationary state π, which is the unique fixed point.

- *Unital Maps:* A CPTP map Φ is called *unital* if it preserves the identity operator, that is, $\Phi(\mathbb{I})=\mathbb{I}$. Unital maps have special properties and include unitary evolution as well as ideal projective measurements. One key property of unital maps is that they cannot decrease the von Neumann entropy of any state ρ [2]

$$S(\rho)\geq S\big(\Phi(\rho)\big), \tag{11.11}$$

  where the von Neumann entropy $S(\rho)=-\mathrm{Tr}(\rho\log_2\rho)$. This property plays a crucial role in the thermodynamic behavior of quantum systems.
- *Combinations of Maps:* Consider a map Φ of the form:

$$\Phi(\rho)=p\rho_0+(1-p)\Phi_0(\rho), \tag{11.12}$$

  where p is a probability $0\leq p\leq 1$, ρ_0 is a fixed density operator, and Φ_0 is another CPTP map. Such maps are strictly contractive, and as a result, they have a unique fixed point, which is ρ_0. The physical interpretation is that with probability p, the state ρ is replaced by ρ_0, and with probability $1-p$, the map Φ_0 is applied.
- *Semigroup Structure:* The set of CPTP maps forms a semigroup under composition. Specifically, if Φ_1 and Φ_2 are two CPTP maps, their composition $\Phi_2\circ\Phi_1$ is also a CPTP map. This is because the composition of two CPTP maps is still trace-preserving and CP. However, the operation is not commutative in general:

$$\Phi_2\circ\Phi_1(\rho)\neq\Phi_1\circ\Phi_2(\rho), \tag{11.13}$$

  for most choices of ρ.

- *Invertibility:* A CPTP map Φ is invertible if and only if there exists another CPTP map Φ^{-1} such that:

$$\Phi^{-1} \circ \Phi = \mathbb{I}. \tag{11.14}$$

In this case, the map Φ represents a reversible process, and an example of such a process is unitary evolution, where $\Phi(\rho) = U\rho U^{\dagger}$ for a unitary operator U.

11.2.4 Quantum States and Density Operators

In quantum mechanics, the state of a physical system is described by a mathematical object known as a *quantum state* [6–8]. There are two primary ways to represent quantum states: pure states and mixed states.

A *pure state* is described by a unit vector $|\psi\rangle$ in a Hilbert space $\mathcal{H}$. The corresponding density operator is given by $\rho = |\psi\rangle\langle\psi|$, which is a rank-one projection operator satisfying $\rho^2 = \rho$ and $(\rho) = 1$.

A *mixed state*, on the other hand, represents a statistical ensemble of pure states and is described by a density operator ρ that is a positive semi-definite, trace-one operator on $\mathcal{H}$. In general, a density operator can be written as:

$$\rho = \sum_i p_i |\psi_i\rangle\langle\psi_i|, \tag{11.15}$$

where $\{p_i\}$ is a probability distribution and $\{|\psi_i\rangle\}$ are pure states.

Density operators possess the following properties:

- *Hermiticity:* $\rho = \rho^{\dagger}$,
- *Positivity:* $\rho \geq 0$,
- *Unit Trace:* $(\rho) = 1$.

The set of all density operators on a Hilbert space $\mathcal{H}$ forms a convex set, denoted as $\mathcal{D}(\mathcal{H})$. The extremal points of this set are the pure states [3].

11.2.5 Quantum Channels

A quantum channel is a mathematical model that describes the physical evolution of quantum states, especially in the presence of noise and interactions with the environment. Mathematically, a quantum channel is a linear map Φ: $\mathcal{B}(\mathcal{H}) \to \mathcal{B}(\mathcal{H}')$ that is CPTP [9,10].

A map Φ is said to be *positive* if it maps positive operators to positive operators. It is *CP* if, for any extension by the identity map $\mathbb{I}_n$ on an auxiliary space $\mathbb{C}^n$, the extended map $\Phi \otimes \mathbb{I}_n$ remains positive. This ensures that the evolution described by Φ is valid even when the system is entangled with an ancillary system [5].

Trace preservation guarantees that the total probability is conserved:

$$(\Phi(\rho)) = (\rho) = 1. \tag{11.16}$$

Any quantum channel admits an operator-sum (or Kraus) representation:

$$\Phi(\rho) = \sum_k K_k \rho K_k^{\dagger}, \tag{11.17}$$

where the Kraus operators $\{K_k\}$ satisfy the completeness relation:

$$\sum_k K_k^\dagger K_k = \mathbb{I}. \tag{11.18}$$

This representation provides a physical interpretation of quantum noise as a probabilistic application of different operations indexed by k [5].

11.3 OPEN QUANTUM SYSTEMS

In practice, quantum systems are never perfectly isolated, they inevitably interact with external degrees of freedom collectively referred to as the environment or reservoir [11]. These interactions lead to non-unitary phenomena such as decoherence, dissipation, and information leakage. The formalism of open quantum systems provides a rigorous framework for describing such dynamics, bridging the idealized unitary evolution of closed systems and the irreversible behavior observed in realistic quantum processes [12–14].

Consider a quantum system S with associated Hilbert space $\mathcal{H}_S$ of dimension d_S, interacting with an environment E described by the Hilbert space $\mathcal{H}_E$, which may be finite- or infinite-dimensional. The total system $S+E$ is assumed to be closed and hence evolves unitarily under a global Hamiltonian H_{SE} acting on the composite space $\mathcal{H}_S \otimes \mathcal{H}_E$,

$$H_{SE} = H_S \otimes \mathbb{I}_E + \mathbb{I}_S \otimes H_E + H_{\text{int}}, \tag{11.19}$$

where H_S and H_E denote the free Hamiltonians of the system and environment, respectively, and H_{int} captures their mutual interaction.

Assuming an initially factorized state $\rho_{SE}(0) = \rho_S(0) \otimes \rho_E(0)$, the evolution of the global system is governed by the unitary operator $U(t) = \exp(-iH_{SE}t)$ (we set $\hbar = 1$ throughout). The reduced state of the system S at time t is then given by tracing over the environment

$$\rho_S(t) = Tr_E\left[U(t)\rho_{SE}(0)U^\dagger(t)\right]. \tag{11.20}$$

This procedure defines a dynamical map Φ_t such that $\rho_S(t) = \Phi_t\left[\rho_S(0)\right]$. Provided that the initial state is separable, the map Φ_t is CPTP, ensuring that the physical requirements of the density operator are preserved [15].

In general, deriving the explicit form of Φ_t is analytically intractable due to the complexity of H_{int} and the properties of $\rho_E(0)$. However, under specific assumptions such as weak coupling and a memoryless (Markovian) environment, it is possible to describe the dynamics using a *quantum master equation* in differential form:

$$\frac{d\rho_S(t)}{dt} = \mathcal{L}\left[\rho_S(t)\right], \tag{11.21}$$

where $\mathcal{L}$ is a superoperator known as the Liouvillian. For time-independent and Markovian dynamics, $\mathcal{L}$ admits the well-known Gorini–Kossakowski–Sudarshan–Lindblad (GKSL) structure [4]

$$\mathcal{L}[\rho] = -i[H,\rho] + \sum_k \gamma_k \left(L_k \rho L_k^\dagger - \frac{1}{2}\{L_k^\dagger L_k, \rho\} \right), \tag{11.22}$$

where H is an effective Hamiltonian (possibly incorporating Lamb shifts), $\{L_k\}$ are Lindblad operators representing environmental interactions, and $\gamma_k \geq 0$ are decay rates. The first term describes unitary (coherent) evolution, while the second encodes the irreversible effects of the environment.

More generally, for non-Markovian dynamics where environmental memory effects become significant, the reduced evolution may be described by integro-differential equations, such as those derived via the Nakajima–Zwanzig projection operator technique:

$$\frac{d\rho_S(t)}{dt} = \int_0^t \mathcal{K}(t,\tau)\rho_S(\tau)d\tau, \tag{11.23}$$

where $\mathcal{K}(t,\tau)$ is a memory kernel superoperator. Under suitable approximations (e.g., Born and Markov), this reduces to a time-local form:

$$\frac{d\rho_S(t)}{dt} = \mathcal{L}(t)[\rho_S(t)], \tag{11.24}$$

with a time-dependent Liouvillian $\mathcal{L}(t)$ generating the dynamics. The corresponding dynamical map is then given by:

$$\Phi_t = \mathcal{T}\exp\left(\int_0^t \mathcal{L}(\tau)d\tau\right), \tag{11.25}$$

where $\mathcal{T}$ denotes the time-ordering operator. Complete positivity of Φ_t imposes structural constraints on the generator $\mathcal{L}(t)$, especially when decay rates become time-dependent or negative [11,15].

Understanding the transition from closed to open quantum system dynamics is fundamental in quantum technologies. While closed systems evolve deterministically and reversibly, open systems exhibit stochastic and often irreversible behavior due to their interactions with uncontrollable environments. Accordingly, control strategies such as dynamical decoupling, quantum error correction, and reservoir engineering have been developed to suppress, correct, or even exploit the effects of openness [11].

11.4 QUANTUM NON-MARKOVIANITY

Let's start by recalling some fundamental notions of open quantum system theory. The temporal evolution of the density matrix, representing the state of an open quantum system, is described by a family of CPTP, denoted Φ_t, called the dynamic map: $\rho_t = \Phi_t\rho_0$, where ρ_0 is the initial state of the system at time $t = 0$. A dynamic map is said to be divisible if it can be expressed as the composition of two CPTP maps: $\Phi_t = \Phi_{t,t_0}\Phi_{t_0,0}$ for all $t_0 \leq t$. Non-divisibility therefore arises when there are instants t_0 for which Φ_{t,t_0} is not CPTP. All the measures of non-Markovianity presented in the following sections have in common that they are based on the non-monotonicity of certain physical quantities, a phenomenon that manifests itself when divisibility is broken. It is important to note that, although non-monotonic evolution implies non-divisibility, the converse is not always true: some non-divisible maps can exhibit monotonic dynamics. So, if we consider divisibility as a criterion for defining Markovianity, the other measures would merely be witnesses to non-Markovianity.

11.4.1 CP- Divisibility

In ref [16] have established an equivalence between Markovian dynamics and infinitesimal divisibility, while introducing a measure that quantifies the deviation from this property. Their method is based on the Choi–Jamiołkowski isomorphism [7] and uses a maximally complicated $|\beta$ state between the system under study and a suitable auxiliary system. A dynamic map $\Phi(t+s,t)$ is CP and preserves the trace (CPTP) if and only if the following condition is satisfied

$$\left[\Phi(t+s,t)\otimes id\right]|\beta\rangle\langle\beta|\geq 0 \tag{11.1}$$

The trace norm of the operator $\left[\Lambda(t+s,t)\otimes id\right]|\beta\rangle\langle\beta|$, then measures the non-CPTP nature of $\Phi(t+s,t)$. More precisely, this norm is 1 if $\Phi(t+s,t)$ is CPTP, and is strictly greater than 1 otherwise, due to the trace preservation property. In practice, we consider the derivative of this quantity, defined by

$$g(t)=\lim_{s\to 0}\frac{\left\|\left[\Phi(t+s,t)\otimes id\right]|\beta\rangle\langle\beta|\right\|_1-1}{s}. \tag{11.2}$$

The condition $g(t)>0$ thus characterizes the non-divisibility of the Φ map, as it indicates a deviation from CPTP behavior.

11.4.2 Trace Distance

Reference [10] introduces a measure of non-Markovianity based on the non-monotonic evolution of the trace distance, interpreted as backflow to the system. This approach is based on the fact that, in a divisible Markov process, the distinguishability between two states decreases continuously, while an increase indicates non-Markovian behavior

$$\sigma\left(t,\rho_0^{1,2}\right)=\frac{d}{dt}D\left(\rho_t^1-\rho_t^2\right)\leq 0 \tag{11.3}$$

where $D(\rho_1,\rho_2)=\frac{1}{2}\|\rho_1-\rho_2\|_1$ is the trace distance between two states ρ_t^1 and ρ_t^2 evolving under the influence of a certain divisible dynamical map $\Phi_t:\rho_0\to\rho_t$. The authors of reference [8] then define as non-Markovian a process for which, for certain time intervals, $\sigma\left(t,\rho_0^{1,2}\right)>0$, that is, information flows back into the system.

The criteria relying on equations (11.2) and (11.3) allow one to define a computable measure of non-Markovianity. A natural quantifier derived from the criterion of Rivas et al, reads [8]

$$N_{RHP}(\Phi_t)=\frac{\int_0^\infty g(t)}{1+\int_0^\infty g(t)} \tag{11.4}$$

The $N_{BLP}(\Phi_t)$ measure evaluates non-Markovianity as the total flow of information back to the system, maximized over all pairs of initial states

$$N_{BLP}(\Phi_t)=\max_{\rho_1,\rho_2}\int_{\frac{dD}{dt}>0}\frac{dD}{dt}dt \tag{11.5}$$

Non-Markovian processes defined according to this approach are always non-divisible. However, the converse is not necessarily true [17,13].

- It has been pointed out that the relationship between the Breuer–Laine–Piilo (BLP) and Rivas–Huelga–Plenio (RHP) non-Markovianity criteria is reminiscent of that between separable states and PPT states in entanglement theory [9].
- Any Markovian dynamics according to the RHP criterion is also Markovian according to the BLP criterion, that is, $N_{RHP}(\Phi_t) \Rightarrow N_{BLP}(\Phi_t)$, but the converse is not necessarily true.
- Comparisons illustrating this asymmetry, particularly in the context of Gaussian channels, have been discussed [18].

11.4.3 Luo, Fu, Song Measure

In reference [19], Luo, Fu, Song (LFS) measure of non-Markovianity is proposed based on the monotonicity property characterizing the time evolution of correlations between the system and an ancilla when the dynamics is divisible. It is interested in total correlations, encompassing both classical and quantum correlations, measured by quantum mutual information, and is defined by

$$I(\rho_{SA}) = S(\rho_S) + S(\rho_A) - S(\rho_{SA}) \tag{11.6}$$

where $\rho_S = \mathrm{Tr}_A(\rho_{SA})$ and $\rho_A = \mathrm{Tr}_S(\rho_{SA})$are the boundary states of the system and the ancilla, respectively, and $S(\rho) = -\mathrm{Tr}\rho\log_2\rho = -\sum_i p_i \log_2 p_i$is the von Neumann entropy of the state ρ with eigenvalues p_i. The von Neumann entropy quantifies the mixing of a state and takes values $0 \le S(\rho) \le \log_2 d$, where d is the dimension of the Hilbert space of the system and the lower and upper bounds correspond to a pure state and a maximally mixed state, respectively. The LFS measure is defined as

$$N_{LFS} = \max_{\rho_{SA(0)}} \int_{\frac{dI}{dt}>0} \frac{dI}{dt}\,dt \tag{11.7}$$

by optimizing over all possible initial states $\rho_{SA}(0)$ with an arbitrary Hilbert space for the ancilla and $\rho_{SA}(t) = (\Phi_t \otimes I)\rho_{SA}(0)$. Considering the exhaustive nature of the original optimization, the authors propose to reduce it as follows:

$$N_{LFS_0} = \int_{\frac{dI}{dt}>0} \frac{dI}{dt}\,dt \tag{11.8}$$

where $\rho_{SA}(t) = (\Phi_t \otimes I)|\psi\rangle\langle\psi|$. and $|\psi\rangle$ is an arbitrary maximally entangled state.

11.4.4 Equivalence of the Measures

Other approaches link non-Markovianity to the non-monotonicity of the quantum Fisher information, which quantifies information flow between the system and environment [16]. Unlike distance-based measures, this method does not directly rely on state-space metrics. The trace distance approach, however, examines the distinguishability of quantum states, connecting divisibility properties to the Helstrom matrix [9]. This framework naturally extends to quantum information theory by viewing dynamical maps as time-dependent channels. For instance, Bylicka et al. [12] analyzed non-Markovianity through the lens of channel capacities – specifically, the entanglement-assisted classical capacity and quantum capacity. The quantum data processing inequality [20] ensures these capacities decrease monotonically under CP-divisible maps; thus, their revival defines non-Markovianity. This connection was further explored by Buscemi and Datta [21].

11.5 OVERVIEW OF QUANTUM TELEPORTATION

Quantum teleportation is a striking quantum mechanical protocol that defies classical intuitions [22]. Unlike the fictional notion of instantaneously moving matter, it enables the transfer of only the quantum information, the exact state of one particle to another, potentially at arbitrarily large distances. This process hinges on the resource of quantum entanglement, a pair of particles is first prepared in a maximally entangled Bell state, with 1 qubit sent to the sender (Alice) and the other to the receiver (Bob). Alice then performs a joint Bell-state measurement on her half of the entangled pair and the qubit whose state she wishes to teleport. This measurement projects both qubits into one of four Bell states and yields 2 classical bits of information. Alice transmits these 2 bits to Bob via a conventional classical channel. Upon receiving them, Bob applies the corresponding Pauli correction (I, X, Z, or XZ) to his qubit, which then assumes exactly the original state of Alice's qubit despite no physical particle having traversed the space between them.

This protocol is of foundational importance in quantum information science, it underpins long-distance quantum communication networks, forms a building block for distributed quantum computing, and offers profound insights into the nonlocal structure of quantum mechanics itself. While quantum teleportation cannot (and does not) transport macroscopic objects or living beings, its ability to faithfully transmit quantum states without moving matter paves the way for advanced quantum technologies and secure communication schemes [23,24].

11.5.1 Basic Scheme of Quantum Teleportation

Alice begins with a qubit A in an unknown state ρ_{in} that she wishes to transmit to Bob. To enable teleportation, Alice and Bob share a preprepared entangled pair of qubits, B and C, supplied by an entanglement source. Alice then performs a joint Bell–state measurement on qubits A and B, projecting them into one of the four Bell states $\{|\psi^{\pm}\rangle, |\phi^{\pm}\rangle\}$. This measurement produces 2 classical bits of information, which Alice sends to Bob via a classical communication channel.

Upon receiving these 2 bits, Bob identifies which Bell outcome occurred. If the result is $|\phi^{-}\rangle$ or $|\psi^{-}\rangle$, no correction is needed and he leaves his qubit C unchanged. For outcomes $|\psi^{+}\rangle$ and $|\phi^{+}\rangle$, Bob applies the corresponding Pauli operator U to qubit C. After this conditional unitary correction, qubit C emerges in the exact same state $\rho_{\text{out}} = \rho_{\text{in}}$ that qubit A originally held, thus completing the teleportation of the quantum state without any physical transfer of the particle itself (see Figure 11.1).

11.5.2 Application for Dephasing Model

We investigate the behavior of the previously discussed quantum criteria under non-Markovian system–environment interactions, explicitly avoiding the Born–Markov approximation. This approximation presumes an infinitesimally short correlation time between the system and its environment, thereby neglecting memory effects [25,26]. We begin by examining a dephasing model subjected to colored noise, with its dissipative evolution governed by the following master equation

$$\dot{\rho}(t) = KL\rho \tag{11.1}$$

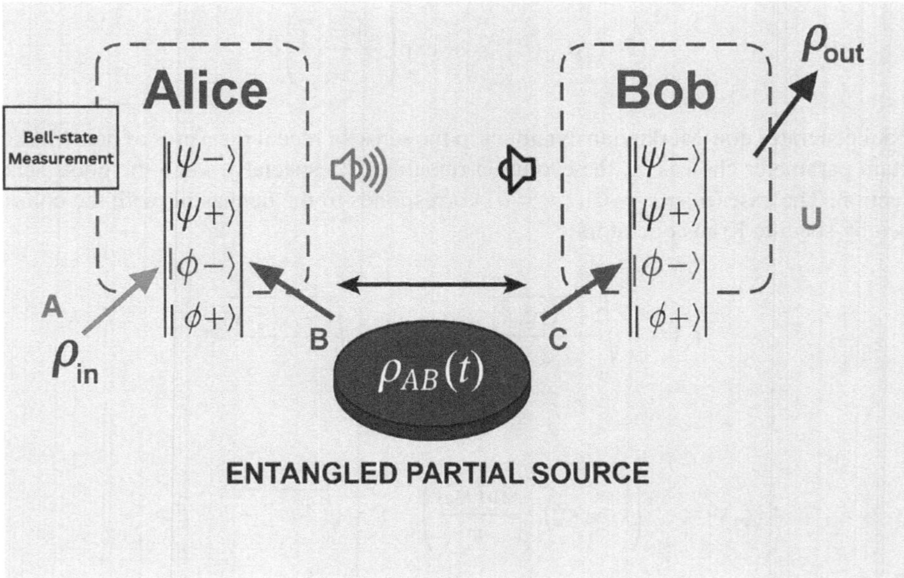

FIGURE 11.1 Quantum teleportation protocol.

The memory effects are modeled by an integral operator K whose kernel $k(t'-t)$ characterizes the nature of the non-Markovian coupling. In the absence of K, the model becomes Markovian. This formalism is suitable for describing a two-level system interacting with a reservoir containing random noise of the telegraph signal type

$$H(t) = \hbar \sum_{i=1}^{3} \Gamma_i(t)\sigma_i \tag{11.2}$$

The variables $\Gamma_i(t)$, modeled as random telegraph signals, translate the stochastic effect of an environment on a two-level system via Pauli matrices σ_i. Their behavior is governed by Poisson laws with a mean determined by τ_i, the inverse of which defines the rate of fluctuation. This formalism is relevant for describing, among other things, the effects of a random phase noise laser field on a two-level atom. Using the von Neumann equation, we obtain the time evolution of the density matrix of the system

$$\rho(t) = \rho(0) - i\int_0^t dk \sum_i \Gamma(k)\left[\sigma_i, \rho(k)\right] \tag{11.3}$$

An effective form of the evolution of the system state is obtained by inserting equation (11.3) into the von Neumann equation and applying stochastic averaging

$$\dot{\rho}(t) = \int_0^t \sum_k \exp\left(-\frac{|t-t'|}{\tau_k}\right) a_k^2 \left[\sigma_k\left[\sigma_k, \rho(t')\right]\right] dt' \tag{11.4}$$

The memory kernel derived from the correlation functions of the random telegraph signal takes the form

$$\langle \Gamma_j(t)\Gamma_k(t')\rangle = a_k^2 \exp\left(-\frac{|t-t'|}{\tau_k}\right)\delta_{jk} \tag{11.5}$$

The model generates non-Markovian dynamics in the sense of recent measures of non-Markovianity. For certain parameter choices a_k, the evolution remains CP, especially when the noise acts in only one direction. The case ($a_1 = a_2 = 0$), ($a_3 = a$) corresponds to the dephasing with the colored noise described by specific Kraus operators

$$F_1(t) = \sqrt{\frac{1+\Lambda(\xi,\kappa)}{2}} I_{2\times 2}, F_2(t) = \sqrt{\frac{1-\Lambda(\xi,\kappa)}{2}} \sigma_3 \tag{11.6}$$

where

$$\Lambda(\xi,\kappa) = e^{-\xi}\left(\cos(\kappa\xi) + \frac{\sin(\kappa\xi)}{\kappa}\right), \quad \kappa = \sqrt{(4a\tau)^2 - 1}, \quad \xi = \frac{1}{2\tau}$$

and the Kraus operators $F_i(t)$ satisfying the normalization condition $\sum_i F_i^\dagger(t) F_i(t) = I$. For bipartite systems, the time evolution of an initial density operator is described as follows

$$\rho_{AB}(t) = A|00\rangle\langle 00| + C(|00\rangle\langle 11| + |11\rangle\langle 00|) + D(|01\rangle\langle 01| + |10\rangle\langle 10|) + D|11\rangle\langle 11| \tag{11.7}$$

with $A = (1-\varsigma)/4 + \varsigma \sin^2\alpha$, $B = (1-\varsigma)/4 + \varsigma\cos^2\alpha$, $C = \varsigma\Lambda(\xi,\kappa)/2\sin 2\alpha$, $D = (1-\varsigma)/4$.

In this scenario, we use a teleportation protocol proposed by Cola and Paris [23], which allows the transfer of a two-qubit system over a single quantum channel. This protocol is based on a pair of entangled qubits, the transmission of 3 classical bits and the introduction of an auxiliary qubit by the receiver (see Figure 11.1). Standard teleportation can be compared to a general depolarization channel [27]. In the non-Markovian dephasing model, we study how the initial state affects the efficiency of the teleportation protocol. In this context, Alice and Bob share a partially complicated state, described by equation (11.7) for the dephasing model. Alice's unknown state is assumed to be of the form $\rho_{in} = |\psi_{in}\rangle\langle\psi_{in}|$, where

$$|\psi_{in}\rangle = \cos\left(\frac{\theta}{2}\right)|10\rangle + e^{-i\varphi}\sin\left(\frac{\theta}{2}\right)|01\rangle \tag{11.8}$$

with $0 \le \theta \le \pi$ and $0 \le \varphi \le 2\pi$. The parameter θ describes an arbitrary quantum state, while φ represents the phase associated with that state. At the end of the teleportation protocol, Bob receives the following state

$$\rho_{out} = \sum_{k,l=\{0,1,2,3\}} p_{kl}(\sigma_k \otimes \sigma_l)\rho_{in}(\sigma_k \otimes \sigma_l) \tag{11.9}$$

where $p_{kl} = Tr[K^k \rho_{ch}] Tr[K^l \rho_{ch}]$, σ_n (with $n = k,l$) are Pauli operators, and normalization is provided by $\sum_{kl} p_{kl} = 1$. The state ρ_{ch} represents the resource shared between Alice and Bob. The different projection operators K^l are defined as follows:

$$K^{0,3} = |\psi^{\pm}\rangle\langle\psi^{\pm}|, \qquad K^{1,2} = |\phi^{\pm}\rangle\langle\phi^{\pm}|$$

$$|\psi^{\pm}\rangle = \frac{1}{\sqrt{2}}(|00\rangle \pm |11\rangle, \qquad |\phi^{\pm}\rangle = \frac{1}{\sqrt{2}}(|01\rangle \pm |10\rangle \tag{11.10}$$

The Bell states presented above (11.10) are perfectly entangled, reflecting the total non-separability between the qubits that make them up. They have been used as quantum channels for the experimental transmission of arbitrary states between two distant parts. However, in the two models studied here, maximum entanglement is not always ensured due to decoherence induced by the interaction with the environment. Nevertheless, our results show that by adjusting the parameters of the reduced density operator it is possible to achieve a significant degree of quantum correlations. To assess the effectiveness of the teleportation protocol, it is appropriate to examine the fidelity between the input state ρ_{in} and the output state ρ_{out}, which measures their similarity. When the input state is pure, fidelity is a relevant tool for quantifying the performance of the teleportation channel. It is defined in references [24,28]. As the specific state to be teleported is generally unknown, it is more relevant to calculate the mean fidelity, which represents the average value of the fidelity associated with the four possible measurement results in the Bell [29] database. This average fidelity is given in references [24,28].

If Alice and Bob share the state $\rho_{AB}(t)$ (Equation (11.10)) with $\rho_{ch}(t) = \rho_{AB}(t)$, which characterizes the non-Markovian dephasing model, then the resulting output state is given by

$$\rho_{out}(t) = \varpi(|00\rangle\langle 00| + |11\rangle\langle 11|) + \Sigma(|10\rangle\langle 01| + |01\rangle\langle 10|) + \Upsilon(|01\rangle\langle 01| + |10\rangle\langle 10|) \tag{11.11}$$

with $\varpi = 2d(a+b), \Sigma = 2c^2 e^{-i\varphi}\sin\theta, \Upsilon = 4d^2\cos^2(\theta/2) + (a+b)\sin^2(\theta/2)$. In accordance, the fidelity of teleportation is as follows

$$F(\rho_{out}) = \left[\left(\frac{1-\varsigma}{4}\right)^2 + \frac{\varsigma^2\Lambda^2(\xi,\kappa)}{2}\sin^2 2\alpha\right]\sin^2\theta + \left(\frac{1+\varsigma}{4}\right)^2(1+\cos^2\theta) \tag{11.12}$$

The average fidelity will then be written as

$$F_{av}(\rho_{out}) = \frac{2}{3}\left(\frac{1+\varsigma}{4}\right)^2 + \frac{1}{3}\left[\left(\frac{1-\varsigma}{4}\right)^2 + \frac{\varsigma^2\Lambda^2(\xi,\kappa)}{2}\sin^2 2\alpha\right] \tag{11.13}$$

Before analyzing the results shown in Figure (11.2), it is worth recalling some previous studies that evaluated the performance of quantum teleportation using mean fidelity. A value of $F_{av} < 2/3$ indicates a failure of the teleportation process, while a result of $F_{av} > 2/3$ suggests an improvement over conventional protocols, indicating a genuine quantum advantage. The maximum value ($F_{av} = 1$) corresponds to perfect teleportation. Figure (11.2) shows the evolution of the average fidelity when the resource state shared between Alice and Bob is derived from the non-Markovian dephasing model described by state (11.7) (see Figure 11.2a, b). In both cases, the state to be teleported is initially prepared in the form (11.8), and different values of purity ς are considered.

The results shown in Figure (11.2a, b) depict the evolution of the average fidelity of the teleported state when the resource state shared between Alice and Bob is derived from the non-Markovian phase shift model. Two specific cases are studied: the Markovian regime (see Figure 11.2a) and the non-Markovian regime (see Figure 11.2b). The study highlights the crucial role of state purity in the success of the teleport protocol, with pure states showing higher average fidelity than mixed states. In the Markovian regime shown in Figure (11.2a), the average fidelity decreases progressively

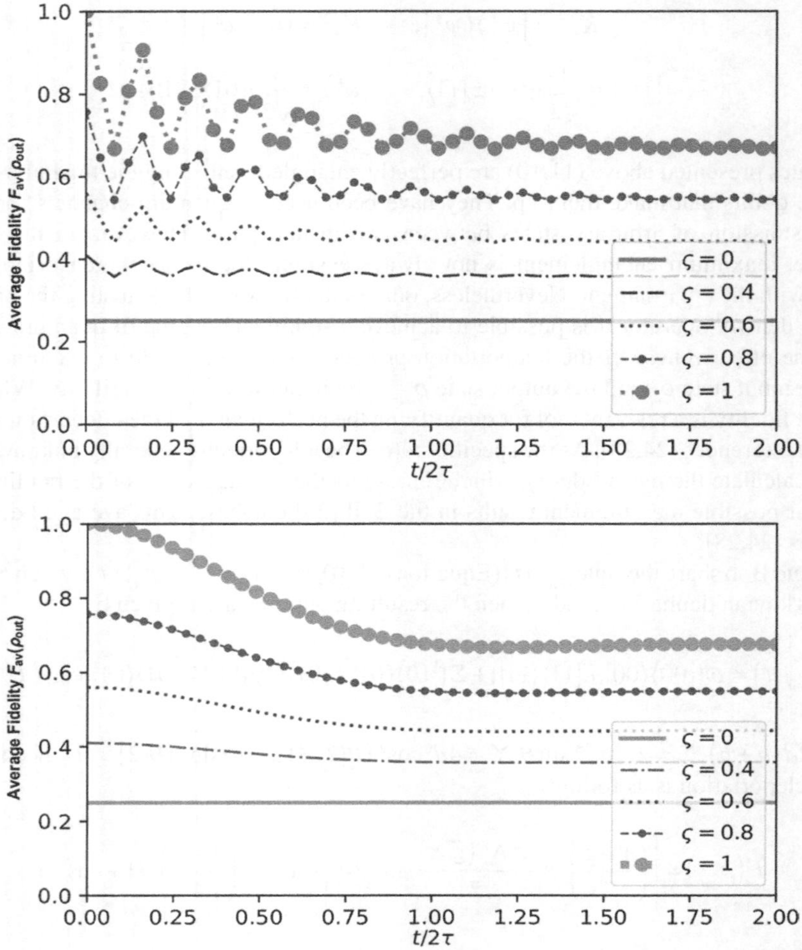

FIGURE 11.2 The average fidelity F_{av} as a function of the temps $t/2\tau$ for different values of the purity ς with $\alpha = \pi/4$; (a) reflects the Markovian regime for a non-Markovian dephasing model with $a = 1$ and $\tau = 0.5$, while (b) for non-Markovian regime with $\tau = 5$.

over time before stabilizing. This behavior indicates a gradual loss of coherence without significant feedback from the environment, characteristic of a memoryless process. Thus, teleportation is still possible, but in an attenuated form. In contrast, Figure (11.2b) shows average fidelity in the non-Markovian regime, with a characteristic parameter $\tau = 5$. This regime shows sharper short-term fluctuations. After an initial fall, the average fidelity temporarily rises to a new peak, albeit lower than the previous one. These oscillations reflect transient transitions between different system states due to the memory of the environment, leading to temporal variations in the mean loyalty.

These results highlight the challenges associated with quantum information transfer in the non-Markovian regime due to memory effects. The system retains information about its previous states, which can affect the quantum entanglement between the particles involved in the teleportation process. Thus, memory effects in the non-Markovian regime introduce additional complexity and fluctuations, making the average fidelity less stable and predictable during quantum teleportation. In contrast, the absence of memory in the Markovian regime guarantees a smoother and more efficient quantum information transfer. Our analysis shows that state purity and the presence or absence of memory effects play a crucial role in the success and efficiency of quantum teleportation. Pure states and Markovian regimes provide more favorable conditions for achieving a high average fidelity of the teleported state.

11.6 CONCLUSION

In this chapter, we have explored the foundational aspects of quantum dynamics, emphasizing the transition from classical to quantum descriptions of physical systems. We began by highlighting the significance of density operators in representing quantum states and their evolution. The use of CPTP maps was shown to be essential in describing valid quantum operations, especially when systems interact with external environments.

The study of open quantum systems reveals the limitations of traditional, Markovian assumptions and underscores the importance of accounting for *memory effects*. These non-Markovian dynamics provide a richer and more accurate framework for understanding realistic quantum processes, where the environment plays a non-negligible role. Recognizing and quantifying such effects are not only of theoretical interest but also central to developing robust and efficient quantum technologies.

Furthermore, the chapter touched on the practical implications of these concepts in quantum information protocols such as teleportation, which highlight the power of entanglement and the utility of CPTP maps in enabling classically impossible tasks. Such protocols demonstrate how foundational principles are directly connected to cutting-edge applications in quantum communication and computation.

Overall, mastering the mathematical and conceptual tools to describe quantum state evolution in both isolated and open systems is critical for anyone seeking to contribute to the rapidly evolving field of quantum science and technology. The ideas presented here lay the groundwork for deeper investigations into quantum control, decoherence, and information processing in the presence of complex environmental interactions.

REFERENCES

[1] M. A. Nielsen and I. L. Chuang, *Quantum Computation and Quantum Information.* 10th Anniversary ed. Cambridge: Cambridge University Press, 2010.

[2] M. M. Wilde, *Quantum Information Theory.* 2nd ed. Cambridge: Cambridge University Press, 2017.

[3] I. Bengtsson and K. Życzkowski. 2017. *Geometry of Quantum States: An Introduction to Quantum Entanglement.* 2nd ed. Cambridge: Cambridge University Press.

[4] V. Gorini, A. Kossakowski, and E. C. G. Sudarshan, "Completely Positive Dynamical Semigroups of N-Level Systems," *Journal of Mathematical Physics* 17 (1976): 821–825.

[5] K. Kraus, *States, Effects, and Operations: Fundamental Notions of Quantum Theory.* Berlin: Springer-Verlag, 1983.

[6] H.-P. Breuer and F. Petruccione, *The Theory of Open Quantum Systems.* New York: Oxford University Press, 2002.

[7] M.-D. Choi, "Completely Positive Linear Maps on Complex Matrices," *Linear Algebra and Its Applications* 10 (1975): 285–290.

[8] Á. Rivas, S. F. Huelga, and M. B. Plenio, "Entanglement and Non-Markovianity of Quantum Evolutions," *Physical Review Letters* 105, no. 5 (2010): 050403. https://doi.org/10.1103/PhysRevLett.105.050403

[9] D. Chruściński, A. Kossakowski, and Á. Rivas, "Measures of Non-Markovianity: Divisibility versus Backflow of Information," *Physical Review A* 83, no. 5 (2011): 052128. https://doi.org/10.1103/PhysRevA.83.052128

[10] H.-P. Breuer, E.-M. Laine, and J. Piilo, "Measure for the Degree of Non-Markovian Behavior of Quantum Processes in Open Systems," *Physical Review Letters* 103, no. 21 (2009): 210401. https://doi.org/10.1103/PhysRevLett.103.210401

[11] H.-P. Breuer and F. Petruccione, *The Theory of Open Quantum Systems.* Oxford: Oxford University Press, 2002.

[12] B. Bylicka, D. Chruściński, and S. Maniscalco, "Non-Markovianity and Reservoir Memory of Quantum Channels: A Quantum Information Theory Perspective," *Scientific Reports* 4 (2014): 5720. https://doi.org/10.1038/srep05720

[13] M. J. W. Hall, J. D. Cresser, L. Li, and E. Andersson, "Canonical Form of Master Equations and Characterization of Non-Markovianity," *Physical Review A* 89 (2014): 042120.

[14] J. Liu, X.-M. Lu, and X. Wang, "Nonunital Non-Markovianity of Quantum Dynamics," *Physicql Review A* 87 (2013): 042103.

[15] Á. Rivas and S. F. Huelga, *Open Quantum Systems: An Introduction.* Berlin: Springer, 2012.

[16] T. Yu and J. Eberly, "Entanglement Evolution in a Non-Markovian Environment," *Optics Communications* 283 (2010): 676–680.

[17] H.-P. Breuer, E.-M. Laine, J. Piilo, and B. Vacchini, "Colloquium: Non-Markovian Dynamics in Open Quantum Systems," *Reviews of Modern Physics* 88 (2016): 021002.

[18] R. Vasile, S. Maniscalco, M. G. A. Paris, H.-P. Breuer, and J. Piilo, "Quantifying Non-Markovianity," *Physical Review A* 84 (2011): 052118.

[19] S. Luo, S. Fu, and H. Song, "Quantifying Non-Markovianity via Correlations," *Physical Review A* 86 (2012): 044101.

[20] M. A. Nielsen and I. L. Chuang, *Quantum Computation and Quantum Information.* Cambridge: Cambridge University Press, 2000.

[21] F. Buscemi and N. Datta, "Equivalence between divisibility and monotonic decrease of information in classical and quantum stochastic processes, " *Phys. Rev. A,* 93 (2016) 012101.

[22] C. H. Bennett, G. Brassard, C. Crépeau, R. Jozsa, A. Peres, and W. K. Wootters, "Teleporting an Unknown Quantum State via Dual Classical and Einstein-Podolsky-Rosen Channels," *Physical Review Letters* 70 (1993): 1895.

[23] M. M. Cola and M. G. A. Paris, "Teleportation of Bipartite States Using a Single Entangled Pair," *Physics Letters A* 337 (2005): 10–16.

[24] M. E. Kirdi, A. Slaoui, H. E. Hadfi, and M. Daoud, "Efficient Quantum Controlled Teleportation of an Arbitrary Three-Qubit State Using Two GHZ Entangled States and One Bell Entangled State," *Journal of Russian Laser* 44 (2023): 121–134.

[25] S. Daffer, K. Wódkiewicz, J. D. Cresser, and J. K. McIver, "Depolarizing Channel as a Completely Positive Map with Memory," *Physical Review A* 70 (2004): 010304.

[26] J. P. Pinto, G. Karpat, and F. F. Fanchini, "Sudden Change of Quantum Discord for a System of Two Qubits," *Physical Review A* 88 (2013): 034304.

[27] G. Bowen and S. Bose, "Teleportation as a Depolarizing Quantum Channel, Relative Entropy, and Classical Capacity," *Physical Review Letters* 87 (2001): 267901.

[28] Y. Dakir, A. Slaoui, A.B.A. Mohamed, R. Ahl Laamara and H. Eleuch, "Quantum Teleportation and Dynamics of Quantum Coherence and Metrological Non-Classical Correlations for Open Two-qubit Systems," *Scientific Reports* 13 (2023): 20526.

[29] M. D. Bowdrey, D. K. Oi, A. J. Short, K. Banaszek, and J. A. Jones, "Fidelity of Single Qubit Maps," *Physics Letters A* 294 (2002): 258–260.

12 Geometric Science of Quantum Systems

Oussama Latifi, Mouhcine Yachi, Brahim Amghar, Rachid Ahl Laamara, and Abdallah Slaoui

12.1 INTRODUCTION

It is well known that most theories of classical physics may be formulated in geometric language. For example, Hamiltonian mechanics is based on symplectic geometry, General Relativity is grounded in Riemannian geometry, and classical Yang–Mills theory utilizes the elegant theory of fiber bundles. These geometric frameworks provide powerful tools and deep insights into the structure and behavior of physical systems.

On the other hand, quantum mechanics has traditionally been formulated through linear operators in Hilbert space, relying on algebraic and functional analytic methods. However, recent advancements have shown that even quantum mechanics can be expressed in an elegant geometric language. It turns out that geometry plays a fundamental role in quantum mechanics, as the space of quantum states is endowed with geometric structures. The geometrical approach to quantum mechanics has been particularly successful in understanding the unitary evolution of quantum states. The states of a quantum system are represented by rays in a complex Hilbert space, leading to a geometrical formulation of the postulates of quantum mechanics. For example, the dynamics of a quantum system can be understood by investigating manifolds that contain all states reachable during the system's evolution. A notable case is the two-level system, or qubit, where the state space can be represented by a two-dimensional (2D) sphere known as the Bloch sphere. The trajectory of quantum evolution between two states is a curve between two points on this sphere. The trajectory of quantum evolution between two states is a curve between two points on this sphere. Similarly, beam splitters can generate entangled coherent states in optical systems, whose geometric and metrological properties have been studied using symplectic structures and quantum Fisher information [1].

Using geometrical methods in quantum mechanics often simplifies the solution of many problems. For instance, the problem of finding the Hamiltonian for a specific quantum evolution can be addressed effectively using geometric approaches.

Additionally, the Zermelo navigation problem and the quantum brachistochrone problem for an arbitrary spin in a magnetic field have been solved using the geometrical properties of rotational manifolds.

Studies [2–4] have shown that the problem of finding the quantum circuit of unitary operations that provides time-optimal evolution in a system of qubits is closely related to finding the minimal distance between two points on a Riemannian metric. This connection between geometry and quantum computing highlights the importance of geometric methods in designing efficient quantum algorithms and circuits.

Berry's discovery of the geometric phase in adiabatic quantum processes has led to significant progress in understanding the geometric nature of quantum-state evolution. Aharonov and Anandan extended Berry's phase to nonadiabatic cyclic evolution by introducing the projective Hilbert space

DOI: 10.1201/9781003621423-14

$\mathcal{H}$, which is the set of rays in the Hilbert space. In their formulation, the geometric phase is the holonomy associated with the parallel transport along a closed curve in $\mathcal{H}$, rather than in the parameter space considered by Berry. This formulation was further advanced by Samuel and Bhandari, and now, the concept of a geometric phase is known to be free not only from the adiabaticity condition but also from the cyclicity or unitarity of the evolution.

The origin of the geometric phase lies in the first-order structure of the overlap, that is, the transition amplitude of normalized states:

$$\langle \psi(\lambda) | \psi(\lambda + d\lambda) \rangle, \tag{12.1}$$

with respect to dλ. Here, λ is a parameter of the curve in $\mathcal{H}$ and is not necessarily identified with time. Works of Anandan and Aharonov considered the second-order structure and found a remarkable relation between uncertainty and the Fubini–Study metric along the evolution curve in $\mathcal{H}$:

$$\mathrm{d}s^2 = 1 - \left| \langle \psi(\lambda) | \psi(\lambda + \mathrm{d}\lambda) \rangle \right|^2 \tag{12.2}$$

They showed that the "velocity" of evolution generated by the operator $A(\lambda)$, depending on λ in general, along the curve is given by:

$$\frac{ds}{d\lambda} = \Delta \mathrm{A}(\lambda) \tag{12.3}$$

where ΔA is the variance of the operator A. This indicates that the uncertainty in quantum mechanics acquires a geometric interpretation, which is an intrinsic characteristic of quantum mechanics itself and independent of specific models.

In conclusion, the geometrical approach to quantum mechanics not only provides a deeper understanding of quantum state evolution but also has significant implications for quantum computing. The geometric phase, in particular, plays a crucial role in the development of efficient quantum algorithms and circuits. This thesis will explore the geometrical description of integrable quantum systems and its relevance to quantum computing, with a focus on the role of geometric phases and the structures underlying quantum mechanics.

12.2 DYNAMICAL ALGEBRAIC STRUCTURE OF QUANTUM SYSTEMS

It is well established that for a quantum system, there is always a well-defined group structure [5]. According to the quantum mechanical axioms, a quantum system is a self-adjoint operator algebra, denoted by g and spanned by the basic physical observables [6, 7]. The algebraic structure is given by commutation relations. Explicitly, the system's Hamiltonian H and the various transition operators $\{\mathrm{A}\}$ can be expressed as functions of a set of the basic operators $\{T_i, i = 1, \ldots, n\}$ of g:

$$H = H(T_i), \quad A = A(T_i) \tag{12.4}$$

This set of operators satisfies a closed commutation relationship,

$$\left[T_i, T_j\right] = \sum_{k-1}^{n} C_y^k T_k \tag{12.5}$$

namely, it spans an algebra where the C_{ij}^k's are the structure constants of g.

Next, we introduce the concept of dynamical group. A dynamical group, denoted by G, is defined by a unitary exponential mapping of g,

$$g \rightarrow G := \left\{ \exp\left(i \sum_i \sigma_i T_i \right) \right\}, \tag{12.6}$$

where unitarity is demanded by quantum mechanics. Mathematically, G is called the covering group of g. According to group representation theory [5], the state space in quantum mechanics $\mathcal{H}$ is a representation space of$g$. Group theory tells us that $\mathcal{H}$ can always be decomposed into a direct sum of the various unitary irreducible subspaces $\xi_\wedge$ of g:

$$\mathcal{H} = \oplus_\wedge d_\wedge \xi_\wedge \tag{12.7}$$

where Λ is a vector which characterizes an irreducible representation (irrep) of g.

When g is a simple Lie algebra, Λ is the highest weight of this irrep [8], d_Λ is the degeneracy of Λ in $\mathcal{H}$.

Different $\mathcal{H}_\Lambda$s cannot mix because the matrix element of any function $f(T_i)$ in $|\psi_\Lambda\rangle$ and $|\psi_{\Lambda'}\rangle (\Lambda \neq \Lambda')$ vanishes.

Therefore, without any loss of generality, we can confine our discussion to a relevant irreducible subspace ξ_Λ of $\mathcal{H}$, and simply denote ξ_Λ as $\mathcal{H}$. Thus, we shall refer to $\mathcal{H}$ as an irrep space of g.

12.3 QUANTUM DYNAMICAL DEGREES OF FREEDOM

The dynamical properties of a quantum system are closely linked to its degrees of freedom, which are determined by the system's group structure. This concept is crucial for understanding both classical and quantum integrability.

Selecting an appropriate basis for the Hilbert space $\mathcal{H}$ is essential for studying quantum dynamics. This choice reflects the system's physical properties and influences our understanding of quantum integrability.

12.3.1 Commuting Operators

For many quantum systems, the operator algebras are Lie algebras. Consider a Lie algebra g with rank l and dimension n. For each subalgebra chain $\{g^1, g^2, \ldots\ldots, g^k\}$, there exists a complete set of commuting operators (CSCO) $\{Q_i^\alpha\}$, where

$$\left[Q_i^\alpha, Q_j^\alpha \right] = 0, \quad i, j = 1, \ldots, d_\alpha \tag{12.8}$$

The total number of operators in the CSCO, d_α is chain-independent and given by:

$$d_\alpha = l + \frac{n-l}{2} = d \tag{12.9}$$

12.3.2 Quantum Numbers and Degeneracy

The eigenstates of the CSCO define the quantum numbers of the system:

$$Q_i^\alpha \left| \gamma^\alpha \right\rangle = \gamma_i^\alpha \left| \gamma^\alpha \right\rangle, \qquad i = 1, \ldots, d \tag{12.10}$$

The CSCO typically includes two classes of quantum numbers:

- Fully degenerate set
- Non-fully degenerate (NFD) set

Operators in the CSCO that do not satisfy the equation:

$$Q_i^{\alpha}\left|\psi\right\rangle = c\left|\psi\right\rangle, \tag{12.11}$$

for all states $|\psi\rangle \in \mathcal{H}$ are called NFD operators. These NFD quantum numbers are crucial for specifying the basis of $\mathcal{H}$. For a compact system S, the ground state $|\psi_0\rangle$ corresponds to the minimal eigenvalue:

$$\left|\psi_0\right\rangle = \left|\Lambda, -\Lambda\right\rangle, \tag{12.12}$$

where $\wedge$ denotes the maximal eigenvalue. In non-compact systems, $|\psi_0\rangle$ represents a lower bound. Given a fixed state $|\psi_0\rangle$, operators $X^{\dagger}$ are considered elementary excitation operators if and only if :

$$X_i^{\dagger}\left|\psi_0\right\rangle \neq \alpha\left|\psi_0\right\rangle, \tag{12.13}$$

for any scalar α. The system's states $|\psi\rangle$ can be generated as:

$$\left|\psi\right\rangle = P\left(X_i^{\dagger}\right)\left|\psi_0\right\rangle, \quad \left|\psi\right\rangle \in S_{\Lambda} \tag{12.14}$$

Here, $P\left(X_i^{\dagger}\right)$ represents a polynomial of $X_i^{\dagger}$. The number of independent elementary excitation operators is related to the system's degrees of freedom. This formulation provides a physical realization of fundamental quantum dynamics.

12.4 GEOMETRIC STRUCTURES OF QUANTUM SYSTEMS

Classical trajectories exist in manifolds with symplectic structures (phase spaces). For a quantum system to exhibit a classical limit, it must generate such a structure. We need to construct a geometry from the Hilbert space that exhibits the necessary symplectic properties.

After the degrees of freedom of quantum dynamics have been defined, it is now possible to construct the manifold associated with the space of quantum states, that is, the quantum phase space, which is equivalent to the phase space in classical mechanics. For this purpose, we define the space of generalized displacement operators as the quotient space G/H, where C is the dynamical group of the quantum system, and H denotes the maximal subgroup of H that leaves the ground state $|\psi_0\rangle$ invariant. Thus, for an integrable quantum system $\left(G, g, \xi_{\wedge}\right)$ with m quantum degrees of freedom, The manifold of quantum states is obtained by a unitary exponential application of the subspace of elementary excitation operators and their complex conjugates $k = \left\{X_i^{\dagger}, X_i, i = 1,2,\ldots,m\right\}$ in the quotient space G/H. This is expressed as:

$$\exp: k \rightarrow G/H \tag{12.15}$$

$$\sum_{i=1}^{m}\left(\eta^i X_i^{\dagger} - \text{h.c}\right) \rightarrow \Omega = \exp\sum_{i=1}^{m}\left(\eta^i X_i^{\dagger} - \text{h.c}\right) \tag{12.16}$$

here, $\eta_i\ (i = 1,\ldots,m)$ are complex parameters, and Ω is a generalized displacement operator, that is, an element of the quantum state manifold G/H. It is also possible to manipulate this geometric manifold in terms of generalized coherent states of the group G. Indeed, there exists a correspondence between the of quantum states manifold $\Omega \in G/H$ and the manifold introduced by generalized coherent states. In other words, by acting with a unitary element $\Omega \in G/H$ on the ground state $|\psi_0\rangle$, we can generate the corresponding coherent state:

$$\Omega \to |\Omega\rangle \equiv \Omega|\psi_0\rangle, \tag{12.17}$$

which shows that the manifold induced by the coherent states is isomorphic to the coset space G/H, while the generalized coherent states offer a continuous basis for spanning $\mathcal{S}$. However, whether G/H is the desired manifold depends on whether or not it is a complex space with a symplectic structure. The complex structure is the fundamental criterion of quantum mechanics, and the symplectic structure is demanded by the existence of a phase space. First, we shall show that G/H is a complex space. We shall confine our attention to the semisimple Lie groups whose g satisfies the Cartan canonical decomposition

$$[\mathrm{k},\mathrm{k}] \subset \mathrm{k}, \quad [\mathrm{k},\mathrm{p}] \subset \mathrm{p}, \quad [\mathrm{p},\mathrm{p}] \subset \mathrm{k} \tag{12.18}$$

the dimensions for k and p are l and m, respectively.

For most applications, G/H will be satisfied by Equation 12.18. In this case, Equation 12.16 can be expressed in terms of $l \times \mathrm{m}$ block matrix representation,

$$\exp\left(\sum_{i=1}^{m}\left(\eta^i X_i^\dagger - \mathrm{h.c.}\right)\right) \to \exp\begin{pmatrix} 0 & \eta \\ \pm\eta^\dagger & 0 \end{pmatrix} = \begin{pmatrix} \sqrt{I \mp \mathrm{X}^\dagger\mathrm{X}} & \mathrm{X} \\ \mp\mathrm{X}^\dagger & \sqrt{I \mp \mathrm{X}^\dagger\mathrm{X}} \end{pmatrix} \tag{12.19}$$

where η's and X's are $l \times \mathrm{m}$ matrices,

$$\mathrm{X} = \eta\frac{\sin\left(\sqrt{\eta^\dagger\eta}\right)}{\sqrt{\eta^\dagger\eta}} \text{ for compact } G, \quad X = \eta\frac{\sinh\left(\sqrt{\eta^\dagger\eta}\right)}{\sqrt{\eta^\dagger\eta}} \text{ for noncompact G} \tag{12.20}$$

and the $+(-)$ sign corresponds to a non-compact (compact) G. It has been shown that a group transformation, that is, $\mathrm{u} = \begin{bmatrix} A & B \\ C & D \end{bmatrix} \in \mathrm{G}$, on G/H is a holomorphic mapping of G/H onto itself,

$$u\Omega\colon Z' = (AZ + B)(CZ + D)^{-1}, \quad Z = X\sqrt{I \mp X^\dagger X} \tag{12.21}$$

Equation 12.21 tells us that G/H is a complex manifold. A complex manifold can be understood in physical terms, namely the wavefunctions should only depend on $\{z\}$ (or $\{z^*\}$), analogous to the dependence on coordinates (or momenta), but it cannot depend on both. Hence, G/H must be a complex space if it is a manifold of the Hilbert space.

Next, it can be shown that G/H has a symplectic structure. Since it is a complex space, there exists at least a Hilbert space of functions $\bar{L}^2$ (G/H) on G/H. Let us denote these functions as $f_1, f_2, \ldots$.. to be any orthonormal basis of $\bar{L}^2(G/H)$. A kernel $\mathrm{K}(z, z^*)$, called the Bergmann kernel, can be defined as follows [9]:

$$\mathrm{K}(\mathrm{z},\mathrm{z}^*) = \sum_n f_n(\mathrm{z}) f_n^*(\mathrm{z}) \tag{12.22}$$

The metric of G/H is then defined by

$$ds^2 = \sum_{ij} g_{ij} dz^i dz^{j*} = \sum_{ij} \left(\frac{\partial^2 ln\, K(z,z^*)}{\partial z^i \partial z^{j*}} \right) dz^i dz^{j*} \tag{12.23}$$

According to differential geometry [10], there exists a closed non-degenerate two-form ω on G/H whose explicit form in complex local coordinate system is

$$\omega = i\hbar \sum_{ij} g_{ij} dz^i \Lambda dz^{j*} \tag{12.24}$$

and the corresponding Poisson bracket is

$$\{f,g\} = \frac{1}{i\hbar} \sum_{ij} g^{ij} \left(\frac{\partial f}{\partial z^i} \frac{\partial g}{\partial z^{j*}} - \frac{\partial g}{\partial z^i} \frac{\partial f}{\partial z^{j*}} \right) \tag{12.25}$$

here f and g are two functions defined on G/H and $g_{ij} g^{jk} = \delta_{ik}$. Thus, the existence of symplectic structure is demonstrated. It should be noted that in Equation 12.24, we have added a factor $i\hbar$. This is because we want to embed G/H into a phase space with physical dimension.

An explicit construction of the symplectic structure is as follows. Using the coherent state theory, a simple and useful realization of an orthogonal basis $\{f_n(z)\}$ is

$$f_\gamma(z) = \langle \gamma \| z \rangle, \tag{12.26}$$

where $\{|\gamma\}$ is a basis of $\mathcal{S}$, and $\|z\rangle$ the unnormalized form of the coherent states $|z\rangle$,

$$|\Omega\rangle = N^{-1/2}(z,z^*) \exp\left(\sum_{i=1}^{m} (z^i X_i^\dagger) \right) |\psi_0\rangle \equiv N^{-1/2}(z,z^*) \|z\rangle \tag{12.27}$$

It can be shown that the Bergmann kernel is the normalization constant:

$$\mathrm{K}(z,z^*) = N(z,z^*) = \langle z \| z \rangle = |\langle \psi_0 | \Omega \rangle|^{-2} \tag{12.28}$$

Hence, the geometrical structure of G/H can be obtained directly from the coherent states. When g is a semisimple Lie algebra and satisfies the decomposition of (12.18), the explicit function of the Bergmann kernel is given as follows [11]:

$$\mathrm{K}(z,z^*) = \det(\mathrm{I} + \mathrm{Z}^\dagger \mathrm{Z})^{\pm \Xi}, \tag{12.29}$$

where the $+(-)$ sign corresponds to the compact (non-compact) G, and Ξ (referred to as the quenching index hereafter) is defined by $h_i |\psi_0\rangle = \mp \Xi |\psi_0\rangle$ or $0, h_i \in k$.

The index Ξ comes from the mapping $G \to G/H$ induced by the fixed point set $|\psi_0\rangle$. This is the geometrical origin of this topological index. Also, each Ξ specifies an irrep of G and represents a quantum system.

By introducing the canonical coordinates (q, p) of G/H,

$$\frac{1}{\sqrt{2h\Xi}} (\mathrm{Q} + i\mathrm{P}) = \mathrm{X}, \tag{12.30}$$

where X is defined by Equation 12.19, the Poisson bracket of Equation 12.25 can be transformed into the standard form [12]

$$\omega = \sum_{i=1}^{m} dp^i \Lambda dq^i, \tag{12.31}$$

$$\{f,g\} = \sum_{i=1}^{m}\left(\frac{\partial f}{\partial q^i}\frac{\partial g}{\partial p^i} - \frac{\partial g}{\partial q^i}\frac{\partial f}{\partial p^i}\right). \tag{12.32}$$

This offers a familiar phase space structure.

As G/H has a complex structure, thus satisfying the demands of quantum mechanics, and has a symplectic structure which allows the development of classical mechanics on it, we shall therefore call G/H a quantum phase space.

12.5 DIRECT CONSTRUCTION OF THE MANIFOLD OF QUANTUM STATES

Now, we will discuss a method of constructing the manifold of quantum states that stems directly from the Hilbert space. Indeed, starting from the fact that the Hilbert space does not correctly describe the states of quantum systems and their evolutions, given that it conflicts with Born's probabilistic interpretation, the space that could be appropriate for describing pure quantum states must necessarily have its elements invariant under $U(1)$ gauge transformation. For a quantum system of $n+1$ independent states, the Hilbert space can be identified with the complex space $\mathbb{C}^{n+1}$. Using the fact that the set of states that are defined up to a phase factor constitutes a single physical state, that is, a class of equivalent states, the physical space accessible for the description of quantum systems is the complex projective space $\mathbb{C}P^n$, also called the projective Hilbert space.

12.5.1 Complex Projective Space

It has been considered that there exists a geometry description of quantum mechanics other than the conventional description in Hilbert space. Here we present the idea that the state space can be viewed as a complex projective space if the number of dimensions of the Hilbert space is finite, and from this point of view we are able to get some principles and dynamical features of conventional quantum mechanics.

A complex projective space $\mathbb{C}P^n$ is defined as follow [10]: it is the set of all the lines that pass through the origin of the $\mathbb{C}^n$ space, with an equivalent relation that every single line is regarded as a point in this multiset; namely, we have an equivalent relation $z \sim cz$ in the $\mathbb{C}^{n+1}$ space

$$\left(z^0, z^1, \ldots, z^n\right) \sim \left(cz^0, cz^1, \ldots, cz^n\right), \forall c \in \mathbb{C} - \{0\} \tag{12.33}$$

These $z^0, z^1, \ldots, z^n$ cannot be zero simultaneously. We can write this equivalent relation

$$[z] = \left[z^0, z^1, \ldots, z^n\right] \tag{12.34}$$

Every $[z]$ represents a point in $\mathbb{C}P^{n+1}$, to be precise, every point in $\mathbb{C}P^n$ has a set of homogeneous coordinates $\left[z^0, z^1, z^k\right]$. Some features of $\mathbb{C}P^n$. Points in $\mathbb{C}P^n$ are equivalent to points in the spherical surface S^{2n+1} of $\mathbb{C}^{n+1}$ space, multiplied with a phase factor of $\{Z^k\} \sim \{e^{i\alpha}Z^k\}$. Therefore,

$$\mathbb{C}P^n \cong S^{2n+1}/U(1) \cong \mathbb{C}^n \cup \mathbb{C}P^{n-1} \tag{12.35}$$

On the other hand, the transitive group of $\mathbb{C}P^n$ is $SU(n+1)$, viz., for every $g \in SU(n+1), p,q \in \mathbb{C}P^n$, there is a map

$$\varphi: SU(n+1) \times \mathbb{C}P^n \rightarrow \mathbb{C}P^n \tag{12.36}$$

$$p \rightarrow \varphi_g(p) = q$$

The isotropic group of $\mathbb{C}P^n$ P^n is the $U(n)$ group, hence

$$\mathbb{C}P^n \cong SU(n+1)/U(n) \tag{12.37}$$

According to the basic principles of quantum mechanics, all the eigenstates of a dynamical operator can form a complete set of basis vectors of a Hilbert space, and any quantum state of a quantum system can be regarded as a vector in this Hilbert space. For example, we have a dynamical operator A, with its eigenequation $A|k\rangle = A_k|k\rangle, A_k$, is the eigenvalue of A, and $|k\rangle$ is the corresponding eigenstate. Then an arbitrary quantum state of a quantum system can be written as

$$|\psi\rangle = \sum_k \langle k|\psi\rangle\langle k| = \sum_k a_k|k\rangle, \tag{12.38}$$

where $ak = \langle k|\psi\rangle$.

We restrict our discussion to the condition that these eigenstates are discrete. We can see the quantum state $|\psi\rangle$ becomes a vector in a Hilbert space, whose basis vectors are the $|k\rangle$s. That is to say, every quantum state $|\psi\rangle$ has a corresponding vector in the Hilbert space, whose coordinates are $(a_1, a_2, \ldots, a_k)$. Usually speaking, the number of the dimensions of this space is infinite, but we only discuss the condition of finite dimensions here.

We know that a state is specified by the direction of the vector, and any length one may assign to the vector is irrelevant. To be exact, if we multiply the vector with an arbitrary complex number c, the resulting state must be the same as the original one apart from the special case that c equals to zero, so that the vector $c|\psi\rangle$ must correspond to the same state as $|\psi\rangle$ does. Therefore, we can define an equivalent relation:

$$(a_1, a_2, \ldots, a_{n+1}) \sim (ca_1, ca_2, \ldots, ca_{n+1}), \forall c \in \mathbb{C} - \{0\}, \tag{12.39}$$

and write this equivalent relation as

$$[a] = [a_1, a_2, \ldots a_{n+1}] \tag{12.40}$$

If we look back at the definition of $\mathbb{C}P^n$, we will realize that we actually defined the homogeneous coordinates of a point in this $\mathbb{C}P^n$. Thus, a quantum state $|\psi\rangle$ actually corresponds to a point in $\mathbb{C}P^n$, with its homogeneous coordinates being $[a1, a2, \ldots\ldots\ldots, a_{n+1}]$. It is important to notice that the complex numbers $z^0, z^1, \ldots, z^n$ in the definition of $\mathbb{C}P^n$ cannot be zero simultaneously. For the state $|\psi\rangle$, if $a_1, a_2, \ldots, a_{n+1}$ all equal to zero, then $|\psi\rangle = 0$, and this does not make any sense. Hence, the demand from the definition of $\mathbb{C}P^n$ that $a_1, a_2, \ldots, a_{n+1}$ cannot be zero simultaneously is satisfied automatically.

Whose elements constitute the set of complex lines $\mathbb{C}P^n$ representing the set of classes of equivalent states under the $U(1)$ gauge transformation. The kth complex line can be represented by

$$[\psi]_k = [\overline{z_0}, \overline{z_1}, \overline{z_2}, \ldots, \overline{z_{k+1}}, \ldots, \overline{z_n}] = \{|\psi\rangle ie^{i\phi}, 0 \leq \phi \leq 2\pi\}, \tag{12.41}$$

where $\overline{z_i} = \frac{\overline{a_i}}{a_k}$ are the local coordinates. Thus, the complex projective space $\mathbb{CP}^n$, that is, the projective Hilbert space, is constructed by $n+1$ open sets, that is, $n+1$ complex lines.

From a topological point of view, we say that it is covered by $n+1$ charts $\mathcal{U}_k$, $k = 0,1,\ldots,n$ such that

$$\mathcal{U}_k : \mathbb{C}^{n+1} \to \mathbb{CP}^n, \tag{12.42}$$

$$\left(\overline{a_0},\overline{a_1},\overline{a_2},\ldots,\overline{a_n}\right) \to k = \left[|\psi\rangle\right]k = \left[\overline{z_0},\overline{z_1},\overline{z_2},\ldots,\overline{z_{k+1}},\ldots,\overline{z_n}\right] \tag{12.43}$$

According to the theory of coherent states, the kth complex line is represented by a coherent state of the dynamic group $SU(n+1)$

$$|\psi\rangle = \left(z_0, z_1, z_2, \ldots, z_{k-1}, z_{k+1}, \ldots, z_n\right)^T, \tag{12.44}$$

where $K(\overline{z}, z^*) = 1 - |\overline{z}k|^2 + \sum_{i=0}^{n} |z_i|^2$ is the Bergmann kernel defined on $\mathbb{CP}^n$. Thus, we arrive at explicating the construction of the manifold of quantum states obeying Born's probabilistic interpretation, that is, verifying the $U(1)$ gauge invariance.

Consequently, it can be considered as the appropriate space to describe quantum dynamics instead of the Hilbert space.

12.5.1.1 Illustrative Case: Electron Spin

To elucidate the preceding concepts, let us examine the spin of an electron, a fundamental quantum property. This example, reminiscent of discussions on spin-1/2 particles in quantum mechanics literature, serves to bridge theory with practical application. Consider the spin eigenstates of a single electron, conventionally denoted as $|+1/2\rangle$ and $|-1/2\rangle$. A general spin state $|\psi\rangle$ of the electron can be expressed as a superposition:

$$|\psi\rangle = \alpha|+1/2\rangle + \beta|-1/2\rangle, \tag{12.45}$$

here, $\alpha = \langle +1/2|\psi\rangle$ and $\beta = \langle -1/2|\psi\rangle$ are complex coefficients.

Reflecting on our earlier discussion, we can interpret this quantum state $|\psi\rangle$ as a vector in a 2D complex space, with (α, β) as its coordinates. An important quantum mechanical principle is that the overall magnitude of this state vector is not physically significant. This leads us to define an equivalence relation:

$$(\alpha, \beta) \sim (c\alpha, c\beta), \qquad \forall c \in \mathbb{C}\backslash\{0\} \tag{12.46}$$

This equivalence allows us to represent the state using homogeneous coordinates $[\alpha, \beta]$, which corresponds to a point in the complex projective space $\mathbb{CP}^1$. Consequently, we can assert that each possible spin state of an electron maps to a unique point in $\mathbb{CP}^1$.

It's worth noting that spin is intimately connected to the $SU(2)$ group in quantum theory. Our earlier derivation showed that $\mathbb{CP}^n \cong SU(n+1)/U(n)$, which in this case yields $\mathbb{CP}^1 \cong SU(2)/U(1)$. This mathematical relationship provides an alternative perspective on the connection between electron spin and the $SU(2)$ group, reinforcing the deep links between quantum mechanics and group theory.

12.6 GEOMETRICAL STRUCTURES ON A MANIFOLD OF QUANTUM STATES

12.6.1 Symplectic Structure

To explore the symplectic structure on $\mathcal{PH}$ as an important geometric structure ensuring the existence of classical dynamics within quantum dynamics, we must first define an interesting geometric quantity called the Hermitian tensor of Dirac. This tensor naturally induces a tensor called the quantum geometric tensor. It is locally defined by [13]

$$Q_{\mu\nu}(\lambda) = \langle D_\mu \psi | D_\nu \psi \rangle \quad (12.47)$$

where $D_\mu = \partial_\mu - i\,\tilde{A}_\mu$ is the covariant derivative on $\mathcal{PH}$. It is interesting to note that the quantum geometric tensor (12.47) is a complex quantity invariant under $U(1)$ gauge. Explicitly, it is written in terms of the evolved quantum state as follows

$$Q_{\mu\nu}(\lambda) = \langle \partial_\mu \psi(\lambda) | \partial_\nu \psi(\lambda) \rangle - \langle \partial_\mu \psi(\lambda) | \psi(\lambda) \rangle \langle \psi(\lambda) | \partial_\nu \psi(\lambda) \rangle \quad (12.48)$$

The quantum geometric tensor decomposes into two essential parts: the first concerns its real part which constitutes the Fubini–Study metric. It can also be expressed in terms of the quantum geometric tensor in the form

$$g_{\mu\nu} = \mathrm{Re}\left(Q_{\mu\nu}\right) = \mathrm{Re}\left(\langle \partial_\mu \psi(\lambda) | \partial_\nu \psi(\lambda) \rangle - \langle \partial_\mu \psi(\lambda) | \psi(\lambda) \rangle \langle \psi(\lambda) | \partial_\nu \psi(\lambda) \rangle\right) \quad (12.49)$$

As for the imaginary part of the quantum geometric tensor called Berry curvature. It is defined as

$$F_{\mu\nu} = -2\,\mathrm{Im}\left(Q_{\mu\nu}\right) = \partial_\mu \tilde{A}_\nu - \partial_\nu \tilde{A}_\mu$$

$$= -2\,\mathrm{Im}\left(\langle \partial_\mu \psi(\lambda) | \partial_\nu \psi(\lambda) \rangle - \langle \partial_\mu \psi(\lambda) | \psi(\lambda) \rangle \langle \psi(\lambda) | \partial_\nu \psi(\lambda) \rangle\right) \quad (12.50)$$

The Berry curvature is antisymmetric $F_{\mu\nu} = -F_{\mu\nu}$ and identifies with the Berry connection $\tilde{A}_\mu$, that is, an emergent quantity of $\mathcal{PH}$, introduced in the Fubini–Study metric to ensure $U(1)$ gauge invariance. On the other hand, the Berry curvature can be intimately linked to the symplectic structure

$$\omega(\lambda) = \omega_{\mu\nu}\, d\lambda^\mu \wedge d\lambda^\nu = -1/2\, F_{\mu\nu}\, d\lambda^\mu \wedge d\lambda^\nu \quad (12.51)$$

This is a closed, non-degenerate two-form differential and invariant under $U(1)$ gauge. It follows that the Poisson structure on $\mathcal{PH}$ is equivalent to the data of a Poisson bi-vector field $\pi \in \chi^2(\mathcal{PH})$ compatible with the following symplectic form [14]

$$\pi = -2\sum_{\mu,\nu=1}^{n'} F^{\mu\nu} \frac{\partial}{\partial \lambda_\mu} \wedge \frac{\partial}{\partial \lambda_\nu} \quad (12.52)$$

It follows that the Poisson bracket of two functions f, g ∈ PH is given in the form

$$\pi(df, df) = \{f, g\} = \frac{1}{i\hbar}\sum_{\mu,\nu=1}^{n} F^{\mu\nu}\left(\frac{\partial f}{\partial \lambda_\mu}\frac{\partial g}{\partial \lambda_\nu} - \frac{\partial g}{\partial \lambda_\mu}\frac{\partial f}{\partial \lambda_\nu}\right) \quad (12.53)$$

Thus, the symplectic structure is now demonstrated.

12.6.2 Riemannian Structure

Consider a quantum system described by the Hamiltonian $\mathrm{H}(\lambda)$, continuously dependent on a set of parameters $\lambda = (\lambda_1, \ldots\ldots, \lambda) \in M$, where M is the parameter space of dimension n'. Let $\| \ \|$ and $(,)$ be the norm and scalar product defined on the Hilbert space of the corresponding parametric system $\mathcal{H}(\lambda)$. The distance between two neighboring states $\psi(\lambda)$ and $\psi(\lambda + \mathrm{d}\lambda)$ is given by [15]:

$$\mathrm{ds}^2 = \|\psi(\lambda+\mathrm{d}\lambda) - \psi(\lambda)\|^2 = \langle \delta\psi | \delta\psi \rangle = (\partial_\mu \psi | \partial_\nu \psi)\, \mathrm{d}\lambda^\mu \mathrm{d}\lambda^\nu \tag{12.54}$$

By writing the Hermitian product in terms of its real and imaginary parts $(\partial_\mu \psi | \partial_\nu \psi) = (\gamma_{\mu\nu} + i\omega_{\mu\nu})$, Equation 12.54 becomes

$$\mathrm{ds}^2 = (\gamma_{\mu\nu} + \mathrm{i}\omega_{\mu\nu})\, \mathrm{d}\lambda^\mu \mathrm{d}\lambda^\nu \tag{12.55}$$

The Hermitian symmetry of the Hilbert space implies that the tensors $\gamma_{\mu\nu}$ and $\omega_{\mu\nu}$ are, respectively, symmetric and antisymmetric. It follows that the geometric distance reduces to

$$\mathrm{ds}^2 = \gamma_{\mu\nu}\, \mathrm{d}\lambda^\mu \mathrm{d}\lambda^\nu \tag{12.56}$$

However, this last metric tensor cannot be the best on the projective Hilbert space, as it is not invariant under the following $\mathrm{U}(1)$ gauge transformation:

$$|\psi(\lambda)\rangle \to |\psi(\lambda)\rangle' = \mathrm{e}^{i\alpha} |\psi(\lambda)\rangle, \quad \alpha \in [0, 2\pi] \tag{12.57}$$

Indeed, the gauge transformation of the metric tensor $\gamma_{\mu\nu}$ gives

$$\gamma'_{\mu\nu} = \gamma_{\mu\nu} - \tilde{\mathrm{A}}_\mu \partial_\nu \alpha - \tilde{\mathrm{A}}_\nu \partial_\mu \alpha + \partial_\mu \alpha \partial_\nu \partial \neq \gamma_{\mu\nu} \tag{12.58}$$

where $\tilde{\mathrm{A}}_\mu(\lambda) = i\langle \psi(\lambda) | \partial_\mu \psi(\lambda) \rangle$ is the Berry connection defined on PH . To maintain the gauge invariance $\mathrm{U}(1)$ of the metric (12.56), the Berry connection must transform under the gauge transformation (12.57) as follows:

$$\tilde{\mathrm{A}}_\mu \to \tilde{\mathrm{A}}'_\mu = \tilde{\mathrm{A}}_\mu - \partial_{\mu\alpha} \tag{12.59}$$

Thus, it results that the Fubini–Study metric takes the form:

$$\mathrm{ds}^2 = \mathrm{g}_{\mu\nu}\, \mathrm{d}\lambda^\mu \mathrm{d}\lambda^\nu = \left(\gamma_{\mu\nu}(\lambda) - (\lambda) \tilde{\mathrm{A}}_\nu(\lambda) \right) \mathrm{d}\lambda^\mu \mathrm{d}\lambda^\nu \tag{12.60}$$

12.7 FUBINI–STUDY METRIC (QUANTUM GEOMETRIC TENSOR)

Fubini–Study metric is the geometric tensor defined on the projective Hilbert space $\mathcal{PH}$. It defines the concept of geodesic distance between quantum states, that is, complex lines. It is a unitary, symmetric metric, defined positively according to the Cauchy–Schwarz inequality, and gauge invariant $\mathrm{U}(1)$.

One of methods to construct the Fubini–Study metric on the manifold of quantum states $\mathcal{PH}$ can be realized by evaluating the transition probability between two neighboring quantum states in the framework of perturbation theory. In other words, this approach considers the Fubini–Study metric as a measure of quantum fluctuations in the neighborhood of each point of the $\mathcal{PH}$ manifold. Indeed, the scalar product between two neighboring states $|\psi(\lambda)\rangle$ and $|\psi(\lambda+\mathrm{d}\lambda)\rangle$ to the second order in λ is written [16]

$$\langle\psi(\lambda)|\psi(\lambda+\mathrm{d}\lambda)\rangle \simeq 1+\mathrm{i}\mathrm{A}\,\tilde{\mathrm{A}}_{\mu}(\lambda)\mathrm{d}\lambda^{\mu}+\frac{1}{2}\langle\psi(\lambda)|\partial_{\mu}\partial_{\nu}\psi(\lambda)\rangle\mathrm{d}\lambda^{\mu}d\lambda^{\nu} \tag{12.61}$$

Using the fact that the terms $\langle\psi|\partial_{\mu}\psi\rangle$ and $\langle\partial_{\mu}\psi|\partial_{\nu}\psi\rangle+\langle\psi|\partial_{\mu}\partial_{\nu}\psi\rangle$ are two pure imaginary numbers. It follows that

$$\mathrm{Re}\langle\psi|\partial_{\mu}\partial_{\nu}\psi\rangle = -\mathrm{Re}\langle\partial_{\mu}\partial_{\nu}|\partial_{\nu}\psi\rangle = -\gamma_{\mu\nu} \tag{12.62}$$

By exploiting Equations 12.61 and 12.62, we arrive at the explicit expression of the Fubini–Study metric:

$$\mathrm{d}s^2 = g_{\mu\nu}(\lambda)d\lambda^{\mu}d\lambda^{\nu} = 2\left(1-\left|\langle\psi(\lambda)|\psi(\lambda+\mathrm{d}\lambda)\rangle\right|^2\right) \tag{12.63}$$

The particularity of this last construction method is that it considers the Fubini–Study metric as a measure of quantum fluctuations in the neighborhood of each point of the manifold of quantum states $\mathcal{PH}$. In the case where the system depends only on time $\mathrm{H}(\lambda=\mathrm{t})$, its dynamics is then governed by the following Schrödinger equation:

$$i\hbar\frac{\mathrm{d}}{\mathrm{dt}}|\psi(\mathrm{t})\rangle = \mathrm{H}(\mathrm{t})|\psi(\mathrm{t})\rangle \tag{12.64}$$

By substituting Equation 12.63 into Equation 12.64, the Fubini–Study metric can be rewritten in the form

$$ds^2 = 2\left(1-\left|\langle\psi(\mathrm{t})|\psi(\mathrm{t}+\mathrm{dt})\rangle\right|^2\right) = \frac{(\Delta E(t))^2}{\hbar}\mathrm{dt} \tag{12.65}$$

where $\Delta\mathrm{E}(\mathrm{t}) = \langle\psi|\mathrm{H}(\mathrm{t})^2|\psi\rangle - \left|\langle\psi|\mathrm{H}(\mathrm{t})|\psi\rangle\right|^2$ is the energy uncertainty for the Hamiltonian $\mathrm{H}(\mathrm{t})$. It follows that the quantum evolution speed can be easily derived from the Fubini–Study metric (12.65). Indeed, we write

$$\mathrm{V} = \frac{ds}{dt} = \Delta E(\mathrm{t}) \tag{12.66}$$

with $\hbar = 1$. Thus, we relate the quantum evolution speed to the energy uncertainty of the system, so that the greater the energy fluctuations, the faster the quantum evolution, and vice versa. This refers to the Aharonov–Anandan theorem [17], according to which energy fluctuations drive the evolution of the quantum system. On the other hand, the evolution speed (12.66) is defined as a function of the

Hamiltonian $H(t)$, which means that it can be measured experimentally, and, similarly, for the evolution time between predefined quantum states. Moreover, the knowledge of the initial state and the form of the Hamiltonian define the manifold of quantum states as well as the evolution speed which are useful for studying numerous problems in quantum information theory.

12.8 GEOMETRIC PHASE

The geometric phases in quantum mechanics have been observed and investigated in numerous contexts, but their full significance was not immediately recognized. In his investigation of polarized light interference, Pancharatnam introduced the first geometric phase in 1956, identifying a phase difference between non-orthogonal polarization states. He showed that the non-transitive character of the "in phase" relationship (which can be seen on the Poincaré sphere) gives this phase its geometric meaning. In 1928, Fock showed that such a phase could be eliminated by a simple redefinition of the initial phase for non-cyclic evolutions, a conclusion that was widely accepted until the 1980s. The turning point came in 1983 when Berry made the astonishing discovery that a quantum system evolving adiabatically around a closed circuit in the space of external parameters acquires, in addition to the familiar dynamic phase, a non-integrable geometric phase that depends only on the geometry of the circuit. This geometric phase confers a profound meaning on quantum physics and suggests a number of observable consequences when understood as the memory of the journey taken.

Despite being revolutionary, Berry's approach was restricted to non-degenerate cyclic evolutions and the adiabatic approximation. These foundations were strengthened by further research conducted by Wilczek and Zee on the non-abelian (degenerate) case, Aharonov and Anandan on the nonadiabatic situation, and Samuel and Bhandari on the possibility of avoiding the assumption of cyclic development. Since then, these advancements have been applied to a wide range of quantum systems, ranging from more complicated systems like three-level quantum systems represented in higher-dimensional spaces, to two-level systems represented on the Poincaré sphere.

12.8.1 Parallel Transport and the Notion of Holonomy

Geometric phases arise due to a phenomenon that can be described as a "global change without local change." To gain an intuitive understanding of geometric phases, it is therefore appropriate to begin with an introduction to the fundamental concept of parallel transport in a purely geometric context. A classic intuitive example of such a situation is the parallel transport of a vector along a loop on a sphere [18]. Consider a surface S (a plane, a sphere, a cone, etc.) and a vector constrained to move everywhere in the tangent plane on the surface. Then, we wish to transport this vector on this surface (without rotating it around the axis normal to the surface), as shown in Figure 12.1.

We are particularly interested in the case where the arrow is transported around a closed circuit $C(1 \rightarrow 2... \rightarrow 6 \rightarrow 7)$. We can encounter two different situations:

1. If the surface is flat, then it is clear that in this case the arrow always remains parallel to its original orientation, and it is therefore unchanged at the end of circuit C.
2. If, however, the surface is curved as in Figure 2.1, the arrow, being constrained to lie in the plane of the local tangent, cannot remain parallel to its original orientation, and at the end of circuit C, we clearly see that it has been rotated by an angle α. This phenomenon is called (holonomy).

The arrow is represented by a unit vector $\mathbf{e}_1$ tangent to the surface. We define another vector $\mathbf{e}_2$ tangent to the surface and which remains perpendicular to $\mathbf{e}_1$. Let $\mathbf{n}(r)$ be another vector that is normal to the surface at point r. These three vectors form an orthonormal frame. While $\mathbf{e}_1$ and $\mathbf{e}_2$ are transported on the surface along a circuit (C), they must rotate with an angular velocity ω which can

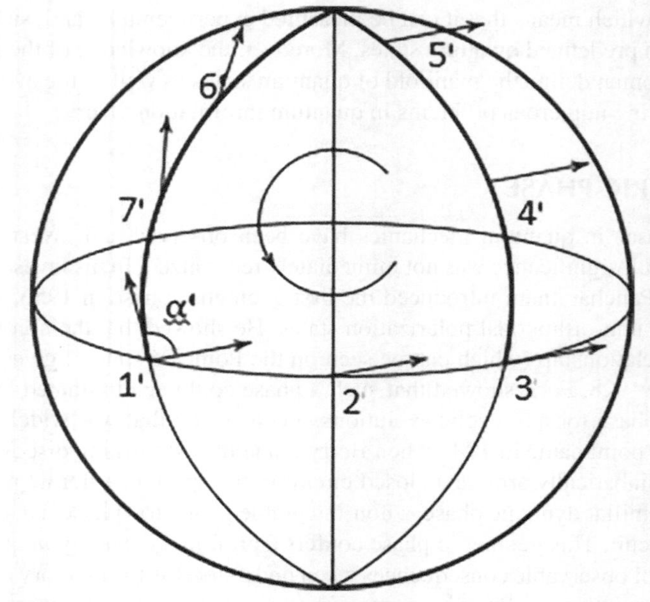

FIGURE 12.1 Parallel transport of a vector along a loop on a sphere.

be determined if the surface is not flat (it is zero if the surface is plate). The equation of motion that describes the vectors $\mathbf{e}_1$ and $\mathbf{e}_2$ is given by:

$$\dot{\mathbf{e}}^r = \omega \times \mathbf{e}^r, \quad r = 1,2 \tag{12.67}$$

which indicates the derivative with respect to time. The angular velocity ω is written in a general form:

$$\omega = a \cdot n + b \cdot \dot{n} + c(n \times \dot{n}) \tag{12.68}$$

To fulfill the condition that $\mathbf{e}_1$ and e_2 remain unitary vectors tangent to the surface, we must have:

$$\mathbf{e}^r \cdot n = 0, \quad r = 1, 2 \tag{12.69}$$

Similarly, parallel transport requires that vectors $\mathbf{e}_1$ and $\mathbf{e}_2$ never rotate around n, that is:

$$\omega \cdot n = 0 \tag{12.70}$$

Replacing Equations 12.69 and 12.70 in Equation 12.68, we finally find that the angular velocity ω is written as:

$$\omega = n \times \dot{n}. \tag{12.71}$$

The law describing the parallel transport of vectors $\mathbf{e}_1$ and $\mathbf{e}_2$ is then:

$$\mathbf{e}^r = (n \times \dot{n}) \times \mathbf{e}^r = -(\mathbf{e}^r \dot{n})n. \tag{12.72}$$

This law can be expressed in another form more appropriate for generalization of parallel transport to the case of quantum mechanics, by defining the complex vector:

$$\phi = \frac{e_1 + i\, e_2}{\sqrt{2}}, \qquad \text{with } \phi^* \dot{\phi} = 1 \tag{12.73}$$

The conditions indicated by Equations 12.67, 12.70, and 12.71 can be expressed in terms of ϕ and its derivative with respect to time ϕ by writing:

$$\phi^* \dot{\phi} = 1/2\left(\mathbf{e}_1^{\cdot} - i\mathbf{e}_2\right)\left(\mathbf{e}_1^{\cdot} - i\mathbf{e}_2^{\cdot}\right) = 1/2\left(\mathbf{e}_1\mathbf{e}_1^{\cdot} + \mathbf{e}_2\mathbf{e}_2^{\cdot} + i\left(\mathbf{e}_1\mathbf{e}_2^{\cdot} - \mathbf{e}_2\mathbf{e}_1^{\cdot}\right)\right) \tag{12.74}$$

The terms $\mathbf{e}_1\mathbf{e}_1^{\cdot}$ and $\mathbf{e}_2\mathbf{e}_2^{\cdot}$ disappear because $\mathbf{e}_1$ and $\mathbf{e}_1^{\cdot}$ are orthogonal (same for $\mathbf{e}_2$ and $\mathbf{e}_2^{\cdot}$). Since $\mathbf{e}_1$ and $\mathbf{e}_2$ are real, the imaginary part of $\phi^* \dot{\phi}$ is $1/2\left(-\mathbf{e}_1\mathbf{e}_2^{\cdot} - \mathbf{e}_2\mathbf{e}_1^{\cdot}\right)$, and this also cancels out because $\mathbf{e}_1$ and $\mathbf{e}_2$ are each perpendicular to the tangent plane containing $\mathbf{e}_1$ and $\mathbf{e}_2$. This leads to the parallel transport law which is written as:

$$\phi^* \dot{\phi} = 0. \tag{12.75}$$

This can also be written in the form:

$$\operatorname{Im}\left(\phi^* \cdot \nabla \phi\right) = 0 \tag{12.76}$$

To express the rotation of unit vectors $\mathbf{e}_1$ and $\mathbf{e}_2$ which move along a closed contour C, we must choose a local orthonormal frame $(\mathbf{n}, \mathbf{u}, \mathbf{v})$ on the surface. The normal unit vector is uniquely determined by the surface, but we have an infinity of possible choices for **u** and **v**. We simply impose that they move on the surface of the sphere under the constraint of parallel transport, which corresponds to a gauge freedom. Once we have made a choice for u, then v is uniquely determined. In an analogous manner to the unit complex vector ϕ, which is a combination of the real unit vectors $\mathbf{e}_1$ and $\mathbf{e}_2$, we introduce the complex vector ψ:

$$\psi = \frac{u + iv}{\sqrt{2}} \tag{12.77}$$

The relationship between parallel transport and the one that is fixed is expressed by

$$\phi(t) = e^{(-i\alpha(t))}\, \psi(t) \tag{12.78}$$

where $\alpha(t)$ is the angle by which $(\mathbf{u}, \mathbf{v})$ must be rotated to coincide with $\left(\mathbf{e}_1, \mathbf{e}_2\right)$. We obtain the equation satisfied by $\alpha(t)$ by inserting Equation 12.78 into 12.76

$$\phi^* \dot{\phi} = -i\dot{\theta}\left(\psi^* \psi + \psi^* \dot{\psi}\right) = 0 \tag{12.79}$$

Since $\psi^* \psi = 1$ and $\psi^* \dot{\psi}$ is imaginary then

$$\dot{\alpha} = \operatorname{Im}\left(\psi^* \dot{\psi}\right) \tag{12.80}$$

If the path C is closed, a simple calculation from differential geometry shows that the accumulated phase $\alpha(C)$ is given by

$$\alpha(C) = \mathrm{Im}\oint_{c} \psi^{*}\, d\psi \tag{12.81}$$

In the case of the sphere, we can choose spherical coordinates such that

$$u = (\cos\theta\cos\varphi, \cos\theta\sin\varphi, -\sin\theta d\theta d\varphi) \quad \text{and} \quad v = (-\sin\varphi,\ \cos\varphi,\ 0).$$

It is easy to show, in this case, that the function to be integrated in Equation 12.78 is $\sin\theta d\theta d\varphi$. Therefore, $\alpha(C)$ is simply the solid angle delimited by the contour C. It is intuitive that the result given by Equation 12.81 does not depend on the choice of the base. If it depended on this base, it would have no physical significance. To see this mathematical independence, we change the base by using the transformation

$$\psi \to \psi e^{i\eta(r)} \tag{12.82}$$

In this case, the complex conjugate ψ^{*} is $\psi^{*}e^{-i\eta(r)}$ and d ψ becomes

$$e^{i\eta(r)} d\psi + i\psi\, e^{i\eta(r)} d\eta(r)$$

This allows us to establish the angle $\alpha(C)$ in the new base

$$\alpha(C) = \mathrm{Im}\oint_{C} \left(\psi^{*} e^{-i\eta(r)} \cdot e^{i\eta(r)} d\psi + i\psi e^{i\eta(r)} d\eta(r)\right) \tag{12.83}$$

$$= \mathrm{Im}\oint_{C} \psi^{*} \cdot d\psi + \mathrm{Im}\oint_{C} i\psi d\eta(r) \tag{12.84}$$

$$= \mathrm{Im}\oint_{C} \psi^{*} \cdot d\psi + \mathrm{Im}\oint_{C} d\eta(r) \tag{12.85}$$

By going from Equations 12.84 to 12.85, we used the fact that $\oint_{C} \psi^{*}\psi$ and $\eta(r)$ is real. The integral of $d\eta(r)$ in Equation 12.78 disappears as it is a scalar that is continuous along the path. Thus, there is no change in the local base at the end of a closed path, namely, the base is the same at the beginning and at the end of the closed path.

12.8.2 Geometric Phase of Berry

In 1984, Berry published his paper [19] in which he considers cyclic evolutions of systems under special, so called adiabatic, conditions. He finds that a cyclic evolution of a wave function yields the original state plus a phase shift, and this phase shift is a sum of a dynamical phase and a geometric (or topological, or Berry) phase shift. Berry points out that the geometrical character of this phase is not negligible. The phase is gauge invariant and therefore cannot be gauged out as was earlier supposed. Many articles have already been written on this subject and, consequently, the so-called Berry phase is now well established, both theoretically as experimentally.

The original point of view of Berry is "dynamical." He starts with a Hamiltonian H that describes the quantum system in question. Further, he considers that the Hamiltonian depends on a multidimensional real parameter *r*, which parametrizes the environment of the system. Then the time evolution of the system is determined by the time-dependent Schrödinger equation

$$\mathrm{H}(\mathrm{r}(\mathrm{t}))|\psi(\mathrm{t})\rangle = \mathrm{i}\hbar\frac{\partial}{\partial \mathrm{t}}|\psi(\mathrm{t})\rangle \tag{12.86}$$

we can choose a basis of eigenstates $|\eta(r(t))\rangle$ corresponding to the energies

$$\mathrm{H}(\mathrm{r}(\mathrm{t}))|\mathrm{n}(\mathrm{r}(\mathrm{t}))\rangle = \mathrm{E}_n(\mathrm{r}(\mathrm{t}))|\mathrm{n}(\mathrm{r}(\mathrm{t}))\rangle \tag{12.87}$$

Now, we assume that the energy spectrum of H is discrete, that the eigenvalues are not degenerated and that no level crossing occurs during the evolution. Moreover, suppose that the environment and therefore $\mathrm{r}(\mathrm{t})$ is adiabatically varied. It means that the changes are slow in time with respect to the characteristic time scale of the system. Then the adiabatic theorem holds and, thus, when the system starts in $\psi(\mathrm{t})\rangle = |\mathrm{n}(0)\rangle$, the system will be over the whole evolution at the nth energy level. However, in general, the state vector gains a phase factor, that is, $|\psi(\mathrm{t})\rangle = e^{i\phi_n}|\mathrm{n}(\mathrm{r}(\mathrm{t}))\rangle$. At first sight, one would guess that the phase factor equals $|\phi_n(\mathrm{t})\rangle = \frac{1}{\hbar}\int E_n(\tau)d\tau$, but the point is that the Schrödinger equation allows an additional phase factor $\gamma_n(\mathrm{t})$, that is, $\phi_n = \theta_n + \gamma_n$. The former is now called the geometric phase, putting this to the Schrödinger equation, we get the following condition for γ_n:

$$\frac{\partial}{\partial t}|\mathrm{n}(\mathrm{x})\rangle + \frac{\mathrm{d}}{\mathrm{dt}}\gamma_{\mathrm{n}}(\mathrm{t})|\mathrm{n}(\mathrm{x})\rangle = 0 \tag{12.88}$$

This could be written in another form:

$$\frac{d}{dt}\gamma_n(t) = \mathrm{i}\left\langle \mathrm{n}(\mathrm{x})\middle|\frac{\partial}{\partial t}\middle|\mathrm{n}(\mathrm{x})\right\rangle = \mathrm{i}\langle n|\nabla \mathrm{n}|\rangle\frac{dx}{dt} \tag{12.89}$$

When we are given a cyclic evolution, described by a closed curve $\mathrm{C}:\mathrm{t}\mapsto \mathrm{x}(\mathrm{t})$ with $\mathrm{x}(\mathrm{T}) = \mathrm{x}(0)$, then the Berry phase for such an evolution is given by

$$\gamma_n(\mathrm{C}) = \mathrm{i}\oint_C \langle \mathrm{n}(\mathrm{x})|\nabla|\mathrm{n}(\mathrm{x})\rangle \mathrm{dx} \tag{12.90}$$

From this, we can see that the Berry phase depends on the geometry of the parameter space (and on the loop C therein). That is why phase of Berry called this phase factor "geometric phase." Now, the geometric phase is used as a universal notion for various generalizations of the original Berry's phase.

12.8.2.1 Illustrative Example

Now we will show how the Berry phase emerges in the concrete example, in a famous and important example where a spin particle occurs in a magnetic field [20].

Consider that the spin $-1/2$ particle is moving in an external magnetic field $\vec{B}$ which rotates adiabatically (slowly) under an angle θ around z-axis as it is depicted in Figure 12.2, where the magnetic field is given by

$$\vec{B}(t) = B_0\begin{pmatrix}\sin\theta\cos(\omega t)\\ \sin\theta\cos(\omega t)\\ \cos\theta\end{pmatrix} \tag{12.91}$$

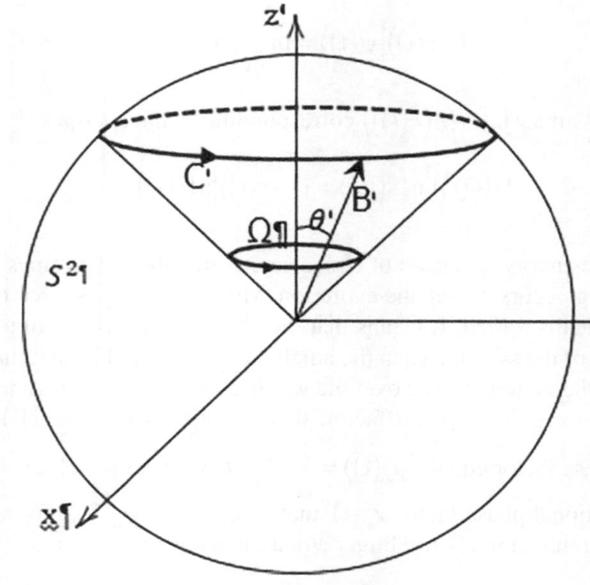

FIGURE 12.2 Spin −1/2 particle in an adiabatically rotating magnetic field $\vec{B}$.

where ω is the angular frequency of the rotation and $B_0 = \left|\vec{B}(t)\right|$. When the field rotates slowly enough and the expected value of the spin was in the direction of field, then the spin of the particle will follow the direction of the field and an eigenstate of the Hamiltonian $\mathrm{H}(0)$ stays for all times t an eigenstate of $\mathrm{H}(\mathrm{t})$. The interaction Hamiltonian for this system in the rest frame is given by $\mathrm{H}(\mathrm{t}) = \mu\vec{B}\vec{\sigma}$ where $\vec{\sigma}$ are Pauli matrices and $\mu = \frac{-e\hbar}{2m}$ is the magnetic dipole moment connected with the spin. When we now use the explicit form of $\vec{\mathrm{B}}$ we get two normalized eigenstates of $\mathrm{H}(\mathrm{t})$

$$\left|n_+(t)\right\rangle = \begin{pmatrix} \cos(\theta/2) \\ e^{i\omega t}\sin(\theta/2) \end{pmatrix}, \qquad \left|n_-(t)\right\rangle = \begin{pmatrix} -\sin(\theta/2) \\ e^{i\omega t}\cos(\theta/2) \end{pmatrix} \tag{12.92}$$

with the corresponding energy eigenvalues $\mathrm{E}_\pm = \pm\mu\mathrm{B}_0$. A calculation of $\nabla\left|\mathrm{n}\pm(\mathrm{t})\right\rangle$ in the spherical coordinates $\mathrm{B}_0, \theta, \phi(\mathrm{t}) = \omega\mathrm{t}$ leads to rather simple expressions

$$\left\langle n_+\middle|\nabla\middle|n_+\right\rangle = \frac{\sin^2(\theta/2)}{B_0\sin\theta}, \qquad \left\langle n_-\middle|\nabla\middle|n_-\right\rangle = \frac{\cos^2(\theta/2)}{B_0\sin\theta} \tag{12.93}$$

The curve C in parameter space (which is now sphere) is given by C: B_0 = const., θ = const., $\phi \in [0, 2\pi]$. Thus, the Berry phase in this example equals

$$\gamma\pm(\mathrm{C}) = \mathrm{i}\oint_{\mathrm{C}}\left\langle n_\pm\middle|\nabla\middle|n_\pm\right\rangle \mathrm{B}_0\sin\theta\, d\phi = -\pi(1\mp\cos\theta) \tag{12.94}$$

This is nothing else but the half of the solid angle enclosed by the path C, and so we get the final expression

$$\gamma\pm(\mathrm{C}) = \mp\frac{1}{2}\Omega(\mathrm{C}) \tag{12.95}$$

We see that whereas the dynamical phase (which is now given by $\pm\mu B_0 T$) depends on the period T of the rotation, the geometrical phase depends only on the special geometry of the problem.

12.8.3 Aharonov–Anandan Phase

In 1987, Aharonov and Anandan [17] considered cyclic evolutions that are not restricted by an adiabatic condition and proposed a generalization of Berry's phase. This generalization is very important because the adiabatic condition is never exactly fulfilled for real evolutions. In the adiabatic approximation, Aharonov and Anandan phase then tends to the Berry phase if the parameters are chosen accordingly.

The appearance of Aharonov and Anandan phase (and other phases and related phenomena) can be explained in terms of quantum mechanics. The crucial role is played by the fact that in quantum physics the physical state of a system is only determined up to a phase. The physical states are therefore in a bijective correspondence with points in the projective Hilbert space, that is, with the pure-state density matrices. However, usually, it is better to compute in Hilbert space and then pass to the projective space. This is because the geometry of Hilbert space is simpler.

According to Aharonov and Anandan, the existence of a phase associated with cyclic evolution, which is universal in the sense that it is the same for the infinite number of possible motions along the curves in the Hilbert space $\mathcal{H}$ which project to a given closed curve in the projective Hilbert space $\mathcal{P}$ of rays of $\mathcal{H}$. Moreover, it is the same for all the possible Hamiltonians $H(t)$ which propagate the state along these curves. The question is, what phase factor $e^{i\phi}$ (which can have observable consequences) that the initial and the final state vector of a cyclic evolution may be related by. Suppose that the normalized state $|\psi(t)\rangle \in \mathcal{H}$ evolves according to the Schrödinger equation

$$H(t)|\psi(t)\rangle = i\hbar\frac{\partial}{\partial t}|\psi(t)\rangle \tag{12.96}$$

such that the evolution is cyclic, that is, $|\psi(T)\rangle = e^{i\phi}.|\psi(0)\rangle$. Let $\Pi : \mathcal{H} \to \mathcal{P}$ be the projection map defined by

$$\Pi(|\psi\rangle) = \{|\tilde{\psi}\rangle : |\tilde{\psi}\rangle = c|\psi\rangle,\ c \in \mathbb{C}\} \tag{12.97}$$

Then $|\psi(t)\rangle$ defines a curve $C^{\sim}:[0,T] \to \mathcal{H}$ with $C \equiv \Pi(C)$ being a closed curve in $\mathcal{P}$. Conversely given any such curve C, we can define a Hamiltonian function $H(t)$ so that the Schrödinger equation is satisfied for the corresponding normalized $|\psi(t)\rangle$. Now define

$$|\psi(t)\rangle = e^{-if(t)}|\psi(t)\rangle \tag{12.98}$$

such that $f(T) - f(0) = \phi$. Then $|\tilde{\psi}(t)\rangle$ is exactly cyclic, that is, $|\tilde{\psi}(T)\rangle = |\tilde{\psi}(0)\rangle$, and from Schrödinger equation,

$$\frac{-df}{dt} - \frac{1}{\hbar}\langle\psi(t)|H|\psi(t)\rangle - \left\langle\tilde{\psi}(t)\left|\frac{d}{dt}\right|\tilde{\psi}(t)\right\rangle. \tag{12.99}$$

Aharonov and Anandan have shown that the accumulated phase Φ_{tot} acquired over the cycle C may be split into a dynamic phase factor ϕ_{dyn} and a geometric phase factor

$$\Phi_{tot} = \phi_{dyn} + \phi_{geo}. \tag{12.100}$$

Hence, if we remove ϕ_{dyn} from the totam phase Φ_{tot} by defining

$$\beta = \phi + 1/\hbar \int \langle \psi(t)|H|\psi(t)\rangle dt, \tag{12.101}$$

it follows from the above equation that

$$\beta = \int_0^T \langle \tilde{\psi} \left| \frac{d}{dt} \right| \tilde{\psi} \rangle dt. \tag{12.102}$$

The geometric phase ϕ_{geo} does not depend upon the exact dynamics of the system but only on the geometric properties of the curve traced out in $\mathcal{P}(\mathcal{H})$. In the case of a two-level system, for example, $\mathcal{P}(\mathcal{H})$ may be identified with the Bloch sphere. For a closed curve on S^2 the geometric phase is given up to a sign by one-half of the solid angle subtended by the cycle C at the origin of the unit sphere

$$\phi_{geo}(C) = \pm 1/2\Omega(C) \tag{12.103}$$

The sign depends in general on the details of the cyclic evolution, for example, the sense in which the cycle is traversed.

12.8.4 Holonomy Interpretations of the Geometric Phase

In the definition of the Berry phase, we considered a Hermitian operator (Hamiltonian of the system) $H(x)$ depending smoothly on a parameter $x \in M$, with an isolated nondegenerate eigenvalue $E_n(x)$depending continuously on x. We now restrict to the unitary evolution for the moment, then the eigenstates $|n(x)\rangle$ of $H(x)$are assumed to be normalized, and it is straightforward that the association $x \to |n(x)\rangle$ defines a principal fiber bundle λ over the parameter space M with fibers

$$L_x = \{|\psi_x\rangle : |\psi_x\rangle = e^{i\varphi}|n(x)\rangle, \varphi \in \mathbb{R}\} \cong U(1) \tag{12.104}$$

The point is that twisting of this $U(1)$-principal bundle $U(1) \to \lambda \to M$ affects the phase of quantum mechanical wave functions.

Consider a closed curve C: $[0,T] \ni t \to x(t) \in M$ in the parameter space $(x(0) = x(T))$ and set $H(t) := H(x(t))$. When the system evolves adiabatically along this loop, the state vector $|\psi(t)\rangle$ of the system which is initially an eigenstate $|\psi(0)\rangle := |n(0)\rangle$ of the Hamiltonian $H(0)$ evolves according to the Schrödinger equation and remains always an eigenstate of $H(T)$ and therefore, after time T, gains a phase factor

$$|\psi(T)\rangle = \exp\left(\frac{-i}{\hbar}\int_0^T E_n\langle(t)\,dt\right).|n(T)\rangle = \exp\left(\frac{-i}{\hbar}\int_0^T E_n(t)\,dt\right).e^{i\gamma(T)}|\psi(0)\rangle \tag{12.105}$$

The Berry's additional phase factor $\gamma(t)$ obviously defines a way of transporting of the basis $|n(t)\rangle$ along C, that is, a lift $C^{\sim} \in \lambda$ of the loop $C \in M$, by assigning $|n(t)\rangle \rightarrow e^{i\gamma(t)}|n(0)\rangle$, that is, a connection ω in the principal fiber bundle. We know that the local expression of the connection (i.e., the gauge potential) is given simply by the $U(1)$ valued one-form

$$A := -\langle n(x)|d|n(x)\rangle = -\left\langle n(x)\left|\frac{\partial}{\partial x^{\mu}}\right|n(x)\right\rangle dx^{\mu} \tag{12.106}$$

In this Berry–Simon (adiabatic) approach, the term $e^{i\gamma(C)} = e^{i\oint_C A}$ is thus nothing else but the element of the local holonomy group $\mathrm{Hol}_{|n(0)\rangle}(\omega)$ based at the point $|n(0)\rangle \in \lambda$ coming from the loop C. Such a holonomy group is (in general) a subgroup of the structure group $U(1)$ of λ. Note that in this non-degenerate case the Lie algebra of the structure group is 1D and therefore $AA = 0$ and we can use the Stoke's theorem to compute the Berry's phase directly, as an integral of the curvature ρ of the connection $\omega : i\gamma(C) = \oint_C A = \iint dA = \iint(\rho - A \wedge A) = \iint \rho$, where $S = \partial C$.

Note also that if we consider a general, nonunitary evolution we get an equivalent description. Instead of λ we have an associated complex line bundle $L = \lambda \times_{U(1)} \mathbb{C}$ with fibers isomorphic to $\mathbb{C}$. The expression of the gauge potential is the same and $e^{i\gamma_n(C)}$ is again an element of the holonomy group of this bundle, which is a subgroup of $GL(L_x) \cong U(1)$.

Aharonov and Anandan realized that instead of considering a closed loop in parameter space, one could consider the closed loop in state space (or equivalently projective Hilbert space) and drop the adiabaticity condition. In the Aharonov–Anandan approach, one can again restrict to the unitary case and consider a $U(1)$-principal bundle η over the projective Hilbert space $\mathcal{P} = \mathbb{C}P^n$ $(n = \dim(H) - 1)$ with fibers

$$\eta_{|\psi\rangle\langle\psi|} = \{e^{i\delta}|\psi\rangle\},\ e^{i\delta} \in U(1) \cong U(1) \tag{12.107}$$

over a point $|\psi\rangle\langle\psi|$ in P. The connection in this Aharonov–Anandan bundle $e^{i\delta}|\psi\,\eta : U(1) \rightarrow \mathcal{H}^* \rightarrow \mathcal{P}$ is given in a quite natural way – the horizontal subspaces are perpendicular to the fibres with respect to the Hilbert space inner product. In the local form, this reads as

$$A = -\langle\psi|d|\psi\rangle, \tag{12.108}$$

which is a $U(1)$-valued one-form (Figure 12.3).

In systems of interacting spins for example, the Aharonov–Anandan phase has been shown to depend non-trivially on entanglement, with geometric phases decreasing linearly as entanglement increases, reflecting the curvature of the underlying quantum state space [21, 22]. This formulation generalizes Berry's phase, removing adiabaticity and cyclicity constraints while retaining its geometric essence.

The most important application of the geometric phase is in quantum computation. The unit of quantum information is called "qubit." A quantum logic gate is then a device that performs a unitary operation on a qubit. A general operation is possible to realize by using one and two-qubit operations, for example, by the Hadamard gate, phase gate, C-NOT gate and controlled phase shift gate.

Geometric phases seem to be good candidates for realizing low-noise quantum computing devices. Because of the dependence only on the geometry of the appropriate space, the geometric phase is an ideal construction for fault-tolerant quantum computation. Nevertheless, it also has several drawbacks, for example, one has to get rid of the dynamical phase. There are many physical realizations of quantum computers and quantum gates.

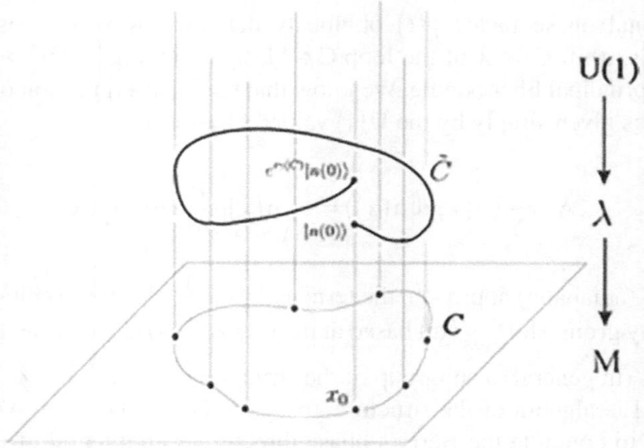

FIGURE 12.3 The fiber bundle for the Berry phase.

12.9 QUANTUM GATES AND BERRY PHASE

Realizations of quantum gates there are many physical realizations of quantum computers and quantum gates. They all have to fulfill four basic requirements:

- There should be a distinct representation of the states $|0\rangle$ and $|1\rangle$.
- They should perform a controllable unitary evolution on the qubits.
- The desired input state should be prepared as the initial qubit state.
- The final state should be measurable to get the final qubit state.

These requirements can be achieved by several constructions. In the following, we mention the most popular physical system with which quantum gates or quantum computation have been tried to realize yet.

12.9.1 Geometric Phases and Nuclear Magnetic Resonance

Zanardi and Rasetti were the first to point out that Berry phases could be used for enabling quantum computation. The idea was picked up, modified, and extended until the first experiments were done [23]. Geometric phases are proper candidates for realizing low noise quantum computing devices. Because of the dependence only on the net area traced out in phase space the geometric phase is an ideal construction for fault-tolerant quantum computation. The phase is achieved by nuclear magnetic resonance (NMR) methods which already were used very early (1987, 1988) to test Berry's phase in experiments [24].

In the experiments, a conditional geometric phase is applied, which means that the state of one spin determines the geometric phase acquired by the other spin. But the adiabatic geometric phase has also several drawbacks. First one is limited in time by the adiabatic condition which has to be removed to take full advantage of the short coherence time of the quantum computer. This can be done by using the more general concept of the Aharonov–Anandan phase, which will not be discussed in this work. The second point is that one wants to get rid of the dynamical phase. This is done by using the so-called spin-echo technique where the adiabatic evolution is applied twice in reversed direction.

However, here additional errors can be produced if the second path is not exactly the reversed of the first path and therefore the dynamical phase does not exactly cancel out.

12.9.1.1 Nuclear Magnetic Resonance

We consider 2 qubit-gates; therefore we start with a system consisting of two spins a and b in magnetic fields $\overrightarrow{B_a}$ and $\overrightarrow{B_b}$ along the z-axis. The Hamiltonian for this system is given by

$$H_0 = \frac{\hbar}{2}\left(\omega_a \sigma_{az} \otimes 1_b + \omega_b 1_a \otimes \sigma_{bz}\right) \tag{12.109}$$

where $\omega_a = \mu\left|\overrightarrow{B_a}\right|$ and $\omega_b = \mu\left|\overrightarrow{B_b}\right|$ denote the characteristic resonance frequencies of spin a and b and $\sigma_{az} = \sigma_{bz} = \sigma_z$. This can be written in the basis of the eigenstates (quantized along the z-axis) of the Hamiltonian $\left\{|\uparrow\uparrow\rangle, |\uparrow\downarrow\rangle, |\downarrow\uparrow\rangle, |\downarrow\downarrow\rangle\right\}_{ab}$ in matrix form are given by

$$H_0 = \frac{\hbar}{2}\begin{pmatrix} \omega_a + \omega_b & 0 & 0 & 0 \\ 0 & \omega_a - \omega_b & 0 & 0 \\ 0 & 0 & -\omega_a + \omega_b & 0 \\ 0 & 0 & 0 & -\omega_a - \omega_b \end{pmatrix} \tag{12.110}$$

where we assume $\omega_a \neq \omega_b$ and $\omega_a > \omega_b$.

When we introduce interactions between the two spins, the Hamiltonian H_0 modifies as follows:

$$H = H_0 + \frac{\hbar}{2} J \sigma_{az} \otimes \sigma_{bz} \tag{12.111}$$

$$= \frac{\hbar}{2}\left(\omega_a \sigma_{az} \otimes 1_b + \omega_b 1_a \otimes \sigma_{bz} + J \sigma_{az} \otimes \sigma_{bz}\right) \tag{12.112}$$

where J denotes an interaction constant. This gives the matrix form

$$H = \frac{\hbar}{2}\begin{pmatrix} \omega_a + \omega_b + \mathrm{J} & 0 & 0 & 0 \\ 0 & \omega_a - \omega_b - \mathrm{J} & 0 & 0 \\ 0 & 0 & -\omega_a + \omega_b - \mathrm{J} & 0 \\ 0 & 0 & 0 & -\omega_a - \omega_b + \mathrm{J} \end{pmatrix} \tag{12.113}$$

Now we add a circularly polarized magnetic field $\overrightarrow{B_z}(t)$ in the xy-plane with the characteristic frequency $\omega_z = \mu|\overrightarrow{B_z}|$, a so-called radio frequency (rf)-pulse, and consider the Hamiltonian H'_a for spin a (tracing out spin b). We obtain the following:

$$H'_a = \frac{\hbar}{2}\left(\omega_a \pm J\right)\sigma_{az} + \frac{\hbar}{2}\omega_z\ \sigma_x(t) \tag{12.114}$$

where the $\pm$sign in front of J depends on the specific state (up or down) of the spin b and $\sigma_x(t) = \begin{pmatrix} 0 & e^{-i\omega t} \\ e^{i\omega t} & 0 \end{pmatrix}$ with the angular frequency ω of the field $\overrightarrow{B_z}(t)$.

When we transform this Hamiltonian into the rotational frame for spin b rotating at speed ω, we get

$$H_a = RHR^{-1} + i\left(\frac{\partial}{\partial t} R\right)R^{-1} \tag{12.115}$$

$$= \frac{\hbar}{2}(\omega_a - \omega)\sigma_z + \frac{\hbar}{2}\omega_a \sigma_x(t) \tag{12.116}$$

$$= \frac{\hbar}{2}(\omega_\pm - \omega)\sigma_z + \frac{\hbar}{2}\omega_a \sigma_x(t) \tag{12.117}$$

where $\omega_\pm = \omega_a \pm J$ and the rotational matrix is given by

$$\mathrm{R} \equiv \mathrm{R}(\mathrm{t}) = e^{-i\omega\frac{\sigma_z}{2}t} \tag{12.118}$$

We see that for $|\omega_1| << |\omega_\pm - \omega|$ the Hamiltonian lies close to the z-axis and for $|\omega_1| >> |\omega_\pm - \omega|$ it lies close to the x-axis. To acquire a Berry phase we want to obtain circular motions which can be imposed by adiabatically varying the phase of the rf-pulse. When the rf-pulse is not swept all the way to resonance but only to some final value ω_f we can implement cones with arbitrary opening angles θ, as shown in Figure 12.4.

The opening angle is given by

$$\cos\theta_\pm = \frac{\omega_z}{\sqrt{(\omega_\pm - \omega)^2 + \omega^2}}, \tag{12.119}$$

which gives for the Berry phase

$$\gamma_\pm = -\pi(1 - \cos\theta) \tag{12.120}$$

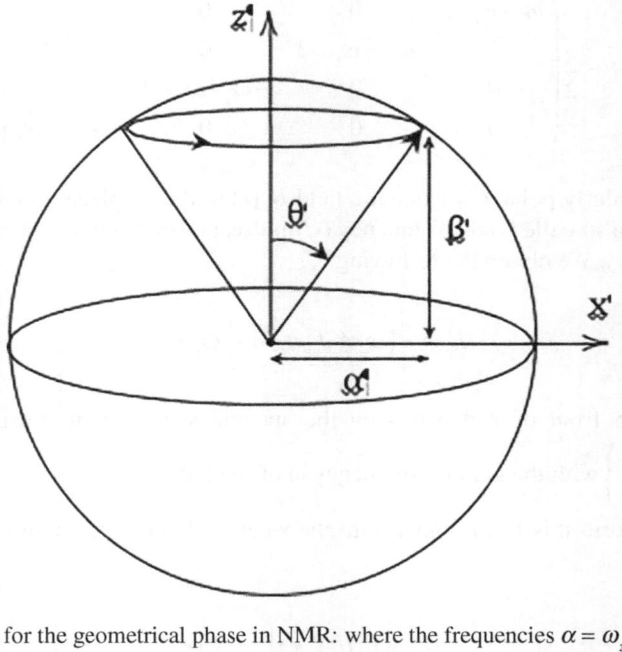

FIGURE 12.4 Setup for the geometrical phase in NMR: where the frequencies $\alpha = \omega_z$ and $\beta = \omega \pm -\omega$.

The Berry phase for spin a depends on the respective state of the spin b, we have applied a conditional Berry phase. The next step is to remove the dynamical phase. This is done by a sequence of operations also known as spin echo approach. First, the procedure to get a geometrical phase is applied to the system. Then the spin states are flipped by a short π pulse and the rotation due to the rf-pulse is repeated but in the other direction. Afterwards a π pulse is applied again to flip the states back to the original direction. After this procedure, the system has acquired the following global phase factors written in matrix form:

$$\left|\psi_f\right\rangle = \begin{pmatrix} e^{2i\Delta\gamma} & 0 & 0 & 0 \\ 0 & e^{-2i\Delta\gamma} & 0 & 0 \\ 0 & 0 & e^{-2i\Delta\gamma} & 0 \\ 0 & 0 & 0 & e^{2i\Delta\gamma} \end{pmatrix} \left|\psi_a\right\rangle \otimes \left|\psi_b\right\rangle \tag{12.121}$$

where $\Delta\gamma = \gamma\gamma-$. This means the acquired geometric phase depends on the respective orientation of the two spins.

12.9.1.1.1 Experimental Realization

The experimental realization of the theory explained above is due to Jones et al. [25]. They used a sample consisting of 100 mg of 99% ^{13}C-labeled $CHCl_3$ dissolved in 0.2 ml of 99.96% $CDCl_3$ $CDCl_3$. The ^1H nucleus are used as spin a and the ^{13}C as spin b. The interaction frequency between both nuclei is given by $J = 209.2Hz$.

The frequency of the rf-pulse was set to $\omega_1 = 221.3$ Hz and the frequency ω was varied between 0 and 774 Hz. The geometric phase was measured for both ^1H resonances. The results are shown in Figure 12.5. We see clearly the splitting of the geometric phases due to the respective spin state.

12.9.2 Topological Quantum Computing

Topological quantum computing (TQC) represents a paradigm shift in quantum information processing by encoding quantum states in topologically protected degrees of freedom. Unlike

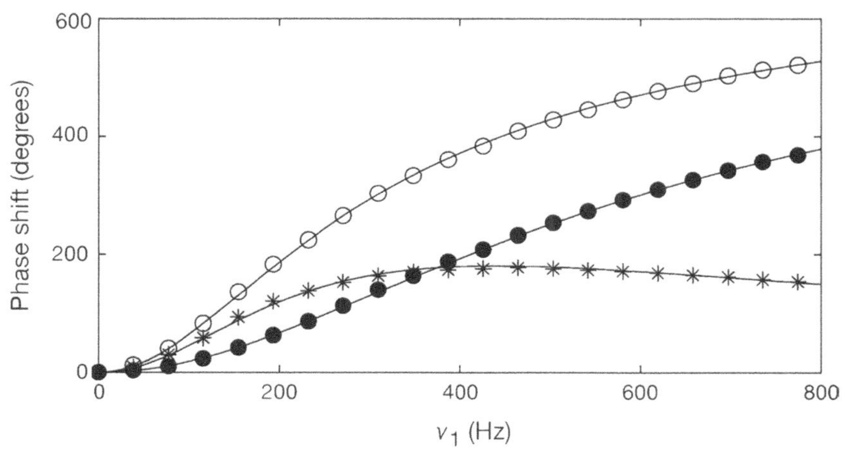

FIGURE 12.5 Experimental results for the conditional Berry phase with NMR: the filled circles show the results for γ^-, the open circles for γ^+ and the stars show $\Delta\gamma$ [25].

conventional qubits, which are susceptible to local noise and decoherence, topological qubits rely on the global geometric properties of entangled quasiparticles, such as Majorana fermions or Fibonacci anyons. These systems leverage non-Abelian geometric phases induced by braiding operations, which are inherently robust against local perturbations.

12.9.2.1 Geometric Foundations of TQC

The robustness of TQC arises from the fact that quantum information is stored in the topological structure of the system's Hilbert space. Consider a 2D topological superconductor hosting Majorana zero modes (MZMs) localized at vortices or defects. Each MZM corresponds to a self-adjoint operator $\gamma_i = \gamma^{\dagger}_i$ satisfying the Clifford algebra:

$$\left\{\gamma_i = \gamma_j\right\} = 2\delta_{ij} \tag{12.122}$$

A pair of MZMs $\left(\gamma_1 = \gamma_2\right)$ forms a conventional fermionic mode $\mathrm{c} = \left(\gamma_1 + i\gamma_2\right)/2$, with the vacuum state $|0\rangle$ and occupied state $|1\rangle = c^{\dagger}|0\rangle|1\rangle$. However, the non-local nature of MZMs ensures that quantum information is stored in the fusion channel of multiple pairs, making it immune to local measurements.

12.9.2.2 Braiding and Non-Abelian Berry Phases

The key operation in TQC is the adiabatic exchange (braiding) of MZMs. When two vortices hosting γ_1 and γ_2 are swapped clockwise, the unitary transformation acting on the ground state subspace is:

$$\mathrm{B}_{12} = \mathrm{e}^{\frac{\pi}{4}\gamma_1\gamma_2} = \frac{1}{\sqrt{2}}\left(1 + \gamma_1\right). \tag{12.123}$$

For four MZMs $\left(\gamma_1, \gamma_2, \gamma_3, \gamma_4\right)$, braiding γ_1 around γ_2 induces a non-Abelian Berry phase proportional to the monodromy matrix of the system. This phase depends solely on the topology of the braid path, not on the details of the trajectory.

Example: Single-Qubit Gate via Braiding

Consider a topological qubit encoded in four MZMs:

$$|\psi\rangle = \alpha|0\rangle_{12} \otimes |0\rangle_{34} + \beta|1\rangle_{12} \otimes |1\rangle_{34} \tag{12.124}$$

where $|0\rangle_{ij}$ and $|1\rangle_{ij}$ represent the fermionic parity states of pairs (i, j). Braiding γ_1 and γ_2 applies the gate:

$$\mathrm{B}_{12}|\psi\rangle = e^{i\frac{\pi}{4}\sigma_z}|\psi\rangle \tag{12.125}$$

where σ_z is the Pauli-Z operator. Similarly, braiding γ_2 around γ_3 generates a Clifford gate equivalent to σx. These operations form a universal gate set when combined with magic-state distillation.

12.9.3 Geometry and Fault Tolerance

The immunity to decoherence in TQC stems from the topological degeneracy of the ground state. Local perturbations cannot distinguish between degenerate states, as they only modify the system's local degrees of freedom. The Berry phase acquired during braiding depends on the global history of the quasiparticles, quantified by the holonomy of the Berry connection over the parameter space M:

$$\gamma_B = \oint_C A d\lambda, A_\mu = \langle i\psi | \partial_\mu \psi \rangle \tag{12.126}$$

where C is a closed path in M (see Equation 12.69). For non-Abelian systems, A becomes a matrix valued connection, and γ_B generalizes to a unitary operator representing the braid group.

12.10 CONCLUSION

This chapter underscores the intrinsic relationship between geometry and quantum mechanics, revealing how geometric frameworks provide profound insights into quantum phenomena. By reformulating quantum dynamics through structures such as symplectic manifolds, the Fubini–Study metric, and projective Hilbert spaces, the discussion demonstrates that quantum state evolution is inherently tied to the geometry of the underlying parameter spaces. Key examples, such as the Berry phase and its non-adiabatic generalization by Aharonov and Anandan, illustrate that quantum phases are not merely dynamical artifacts but encode geometric and topological properties of the system's evolution.

The practical implications of these geometric principles are particularly evident in quantum computing. Geometric phases, resistant to local noise due to their dependence on global path properties, offer a promising route for designing fault- tolerant quantum gates. Experimental implementations, such as NMR-based conditional geometric phases and TQC with non-Abelian anyons, highlight how geometric robustness can mitigate decoherence a critical challenge in quantum technologies.

Ultimately, the geometric perspective bridges abstract mathematical formalism with physical intuition, offering a unified framework to interpret quantum mechanics. As quantum systems grow in complexity, these geometric tools will remain indispensable for both theoretical exploration and the development of scalable quantum technologies.

REFERENCES

[1] A. Slaoui, B. Amghar, and R. A. Laamara. Interferometric phase estimation and quantum resource dynamics in bell coherent-state superpositions generated via a unitary beam splitter. *Journal of the Optical Society of America B*, 40(8):2013–2027, Aug 2023.

[2] M. A. Nielsen, M. R. Dowling, M. Gu, and A. C. Doherty. Quantum computation as geometry. *Science*, 311(5764), 1133–1135, 2006.

[3] M. A. Nielsen. A geometric approach to quantum circuit lower bounds. *Quantum Information & Computation*, 6(3), 213–262, 2006.

[4] N. Khaneja, S. J. Glaser, and R. Brockett. Sub-Riemannian geometry and time optimal control of three spin systems: Quantum gates and coherence transfer. *Physical Review A*, 65(3):032301, 2002.

[5] H. Weyl. *The Theory of Groups and Quantum Mechanics*. Dover, New York, 1950. Translated by H. P. Robertson.

[6] A. Bohm. *Quantum Mechanics: Foundations and Applications*. Springer-Verlag, Berlin, 2nd ed., 1986.

[7] J. von Neumann. *Mathematical Foundations of Quantum Mechanics*. Springer-Verlag, Berlin, 1932.

[8] B. G. Wybourne. *Classical Groups for Physicists*. Wiley, New York, 1974.

[9] L. K. Hua. *Harmonic Analysis of Functions of Several Complex Variables in the Classical Domains*. American Mathematical Society, Providence, RI, 1963.

[10] B.-Y. Hou and B.-Y. Hou. *Differential Geometry for Physicists*. World Scientific Publishing, Singapore, 1997.

[11] W. M. Zhang, D. H. Feng, and R. Gilmore. Coherent states: theory and some applications. *Reviews of Modern Physics*, 62:867, 1990.

[12] V. I. Arnold. *Mathematical Methods of Classical Mechanics*. Springer-Verlag, Berlin, 1978.

[13] A. Botero. Modewise entanglement of Gaussian states. *Journal of Mathematical Physics*, 44(12): 5279–5295, 2003.

[14] M. Boucetta. *Introduction to Poisson Geometry*. School of Geometry and Dynamic Systems, Houari Boumediène University of Science and Technology, Algiers, 2007.

[15] A. Dey, S. Paul, P. Roy, and T. Sarkar. Information geometry and quantum phase transitions in the Dicke model. arXiv preprint, 2016.

[16] R. Cheng. Quantum geometric tensor (Fubini-study metric) in simple quantum system: A pedagogical introduction. arXiv:1012.1337v2, 2013.

[17] J. Anandan and Y. Aharonov. Geometry of quantum evolution. *Physical Review Letters*, 65:1697, 1990.

[18] B. Goss Levi. Superconductivity: Hot topics cool down. *Physics Today*, 46:17, 1993.

[19] M. V. Berry. Quantal phase factors accompanying adiabatic changes. *Proceedings of the Royal Society of London*, 392:45–57, 1984.

[20] K. Durstberger. *Geometric Phases in Quantum Theory*, University of Vienna, Vienna, Austria, 2002.

[21] B. Amghar, A. Slaoui, J. Elfakir, and M. Daoud. Geometrical, topological and dynamical description of N interacting spin-s under long-range Ising model and their interplay with quantum entanglement. *Physical Review A, 107,* 032402, 2023.

[22] J. Elfakir, B. Amghar, A. Slaoui, and M. Daoud. Complementarity between quantum entanglement and geometric and dynamical appearances in N spin-1/2 system under all-range Ising model. *The European Physical Journal Plus*, 139:771,2024.

[23] J. A. Jones, V. Vedral, A. Ekert, and G. Castagnoli. Geometric quantum computation using nuclear magnetic resonance. *Nature*, 403:869–871, 2000.

[24] D. Suter, G. C. Chingas, R. A. Harris, and A. Pines. Berry's phase in magnetic resonance. *Molecular Physics*, 61:1327–1340, 1987.

[25] J. A. Jones, V. Vedral, A. Ekert, and G. Castagnoli. Geometric quantum computation using nuclear magnetic resonance. *Nature*, 403:869–871, 2000.

13 Quantum Cryptography
Reinventing Secure Communication

Youssef Oulouda, Abdallah Slaoui, and Ahmed A. Abd El-Latif

13.1 INTRODUCTION

Quantum cryptography represents a fundamental domain within quantum information theory that has undergone practical implementation [1, 2], driven by the imperative need to secure personal data and communications technology. Offering an unparalleled level of security in contrast to classical cryptography, quantum cryptography's robustness lies in its foundation on the principles of quantum mechanics. Two core principles of quantum mechanics form the basis of quantum cryptography: the measurement postulate [3] and the non-cloning theorem [4]. The measurement postulate asserts that any attempt to measure a quantum state inevitably disturbs its original form, effectively exposing the potential presence of an eavesdropper [5]. This disturbance serves as an alert to authorized parties about the potential intervention of an eavesdropper on the communication channel. And, thus, follow the procedure adopted for that case. On the other hand, the non-cloning theorem prevents any eavesdropper to duplicate an arbitrary quantum state. These principles underline the strength of quantum cryptography over classical one. The BB84 protocol [6] stands as one of the pioneering and successful protocols in the realm of quantum cryptography, marking a significant leap in the field by actualizing quantum communication and transitioning it from theory to practice. Following this groundbreaking discovery, researchers have been intrigued by the possibilities offered by protocols based on BB84. For instance, protocols like the six-state protocol [7], Eckert 91 [8], B92 [9], among others, emerged from the BB84 foundation. However, despite these notable strides and advancements in the field, quantum cryptography encounters several challenges that impede its widespread adoption in global technologies. Implementing such protocols requires expensive and scarce devices [10]. Additionally, these protocols often suffer from information loss during processing, not to mention vulnerabilities to eavesdropping strategies and channel noise. The aforementioned protocols are categorized under Discrete Variables Quantum Key Distribution (DV-QKD) [11], wherein the examination and design are based on discrete variables, like photon polarization or phase, electron spin. Continuous Variables Quantum Key Distribution (CV-QKD) [12] offers a new approach to address the challenges faced by DV-QKD protocols. CV-QKD protocols utilize readily available devices accessible to anyone, such as laser sources, beam splitters, and others, making them more practical and widespread. In this chapter, our aim is to present quantum cryptography as a means to achieve secure communication. The first section highlights significant milestones achieved in the birth of quantum cryptography. In the second section, we introduce classical cryptography and outline its limitations. The third section covers basic concepts of quantum cryptography and detection schemes, including a definition of the differences between DV-QKD and CV-QKD. We provide explanations of various protocols and eavesdropping strategies, analyze the advantages and disadvantages of quantum cryptography, explore its application fields, draw a comparison between classical and quantum cryptography, and conclude accordingly.

DOI: 10.1201/9781003621423-15

13.2 CRITICAL TURNING POINTS IN THE HISTORY OF QUANTUM CRYPTOGRAPHY

1917: Gilbert S. Vernam [13] devised the one-time pad, a revolutionary machine generating non-repeating, seemingly random character sequences. This method utilized an encryption key matching the message length, ensuring unique and never-reused keys, standing as an empirically proven secure communication method due to perfect secrecy.

1940s: Claude Shannon [14] laid the groundwork for cryptographic theory, emphasizing the relationship between uncertainty in encoded messages and cryptographic keys' inherent uncertainty. This established the basis for information-theoretic security.

Early 1970s: Pioneers like Whitfield Diffie, Martin Hellman, Ralph Merkle, Ron Rivest, Adil Shamir, Leonard Adleman, James Ellis, Clifford Cocks, and Malcolm Williamson [15, 16] introduced asymmetric or public-key cryptography, revolutionizing secure communication by using distinct keys for encryption and decryption.

1970: Stephen Wiesner's paper [17], "Conjugate Coding," presented the theoretical framework using quantum mechanics to create counterfeiting-resistant banknotes and introduced the concept of a multiplexing channel.

1979: Charles H. Bennett learned about Wiesner's work through a close association. In collaboration with Gilles Brassard, they combined public-key cryptography with Wiesner's quantum framework, leading to the Crypto 82 paper and coining the term "quantum cryptography."

1984: Bennett and Brassard redirected their efforts in quantum cryptography after recognizing that leveraging the quantum channel to securely transmit a random key was simpler than transmitting the actual message. This marked the birth of Quantum Key Distribution (QKD), acknowledged in an information-theory conference in 1983, albeit with limited publication.

1984 (Later): Brassard published the comprehensive description of their QKD scheme, known as the BB84 QKD scheme [6], at the 1984 IEEE International Conference on Computers, Systems, and Signal Processing in India, overcoming challenges in getting such results published at the time.

Quantum Cryptography Evolution: The period witnessed skepticism regarding the feasibility of quantum cryptography due to technological limitations. However, breakthroughs in understanding photon behavior and its implications for secure communication initiated a shift in perspective. Bennett and Brassard's efforts centered on transmitting quantum states through a quantum channel, culminating in the birth of QKD [18].

13.3 CLASSICAL CRYPTOGRAPHY AND BOUNDARIES

13.3.1 Classical Cryptography

Classical cryptography encompasses three pivotal components within this form of communication:

- *The Message:* Representing the text or understandable information intended to be converted into an incomprehensible form.
- *The Legitimate Communicators:* Comprising the legitimate parties interested in communication or message sharing, potentially involving more than two parties and segregated into sender and receiver roles.
- *The Transmission Channel:* Serving as the medium responsible for conveying the message from sender to receiver, encompassing both airborne and physical mediums (such as coaxial cable or optical fiber).
- *The Encryption Key:* Responsible for converting clear text into an incomprehensible format.

- *The Decryption Key:* Responsible for transforming incomprehensible or encrypted text back into clear and understandable content.
- *Cryptanalysis:* An unauthorized facet of communication that aims to uncover the shared message between the legitimate communicators. It utilizes sophisticated attack strategies to extract information covertly. Two distinctive forms of classical cryptography are identified: symmetric key cryptography and asymmetric key cryptography.
- *Symmetric Key Cryptography:* Also referred to as private key cryptography. In this method, a shared secret key is maintained between the sender and the receiver, remaining confidential. This single key serves the dual purpose of encrypting and decrypting messages. Its symmetrical nature stems from using the same key for both encryption and decryption. Algorithms such as Advanced Encryption Standard (AES) exemplify symmetric key cryptography.
- *Asymmetric Key Cryptography:* Also referred to as public key cryptography, this strategy utilizes two distinct keys: a public key and a private key. The public key is openly available and is used for encryption, while the private key, safeguarded by the owner, is employed for decryption. Cryptographic systems like RSA [16] and Elliptic Curve Cryptography (ECC) [19] are examples of asymmetric key cryptography.

Both symmetric and asymmetric key cryptography possess unique advantages and applications. Symmetric key cryptography is generally faster and demands less computational power. However, it requires a secure key exchange between communicating parties. Asymmetric key cryptography resolves the key exchange issue by employing a public key for encryption. Nevertheless, it tends to be slower due to the complexity of the algorithms involved.

13.3.2 Limitations of Classical Cyptography

In the early 1980s, scientific interest surged in crafting a quantum computer, aspiring to tackle complex systems beyond classical computing capabilities. Feynman suggested simulating classical physics systems via quantum computers [20, 21]. Expectations soared for quantum computing to handle computations that exceeded classical computer capabilities. Grover's algorithm [22] seeks elements within lists, while Peter Shor in 1999 [23] proposed an algorithm which factors integers, surpassing classical algorithms like General Number Field Sieve (GNFS) due to quantum Fourier transformation, as illustrated in Figure 13.1.

While GNFS takes billions of years to factorize a 2048-bit RSA encryption key, Shor's algorithm accomplishes this in seconds to a month, demonstrating quantum computing's power. Yet, Shor's algorithm encounters notable challenges due to high error rates when attempting to factorize large integers using quantum computers. Its maximum factorized integer is 21 [25]. Despite this, RSA encryption's future security might be compromised, today's encrypted messages might become susceptible to decryption in the future if RSA's security is breached. To prevent this from happening, concerted efforts have been directed toward the development of Quantum Cryptography, spearheading what is known as Post-Quantum Cryptography. This initiative was instigated by the National Institute of Standards and Technology (NIST) in 2016 [26]. NIST proposed developing more complex math equations or those posing longer-solving times, even for quantum computers. Physical constraints were leveraged by using quantum mechanics to create secure protocols. Additionally, alternative encryption systems resistant to Shor's algorithm have been proposed, including McEliece's Code-Based Encryption, Quadratic Multivariable Equations Encryption, NTRU Network-Based Encryption, and Merkle's Hash-Based Encryption. Shor's algorithm is the primary threat to cryptographic systems. However, a possibility exists wherein a more potent quantum algorithm may arise in the future, potentially compromising the encryption systems mentioned earlier, even those classified as secure by NIST.

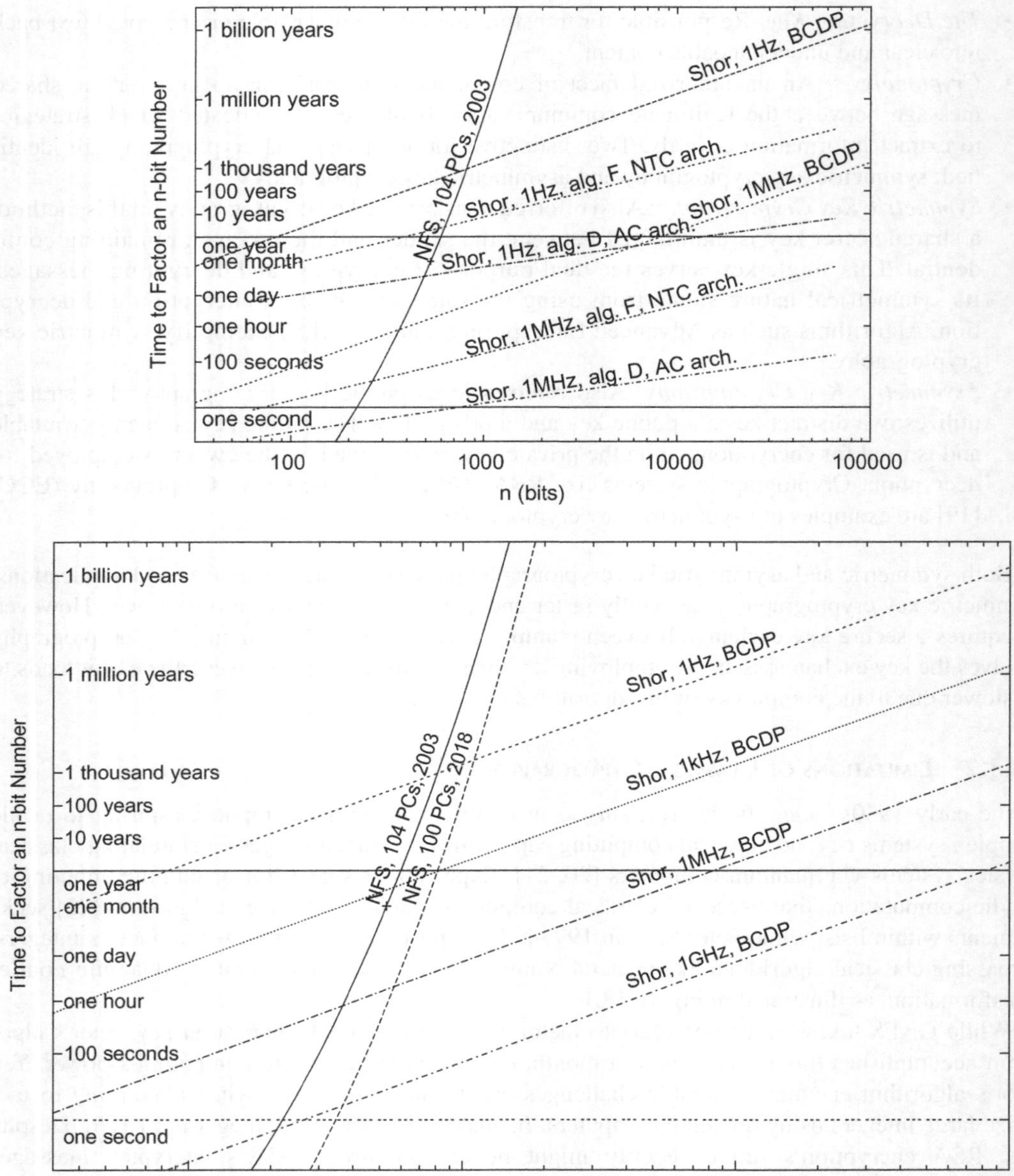

FIGURE 13.1 Power of the SHOR algorithm versus the GNFS protocol using binary-continuous double-precision modular exponentiation for various factorization durations [24].

13.4 QUANTUM CRYPTOGRAPHY

In this section, we will explain the basic ideas in quantum information that give us an idea to understand how quantum cryptography works. Quantum information theory is based on the core concept of superposition, which means things can be in many states at once until we check them. Quantum cryptography, a part of this theory, aims to change how we keep information secure. It uses special features of quantum mechanics like superposition and entanglement to create new and secure ways of communicating. Unlike classical cryptography based on mathematical complexity, quantum cryptography uses these special features to protect data. It is a big step forward in how we share information securely.

13.4.1 Basic Concepts

13.4.1.1 Density Operator

The quantum systems are always represented by quantum state vectors describing one of the properties of a given system, such as the wave function $|\psi\rangle$, the polarization state of light $|\uparrow\rangle/|\rightarrow\rangle$' the spin state of an electron, $|+1.2\rangle/|-1/2\rangle$, etc. Even the postulates of quantum mechanics are formulated in the language of these vectors known as kets. There exists an equivalent substitute for these vectors that offers more advantages, called the density operator or density matrix [27]. This operator provides a comprehensive description of all the possible states of the quantum system for a given property. We define the density operator of a system described by the state $|\psi\rangle$

$$\rho = |\psi\rangle\langle\psi| \tag{13.1}$$

The density operator satisfies two conditions:

- Trace condition: The trace of the density operator is always equal to 1, i.e., $tr(\rho) = 1$
- Positivity trace: the density operator is always positive and satisfies fs $\langle\rho\rangle = 1$

Here are some examples of density operators.

⋆ *For a Qubit:* A qubit is defined by

$$|q\rangle = \alpha|0\rangle + \beta|1\rangle \tag{13.2}$$

where $\{\alpha, \beta\} \in$ C2 are complex numbers satisfying $\langle q|q\rangle = 1 \Rightarrow |\alpha|^2 + |\beta|^2 = 1$.

$$\begin{aligned} \rho &= |q\rangle\langle q| \\ &= (\alpha|0\rangle + \beta|1\rangle)((\langle 0|\alpha^* + \langle 1|\beta^*) \\ &= \begin{pmatrix} |\alpha|^2 & \alpha^*\beta \\ \alpha\beta^* & |\beta|^2 \end{pmatrix} \end{aligned} \tag{13.3}$$

So, the density operator in $\{|0\rangle, |1\rangle\}$ basis is given by

$$\rho = \begin{pmatrix} |\alpha|^2 & \alpha^*\beta \\ \alpha\beta^* & |\beta|^2 \end{pmatrix} \tag{13.4}$$

we can represent this density operator in $\{|+\rangle, |-\rangle\}$ basis where

$$|+\rangle = \frac{1}{\sqrt{2}}(|0\rangle + \langle 1|) \tag{13.5}$$

$$|-\rangle = \frac{1}{\sqrt{2}}(|0\rangle - |1\rangle) \tag{13.6}$$

We get the following after the calculation:

$$\rho = 2\begin{pmatrix} |\alpha+\beta|^2 & (\alpha+\beta)*(\alpha-\beta) \\ (\alpha-\beta)*(\alpha+\beta) & |\alpha-\beta|^2 \end{pmatrix} \tag{13.7}$$

In classical information theory, measurement refers to extracting information about a physical quantity, such as distance, temperature, pressure, volume, energy, or mere observation. Crucially, this process does not affect or change the system's state. However, in quantum mechanics, measurement shares a similar definition but distinctly influences the system's state, even if it is just an observation [3]. This property, while initially seen as a disadvantage due to the disturbance it causes, actually serves as a valuable advantage in certain contexts, particularly in quantum communication and cryptography.

In the context of our study, if two legitimate users are exchanging information and an unauthorized third party attempts to eavesdrop or intercept the communication, the act of measuring or observing the quantum information disturbs the state of the quantum system. This disturbance can be detected by the legitimate users, serving as an indication that an unauthorized party is trying to intercept the communication.

13.4.1.2 No-Cloning Theorem

The non-cloning theorem [4] is assumed to be the second fundamental principle of quantum cryptography after the measurement postulate. This is due to the impossibility of duplicating random quantum state, which is a trivial operation in classical information theory, as it is possible to make copies of a bit or a set of bits without difficulty and without causing errors or leaving a trace. This theory arises from the intrinsic properties of quantum states. To demonstrate this theorem, considering a quantum state $|\psi_1\rangle$ representing the original state we wish to duplicate onto the state $|\phi\rangle$ representing a blank sheet, the cloning transformation of the state can be described by a specific quantum operation:

$$|\psi_1\rangle \otimes |\phi\rangle = |\psi_1\rangle \otimes |\psi_1\rangle \tag{13.8}$$

This transformation is governed by a unitary operator U, which does not need to be specified:

$$U|\psi_1\rangle \otimes |\phi\rangle = |\psi_1\rangle \otimes |\psi_1\rangle \tag{13.9}$$

U must be independent of $|\psi_1\rangle$ as it is arbitrary by assumption. If the state $|\psi_1\rangle$ is known, the preparation procedure will be known, and if we want to duplicate a second original state $|\psi_2\rangle$, we will have

$$U|\psi_2\rangle \otimes |\phi\rangle = |\psi_2\rangle \otimes |\psi_2\rangle \tag{13.10}$$

If we calculate the inner products (13.9) and (13.10),

$$N = \langle\phi| \otimes \langle\psi_1|U^{\dagger}U|\psi_2\rangle \otimes |\phi\rangle \tag{13.11}$$

Equation 13.15 has two possible ways to be calculated

$$N = \langle \phi | \otimes \langle \psi_1 | I | \psi_2 \rangle \otimes | \phi \rangle$$
$$= \langle \psi_1 | \psi_2 \rangle$$
$$N = (\langle \psi_1 | \otimes \langle \psi_1 |)(| \psi_2 \rangle \otimes | \psi_2 \rangle)$$
$$= (\langle \psi_1 | \psi_2 \rangle)^2 \quad (13.12)$$

We observe from Equation 13.12 that the condition $\langle \psi_1 | \psi_2 \rangle = (\langle \psi_1 | \psi_2 \rangle)^2$ implies that the cloning operation is possible only if $| \psi_1 \rangle \equiv | \psi_2 \rangle$ or $| \psi_1 \rangle \perp | \psi_2 \rangle$. In other words, the cloning operation can only occur if the original state and the cloning state are identical or orthogonal. It should be noted that the superposition of the two states cannot be duplicated.

13.4.1.3 Quadratic Operator

The quadratic operator [28] is an important tool in quantum cryptography, especially in continuous variables quantum key distribution CV-QKD. It is used to modulate the secret key on the properties of light (phase, amplitude, etc). Beginning with the classical quadratic operator definition, an exploration of the electromagnetic field can be initiated, followed by its quantum analogue associated with the quantum harmonic oscillator. The electromagnetic field can be expressed in terms of its amplitude, E_0, as follows:

$$E(t) = E_0 \cos(\omega t + \phi)$$
$$= E_0 \cos(\omega t)\cos(\phi) - E_0 \sin(\omega t)\sin(\phi)$$
$$= Z_1 \cos(\omega t) + Z_1 \sin(\omega t) \quad (13.13)$$

This can be expressed as a cosine oscillation with amplitude Z_1 and a sine oscillation with amplitude Z_2. The variables Z_1 and Z_2 are termed the quadratures of the electromagnetic field oscillating with a phase shift of π. Another approach to define this classical field is by utilizing the phase representation:

$$a(t) = E_0 e^{-i\phi} e^{-i\omega t}$$
$$= a e^{-i\omega t} \quad (13.14)$$

With $a = E_0 e^{-i\varphi}$ representing the phase amplitude, the phase amplitude "a" can be expressed in terms of Z_1 and Z_2:

$$a = Z_1 + iZ_2 \quad (13.15)$$

where,

$$Z_1 = Re(a) = \frac{1}{2}(a + a^*) \tag{13.16}$$

$$Z_2 = Im(a) = \frac{1}{2i}(a - a^*) \tag{13.17}$$

Now, moving to the quantum analogy of these classical quadratic operators in terms of the creation and annihilation operators $\hat{a}$ and $\hat{a}^\dagger$ that satisfy the commutation relation $[\hat{a}, \hat{a}^\dagger] = 1$

$$[\hat{Z}_1, \hat{Z}_2] = i/2 \tag{13.18}$$

$\hat{Z}_1$ and $\hat{Z}_2$ do not commute as $\hat{x}$ and $\hat{p}$ do not commute, and they are Hermitian operators.

13.4.2 Detection Technique

13.4.2.1 Photon Detection

Photon detection [29] is a critical element in QKD, involving the direct measurement of individual photons, which are the smallest units of light. In QKD protocols like BB84, photon detectors serve as indispensable instruments for accurately registering, identifying, and quantifying these photons. They play a pivotal role in extracting quantum information encoded in photons, ensuring the secure exchange of cryptographic keys between parties by preserving the quantum state of transmitted photons. However, the photon detection technique has several disadvantages:

- *Poor Efficiency:* Photon detectors may suffer from limitations in efficiently capturing individual photons.
- *Dark Counts:* These detectors can generate signals, termed dark counts, even in the absence of incident photons, contributing to background noise.
- *Sensitivity to Channel Noise:* Photon detectors can be sensitive to noise channel or environmental conditions like temperature changes or electromagnetic interference.
- *Expensive:* High-performance photon detectors can be expensive, posing challenges for their widespread adoption and implementation in QKD systems.

13.4.2.2 Homodyne Detection

Homodyne detection [30] is an essential technique in continuous variable quantum cryptography and communication systems in general [31], enabling the measurement of a quadrature measurement (Figure 13.2). This method combines the incoming signal with a local oscillator beam of the same frequency using a beam splitter; the resulting signal is sent to a detector. The key aspect lies in the manipulation of the local oscillator's phase, which allows for precise extraction of information pertaining to the phase and amplitude of the original signal. This approach facilitates the identification of the light signal, helping to extract encrypted information within it.

The photodetectors, D1 and D2, measure the photocurrent at outputs 1 and 2, respectively. These photocurrents are directly proportional to the intensity of light. Moreover, the intensity of light is in turn proportional to the number operators of outputs 1 and 2, as

$$\hat{n}_1 = \hat{b}_1 b_1^\dagger \tag{13.19}$$

$$\hat{n}_2 = \hat{b}_2 b_2^\dagger \tag{13.20}$$

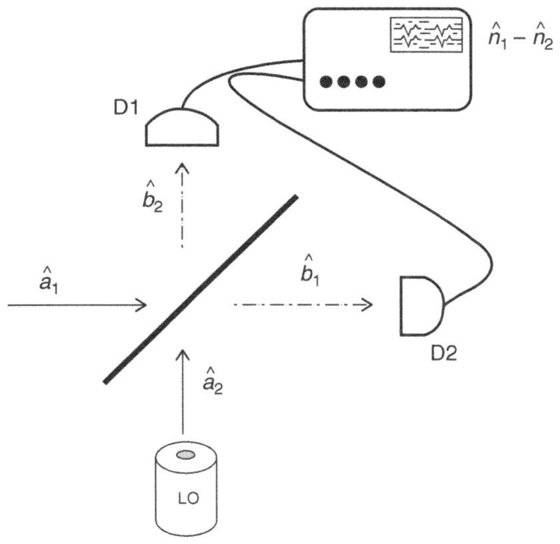

FIGURE 13.2 Homodyne detection scheme.

The average value of the photocurrent is proportional to the input operators of the beam splitter:

$$\hat{b}_1 = T\hat{a}_2 + R\hat{a}_1 \tag{13.21}$$

$$\hat{b}_2 = T\hat{a}_1 + R\hat{a}_2 \tag{13.22}$$

By substituting Equation 13.24 into Equations 13.23, we find

$$\hat{n}_1 = \frac{1}{2}\left(\hat{a}_2^\dagger\hat{a}_2 + \hat{a}_1^\dagger\hat{a}_1 + i\hat{a}_2^\dagger\hat{a}_1 - i\hat{a}_1^\dagger\hat{a}_1\right) \tag{13.23}$$

$$\hat{n}_2 = \frac{1}{2}\left(\hat{a}_2^\dagger\hat{a}_2 + \hat{a}_1^\dagger\hat{a}_1 - i\hat{a}_2^\dagger\hat{a}_1 + i\hat{a}_1^\dagger\hat{a}_1\right) \tag{13.24}$$

If we calculate the difference of photocurrents between outputs 1 and 2, we will obtain

$$\hat{n}_1 - \hat{n}_2 = i\left(\hat{a}_2^\dagger\hat{a}_1 - \hat{a}_1^\dagger\hat{a}_2\right) \tag{13.25}$$

This technique of detection is extensively employed in continuous variables QKD protocols.

13.4.2.3 Heterodyne Detection

This technique is similar to homodyne detection; it utilizes a local oscillator, employing two different frequencies: one for the signal and another for the oscillator. These frequencies are combined, generating an interference pattern that is subsequently examined by a detector. Heterodyne detection offers simultaneous measurement of amplitude and phase information. It is often used in high-resolution spectroscopy and communication systems that require precise signal characterization. Heterodyne detection is therefore most useful for detecting weak signals that are coherent with a locally produced source.

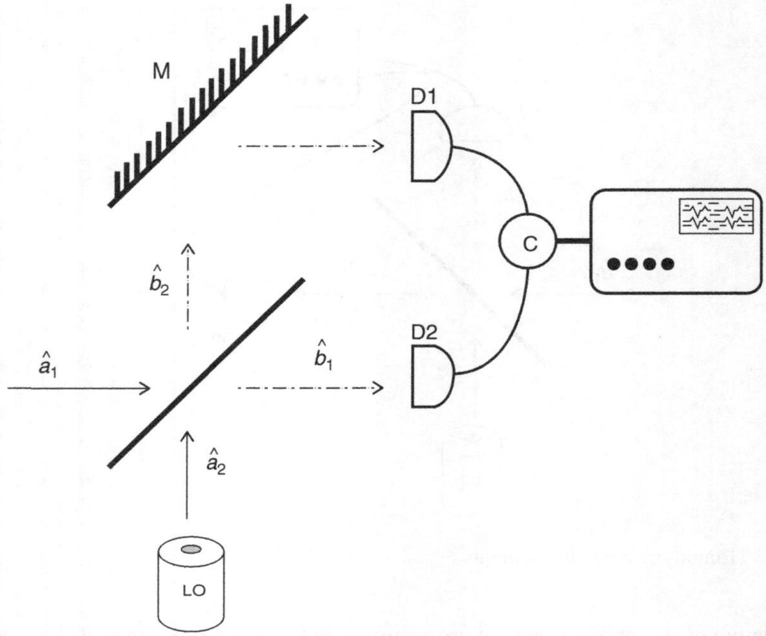

FIGURE 13.3 Heterodyne detection scheme.

Figure 13.3 demonstrates how a basic optical heterodyne detection is organized. It involves merging the signal and local-oscillator beams using a beam splitter. The two resulting beams are measured by two detectors, each receiving half of the signal power along with half of the local oscillator's power. To optimize light usage, two detectors are utilized. The beams are superimposed onto the detectors, aligning their electric fields to combine. For more information about this technique, one should consult the references [30, 32, 33].

13.4.3 Quantum Key Distribution

The core principle of QKD is to securely share an encryption key between two authorized communicators, leveraging the laws of quantum mechanics. The security of this communication is ensured by the measurement postulate, which detects any interception attempts by a potential eavesdropper who might interact with quantum states to obtain information. These interactions introduce disturbance on quantum state, detected by the communicators during channel parameter estimation. Analysis of these estimations indicates the potential presence of an eavesdropper. Additionally, communication security is guaranteed by the previously defined non-cloning theorem, preventing an attacker from replicating information exchanged over the channel. Two main families of Quantum Key Distribution protocols exist: discrete-variable QKD (DV-QKD) and continuous-variable QKD (CV-QKD). DV-QKD protocols use discrete variables such as photon phase or polarization to encode the encryption key, while CV-QKD protocols utilize continuous variables like electromagnetic field quadratures for encoding. Table 13.1 summarizes key differences between these two protocol types.

13.4.3.1 BB84

The BB84 protocol [6] stands as one of the earliest quantum key distribution methods that has been successfully tested in experiments. Named after its inventors Charles Benett and Gilles Brassad in 1984, it operates by encoding bits through light polarization. The protocol follows these key steps:

TABLE 13.1
Classification of Some Important Characteristics of CV-QKD and DV-QKD Protocols

	CV-QKD	DV-QKD
Type of variables	EM-field, amplitude, phase	phase/ photon polarization
Emission	laser sources	Single-photon source
Detection	coherent detection homodyne/ heterodyne	avalanche photodiode
Range	202.81 km on optic fibr 23 km on air	100 km
Rate	10 kb/s	1Mb/s

Firstly, Alice prepares two encoding bases: B_1, the computational basis, and B_2, the conjugate basis:

$$B_1 = \{|0\rangle;|1\rangle\} \tag{13.26}$$

$$B_2 = \{|+\rangle;|-\rangle\} \tag{13.27}$$

where

$$|+\rangle = \frac{1}{\sqrt{2}}(|0\rangle + |1\rangle) \tag{13.28}$$

$$|-\rangle = \frac{1}{\sqrt{2}}(|0\rangle - |1\rangle) \tag{13.29}$$

Subsequently, Alice generates two random bit sequences of equal length. The first sequence indicates the order of bases she will use to encode each bit in the second sequence, which denotes the bit chain to be encoded onto the polarization states of each base.

$$Sequence\ 1: \begin{cases} B_1 \rightarrow "0" \\ B_2 \rightarrow "1" \end{cases} \tag{13.30}$$

$$Sequence\ 2: \begin{cases} |0\rangle;|+\rangle \rightarrow "0" \\ |1\rangle;|-\rangle \rightarrow "1" \end{cases} \tag{13.31}$$

Next, Alice encodes her bit sequence onto the four polarization states of bases B_1 and B_2, thereby transmitting them to Bob through a quantum channel. On Bob's side, he randomly selects either B_1 or B_2 to measure each signal received 4. Once all qubits are transmitted, Alice publicly declares her base sequence. Then, Bob and Alice retain only the measurements made in the correct base, halving the bit sequence compared to Alice's original.

An important consideration is that an eavesdropper (commonly known as Eve) could potentially eavesdrop on the public channel. Additionally, Alice and Bob allocate a portion of their encryption key within the channel to estimate the error rate. If the error rate exceeds a certain threshold, it

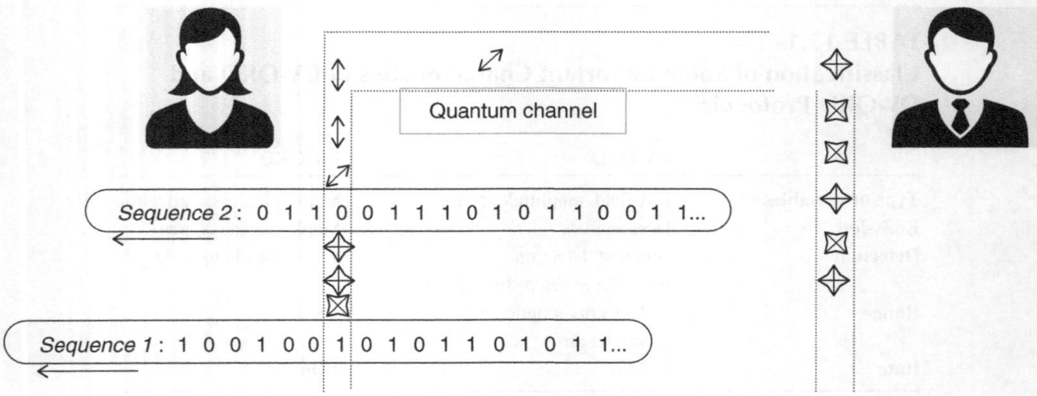

FIGURE 13.4 Discrete variable BB84 protocol.

suggests the presence of an eavesdropper. In such a case, Alice and Bob have the choice to restart the protocol or implement error correction techniques.

13.4.3.2 Six-State Protocol

The six-state protocol [7], an extension of the BB84 protocol (Figure 13.4), introduces an extra third basis to the transmission process, enriching the complexity of quantum communication:

$$|+i\rangle = \frac{1}{\sqrt{2}}\left(|0\rangle + i|1\rangle\right) \tag{13.32}$$

$$|-i\rangle = \frac{1}{\sqrt{2}}\left(|0\rangle - i|1\rangle\right) \tag{13.33}$$

This augmentation is intended to bolster the security of the communication channel by intensifying the difficulty of intercepting communications and amplifying the errors incurred by any potential eavesdropper. While this enhancement enhances the protocol's security, it results in a trade-off: the legitimate measurement outcomes in the correct basis now significantly condense Alice's original sequence by a factor of three, potentially impacting the efficiency of data transmission.

13.4.3.3 B92

Bennett demonstrated that neither four nor six states are necessary for secure QKD; only two non-orthogonal states are sufficient [9]. His protocol, famously termed the B92 protocol, received considerable attention within the community for its fundamental contributions to understanding the security of QKD schemes (Figure 13.5). Initially designed for DV-QKD, it has been adapted for CV-QKD for the security reasons (Figures 13.6 and 13.7). They found out that CV-QKD B92 is an enhanced version that offers more advantage than its DV counterpart.

The situation is quite different from the BB84 protocol. In this protocol, Alice uses two states, 0 and +, from the computational and conjugate bases, respectively. This choice is made for the non-orthogonality condition. Alice encodes her bit sequence in these two states as follows: The situation is significantly different from the BB84 protocol. Within this protocol, Alice employs two distinct states, 0 and +, derived from the computational and conjugate bases 28, respectively, chosen specifically to fulfil the non-orthogonality requirement. Alice then proceeds to encode and send her bit sequence using these two states in the following manner:

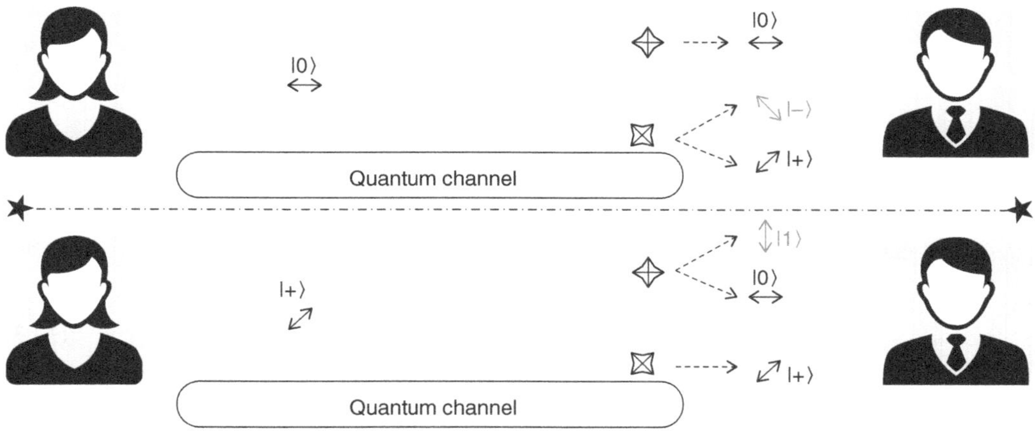

FIGURE 13.5 Discrete variable B92 protocol.

$$\begin{cases} |-\rangle; \rightarrow "0" \\ |+\rangle; \rightarrow "1" \end{cases} \tag{13.34}$$

Bob, on his end, randomly measures each signal in either the computational basis or the conjugate basis, obtaining his results as follows:

$$\begin{cases} |-\rangle; \rightarrow "0" \\ |1\rangle; \rightarrow "1" \\ else; \rightarrow inconclusive\,results \end{cases} \tag{13.35}$$

If Bob measures and finds the state $|-\rangle$, he sets the bit as "0." This decision is made because this particular state is not used in the protocol, and this state cannot be obtained if the state $|0\rangle$ was not sent by Alice. Similarly, if Bob measures and finds the state $|1\rangle$, he sets the bit as "1." This choice is made because the state $|1\rangle$ is not utilized in this protocol, and this state cannot be obtained if the state $|+\rangle$ was not transmitted by Alice 5.

13.4.3.4 Continuous Variables Protocols

CV protocols in QKD refer to protocols that utilize continuous parameters, like the amplitude or phase, for encoding and decoding information. These protocols often utilize the quadrature phase amplitudes of light to encode information. Gaussian modulation, coherent states, and homodyne detection are frequently employed techniques in CV-QKD protocols. Continuous-variable BB84 is an example of CV protocols that has been studied in the literature [34]. In the CV-BB84 protocol, Alice prepares a coherent state, characterized by the parameter α [12], which represents a good example of a laser source. We assume α is a real number. The continuous-variable QKD protocol BB84, similarly to DV-BB84 discussed in Section 13.4.3, necessitates two bases, each comprising two states achieved by altering the pulse phases $\alpha = |\alpha| e^{i\phi}$ as follows:

$$B_1 = \begin{cases} \phi = 0 & |\alpha\rangle \\ \phi = \pi & |-\alpha\rangle \end{cases} \tag{13.36}$$

$$B_2 = \begin{cases} \phi = \frac{\pi}{2} & |\alpha\rangle \\ \phi = \frac{3\pi}{2} & |-\alpha\rangle \end{cases} \quad (13.37)$$

Once the encoding bases B_1 and B_2 are established, Alice generates a long random bit series of "0"s and "1"s. She encodes her bit sequence randomly using the four states and sends these encoded states via a quantum channel to Bob:

$$"0" = \begin{cases} |-\alpha \\ |-i\alpha \end{cases}, \quad "1" = \begin{cases} |\alpha \\ |+i\alpha \end{cases} \quad (13.38)$$

Bob, on his part, prepares to receive Alice's pulses with his two measurement quadratures $\hat{z}_1$ and $\hat{z}_2$. These quadratures are defined by the relations $\hat{a} = \hat{z}_1 + i\hat{z}_2$ and $\hat{a}^\dagger = \hat{z}_1 + i\hat{z}_2$ where $\left[\hat{z}_1, \hat{z}_2\right] = \frac{i}{2}$. Employing homodyne detection 2, he randomly chooses a measurement quadrature $\hat{z}_1$ or $\hat{z}_2$ for each received signal.

Following the transmission of numerous pulses, Alice and Bob engage in reconciliation via a classical channel. During this phase, Alice discloses the bases used for each transmitted bit, and Bob confirms whether he measured in the correct base or not. The bits measured in the right base form Alice's encryption key. Bob retains the bits measured in the correct base, known as the correct data, discarding those measured incorrectly. Afterward, using his post-selection threshold z_c, Bob further filters the retained data, classifying bits based on specific thresholds:

$$\begin{cases} z_{1,2} < -z_c \rightarrow "0" \\ z_{1,2} > z_c \rightarrow "1" \\ else \rightarrow ignore \end{cases} \quad (13.39)$$

This filtering process involves disregarding data within the interval $\left]-z_c, z_c\right[$, representing the uncertainty range where distinguishing between "0" and "1" becomes ambiguous. The post-selection threshold is an important tool in CV-QKD protocols [12].

13.5 EAVESDROPPING

There is no perfect communication during the exchange of information. Despite the robust encryption methods employed, various factors can contribute to the communication. Environmental factors, such as interference or signal attenuation due to channel noise, are among these factors. Additionally, the presence of an eavesdropper poses a significant threat, as it cannot only corrupt sensitive information but also exploit it against legitimate users. This compels researchers to invest considerable effort in proposing new protocols and fortifying communication channels to secure information against potential threats. In this section, we aim to define several attack strategies employed by eavesdroppers.

13.5.1 Intercept Resend Attack

The intercept resend attack is a well-known strategy extensively investigated in various studies, DV-QKD or CV-QKD [34–36]. In DV-QKD Eve, the potential eavesdropper aims to capture photons

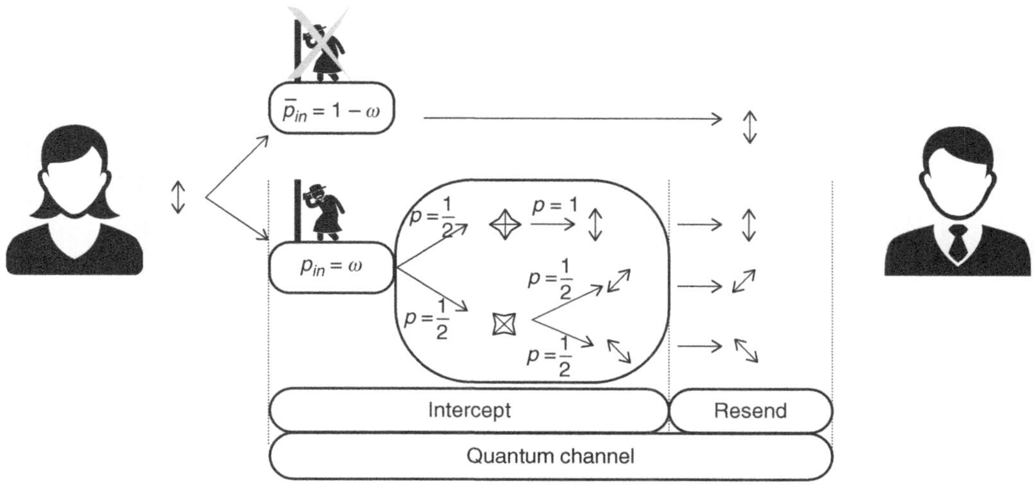

FIGURE 13.6 Intercept resend attack scheme in DV-QKD.

transmitted by Alice, measure them, and then resend them onto Bob (Figure 13.6). However, in the BB84 protocol, Alice uses a random basis to encode each photon, preventing Eve from predicting which basis Alice will employ for each intercepted photon Figure 13.4. Consequently, this situation leads Eve to similarly select her bases randomly. In 50% of cases, she selects the wrong bases.

Suppose Alice transmits the encrypted bit "1" using the computational basis, signifying the photon is in the state $|1\rangle$. Eve faces two scenarios: intercepting or not intercepting the photon, with complementary probabilities of ω and $(1-\omega)$,respectively.

If Eve intercepts the photon, two possibilities arise. She might select the correct or incorrect basis for measurement, each with a probability of 1/2. If she chooses the correct basis (which is, in our case, the computational basis), Eve determines Alice's value "1 precisely" with a perfect projection. She can then resend it to Bob using the same basis.

However, if Eve chooses the wrong basis (the conjugate basis 28), she will obtain either "0" or "1"as a result of her measurement, each with an equal probability 1/2, depending on the following projection:

$$|\langle + \mid 1\rangle|^2 = \left|\frac{1}{\sqrt{2}}(\langle 0 \mid 1\rangle + \langle 1 \mid 1\rangle\right|^2 = |\langle - \mid 1\rangle|^2 = \frac{1}{2} \tag{13.40}$$

On the contrary, if Eve does not intercept the photon sent by Alice, the photon safely reaches Bob. In this scenario, Eve randomly chooses either "0" or "1" because of her measurement, each with a probability of 1/2.

In CV-QKD, the scenario differs significantly (Figure 13.7). Here, Eve employs a 50:50 beam splitter to divide Alice's state, after which she performs simultaneous measurements on both outputs of the beam splitter. Eve employs $\hat{x}_1$ and $\hat{x}_2$ with homodyne detection to measure the two emerging beams from the splitter. Subsequently, she compares her measurement outcomes x_1 and x_2 using the following criterion:

$$\begin{cases} |\alpha\rangle & si \;\; x_1 \geq |x_2| \\ |i\alpha\rangle & si \;\; x_2 \geq |x_1| \\ |-\alpha\rangle & si \;\; -x_1 \geq |x_2| \\ |-i\alpha\rangle & si \;\; -x_2 \geq |x_1| \end{cases} \tag{3.41}$$

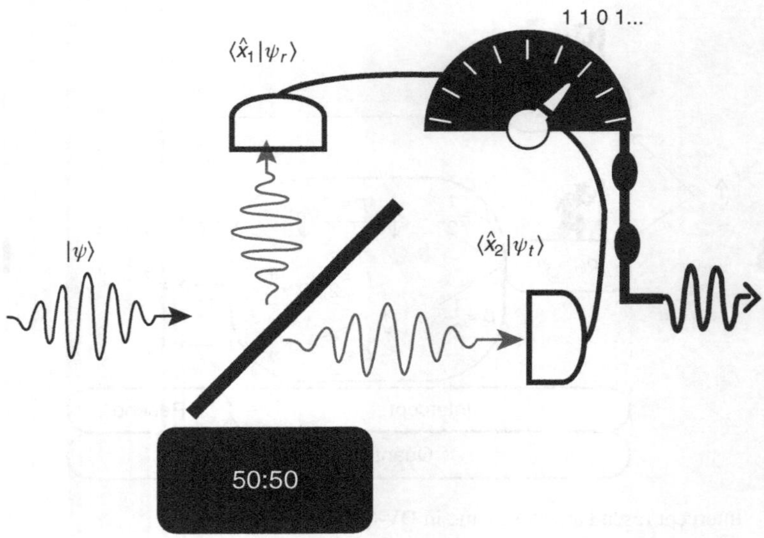

FIGURE 13.7 Intercept resend attack scheme in CV-QKD.

This criterion arose from representing the probability distribution of all states, taking into consideration the projection in both the correct and incorrect bases. For more details about this attack strategy, one can consult the references [37-38].

13.5.2 Superior Channel Attack

In this attack, Eve is assumed to have a perfect quantum channel and a quantum memory capable of storing light states until the reconciliation step [34]. Eve stores the reflected part of the original signal emerging from the 50:50 beam splitter. This portion of the signal would typically be lost in the quantum channel during the transmission of the encryption key, while the transmitted part remains intact through her own perfect channel to Bob. Subsequently, she waits for the reconciliation step to determine the exact bases to use for each stored state.

Unfortunately, this type of attack cannot be detected by legitimate users due to its indistinguishability from ordinary channel noise (Figure 13.8).

13.5.3 Gravitational Attack

In the BB84 protocol, Alice and Bob establish a private key using the principles of quantum mechanics (as outlined in Section 13.4.3), relying on the laws of physics to secure their communication. However, the impact of gravitational laws is often overlooked. In a "gravitational attack," [37] the eavesdropper splits the quantum state $|\psi\rangle$ using a beam splitter, resulting in the state: $|\psi\rangle = |\psi\rangle(x_1) + |\psi\rangle(x_2)$. Here, $|\psi(x_1)\rangle$ is measured in one output arm of the beam splitter using the computational basis, while $|\psi(x_2)\rangle$ is measured in the other output arm using the conjugate basis 28. Depending on the obtained measurement results s where $(s = 0, s = 1, s = +, \text{or } s = ?)$,a macroscopic test mass minitially positioned at $x(1)$ is automatically relocated to one of four separate spatial positions: $x(0), x(1), x(+), x(\ )$ Subsequently, Eve performs experimental measurements to determine the gravitational field around these four positions, with the help of the Cavendish setup (Figure 13.9).

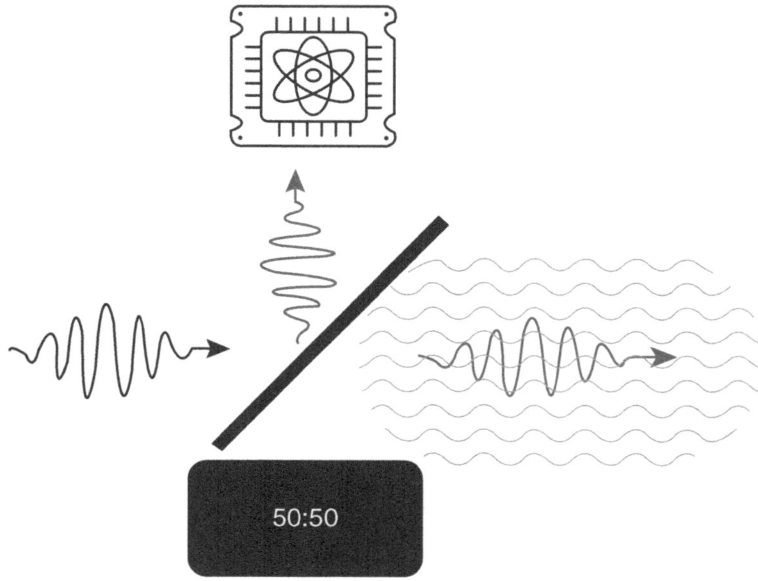

FIGURE 13.8 Superior channel attack scheme.

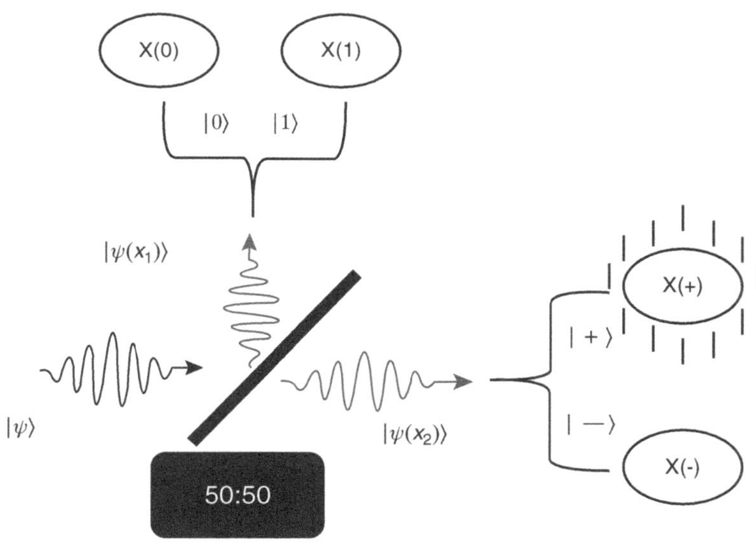

FIGURE 13.9 Gravitational attack scheme.

13.6 ADVANTAGES, DISADVANTAGES, AND APPLICATION OF QUANTUM CRYPTOGRAPHY

In this section, we will explore several advantages and disadvantages associated with quantum cryptosystems. Additionally, we will categorize and discuss various application fields where these systems find relevance [38].

13.6.1 Advantages of Quantum Cryptography

Ease of Use: Quantum cryptography offers user-friendly applications and protocols, making it accessible for secure communication.

Security: Utilizes the principles of quantum mechanics, rendering it theoretically impenetrable and highly secure.

Efficient Resource Usage: Generally, requires fewer resources compared to classical cryptographic methods.

13.6.2 Disadvantages of Quantum Cryptography

Limited Range: Presently constrained to a range of approximately 90 miles, limiting long-distance applications.

Job Displacement: Widespread adoption might replace numerous jobs and impact existing physical resources.

13.6.3 Applications of Quantum Cryptography

Data Encryption: Securing sensitive data against unauthorized access and maintaining confidentiality.

Digital Signatures: Ensuring the authenticity and integrity of digital documents, transactions, or identities.

Secure Communication with Space: Enabling secure data exchange between satellites and ground stations for space missions and communications.

Quantum Internet: Forming the backbone for an ultra-secure, quantum-based internet infrastructure, offering enhanced security in data transmission.

Key Agreement: Facilitating the secure exchange of cryptographic keys between multiple parties for secure communication.

Advanced Power Grids: Providing highly secure communication and control in power grid systems, ensuring protection against cyber threats.

Ultra-Secure Voting: Implementing tamper-proof voting systems, ensuring secure and verifiable election processes.

Analysis and Prediction of DNA Structure: Leveraging secure communication to analyze and predict complex DNA structures for medical research and treatment.

Functional Brain Analysis: Enabling secure communication for in-depth research and understanding of brain functionalities, aiding in neuroscience.

Secure Financial Transactions: Ensuring secure and tamper-proof financial transactions, protecting sensitive information in banking and finance.

Enhanced Healthcare Security: Facilitating secure communication for confidential patient data and healthcare records.

Smart City Infrastructure: Implementing secure communication protocols in smart city networks to safeguard critical data and infrastructure.

13.7 CLASSICAL V/S QUANTUM CRYPTOGRAPHY

Quantum cryptography fundamentally differs from classical cryptography across multiple dimensions [18]:

- ⋆ *Fundamental Dimension:* From a fundamental standpoint, classical cryptography relies on classical physics laws to safeguard the transmission of both encrypted messages and cipher keys. Under Shannon's classical information theory, legitimate users leverage macroscopic

TABLE 13.2
Comparing Key Features of Classical Cryptography and Quantum Cryptography

Features	Quantum Cryptography	Classical Cryptography
Basis	Quantum mechanics	Mathematical computation
Security level	Unconditional security	Limited security
Unity	Quantum bit (qubit)	Classical bit (bit 0/1)
Cost	Expensive	Almost zero
Requirements	Dedicated hardware and communication lines	Software and portable
Communication range	202.81 km on optic fiber	Million of kilometers
Bit rate	1 Mb/s	Depend on computing power

properties of physics for the transmission process. On the contrary, quantum cryptography relies on the properties of quantum mechanics to ensure the security of the cipher key transmission. It operates as a branch of quantum information theory, where legitimate users utilize microscopic properties of physics for the key transmission. The key distinction lies in classical cryptography which transmits both the encrypted message and the key, while quantum cryptography exclusively transmits the key.

- *Application Dimensions:* Classical cryptography finds wide application in everyday life, encompassing various communication means and serving as a foundational security component for internet usage. It is a cornerstone for ensuring the safety of banking transactions, relying on robust mathematical algorithms. However, its security weakens with the progress of quantum cryptography, which is utilized across diverse fields, as previously mentioned in section 0.4.5. Nonetheless, the implementation of its algorithms is not simple.
- *Commercial Dimensions:* In the commercial domain, quantum cryptography requires expensive and rare tools to implement a QKD protocol. The transmission and detection tools needed for high-performance encryption in quantum cryptography are costly. On the other hand, classical cryptography can be effortlessly implemented through software at almost zero cost for consumers. Moreover, the transmission and detection tools for classical cryptography are readily available anywhere and at any time.
- *Technological Dimensions:* Quantum cryptography, despite its potential, has not reached the technological advancement of classical cryptography due to equipment limitations and sensitivity to channel noise, resulting in key loss. Conversely, classical cryptography boasts of high security and transmission at long distances, securing crypto coins, but remains vulnerable to future quantum computers. Quantum cryptography excels in smaller distances with unconditional security based on quantum principles, yet faces limitations in scalability. While classical cryptography demonstrates broader applications and robustness, quantum cryptography, constrained by technology and quantum system sensitivities, shows promise in specific scenarios at a more limited scale (Table 13.2).

13.8 CONCLUSION

In this chapter, we have explored quantum cryptography and its crucial role in providing an unconditioned level of security in communication between parties. We have traced its history, giving credit to Wiesner for sparking the development of quantum cryptography. We started by explaining the two main types of classical cryptography – private and public – to help draw comparisons with quantum cryptography. We also talked about Shor's algorithm, which poses a serious threat to classical cryptography and is one of the first quantum-based algorithms. We have described quantum cryptography as a better alternative that relies on quantum mechanics principles, such as the measurement postulate, which reveals any act of measurement and gives the possibility to detect any intruder within the communication, and the non-cloning theorem which prevent eavesdroppers from copying a quantum state exchanged between legitimate users. These principles form the foundation of its robust security features.

We have outlined several detection techniques utilized in quantum key distribution protocols, including the use of photon detection, homodyne and heterodyne detection as prominent example. We have introduced several well-known QKD protocols, namely BB84, the six-state protocol, and B91. Additionally, we have discussed some eavesdropping strategies, such as the intercept-resend attack, highlighting potential threats in securing quantum communication. We have explored several advantages of quantum cryptography, primarily focusing on its superior security and efficiency. However, we have also acknowledged its limitations, notably its restricted operational range. Furthermore, we have highlighted its diverse applications in various fields, including data encryption and digital signatures.

In the final section, we conducted a comprehensive comparison between classical and quantum cryptography across multiple dimensions, pointing out their differences. This comparison emphasized that quantum cryptography excels in nearly every aspect, highlighting its superiority as a secure system in multiple dimensions. To sum up, quantum cryptography is a big deal when it comes to secure communication. It shows how integrating quantum physics into information security can lead to unbreakable encryption methods.

REFERENCES

[1] Amer, Omar, Vaibhav Garg, and Walter O. Krawec. "An introduction to practical quantum key distribution." *IEEE Aerospace and Electronic Systems Magazine* 36.3 (2021): 30–55.

[2] Hughes, Richard J., George L. Morgan, and C. Glen Peterson. "Practical quantum key distribution over a 48-km optical fiber network." *arXiv preprint quant-ph/9904038* (1999).

[3] Shankar, Ramamurti. *Principles of quantum mechanics*. Springer Science & Business Media (2012).

[4] Wootters, William K., and Wojciech H. Zurek. "A single quantum cannot be cloned." *Nature* 299.5886 (1982): 802–803.

[5] Lee, Chankyun, Ilkwon Sohn, and Wonhyuk Lee. "Eavesdropping detection in BB84 quantum key distribution protocols." *IEEE Transactions on Network and Service Management* 19.3 (2022): 2689–2701.

[6] Bennett, C. H., Brassard, G., "Quantum cryptography: Public key distribution and coin tossing." *Theoretical Computer Science* 560 (2014): 7–11.

[7] Bruβ, Dagmar. "Optimal eavesdropping in quantum cryptography with six states." *Physical Review Letters* 81.14 (1998): 3018.

[8] Ekert, Artur K. "Quantum cryptography based on Bell's theorem." *Physical Review Letters* 67.6 (1991): 661.

[9] Bennett, Charles H. "Quantum cryptography using any two nonorthogonal states." *Physical Review Letters* 68.21 (1992): 3121.

[10] Tosi, Alberto, Franco Zappa, and Sergio Cova. "Single-photon detectors for practical quantum cryptography." *Electro-Optical Remote Sensing, Photonic Technologies, and Applications VI*. Vol. 8542. SPIE (2012).

[11] Gao, Fangzhou. "Practical analysis of discrete variable quantum key distribution." *2020 IEEE 2nd International Conference on Circuits and Systems (ICCS)*. IEEE (2020).

[12] Pirandola, Stefano, et al. "Continuous-variable quantum cryptography using two-way quantum communication." *Nature Physics* 4.9 (2008): 726–730.

[13] Vernam, Gilbert S. "Cipher printing telegraph systems: For secret wire and radio telegraphic communications." *Journal of the AIEE* 45.2 (1926): 109–115.

[14] Shannon, Claude Elwood. "A mathematical theory of communication." *The Bell System Technical Journal* 27.3 (1948): 379–423.

[15] Diffie, Whitfield, and Martin E. Hellman. "Multiuser cryptographic techniques." *Proceedings of the June 7–10, 1976, National Computer Conference and Exposition* (1976).

[16] Rivest, Ronald L., Adi Shamir, and Leonard M. Adleman. *On digital signatures and public-key cryptosystems*. Massachusetts Institute of Technology, Laboratory for Computer Science (1977).
[17] Wiesner, Stephen. "Conjugate coding." *ACM Sigact News* 15.1 (1983): 78–88.
[18] Goyal, Aakash, Sapna Aggarwal, and Aanchal Jain. "Quantum cryptography & its comparison with classical cryptography: A review paper." In *5th IEEE International Conference on Advanced Computing & Communication Technologies [ICACCT-2011]*, (2011) 428–432.
[19] Koblitz, Neal. "Elliptic curve cryptosystems." *Mathematics of Computation* 48.177 (1987): 203–209.
[20] Feynman, Richard P. "Quantum mechanical computers." *Optics News* 11.2 (1985): 11–20.
[21] Benioff, Paul. "The computer as a physical system: A microscopic quantum mechanical Hamiltonian model of computers as represented by Turing machines." *Journal of Statistical Physics* 22 (1980): 563–591.
[22] Grover, Lov K. "A fast quantum mechanical algorithm for database search." In *Proceedings of the twenty-eighth annual ACM symposium on Theory of computing* (1996) 212-219.
[23] Shor, Peter W. "Polynomial-time algorithms for prime factorization and discrete logarithms on a quantum computer," *SIAM Review* 41.2 (1999): 303–332.
[24] Van Meter, R., K.M. Itoh, T.D. Ladd. "Architecture-dependent execution time of Shor's algorithm." *Controllable Quantum States: Mesoscopic Superconductivity and Spintronics (MS+ S2006)*. 183 (2008).
[25] Martin-Lopez, Enrique, et al. "Experimental realization of Shor's quantum factoring algorithm using qubit recycling." *Nature Photonics* 6.11 (2012): 773–776.
[26] Chen, Lily, et al. *Report on post-quantum cryptography*. Vol. 12. US Department of Commerce, National Institute of Standards and Technology, (2016) https://nvlpubs.nist.gov.
[27] Fano, Ugo. "Description of states in quantum mechanics by density matrix and operator techniques." *Reviews of Modern Physics* 29.1 (1957): 74.
[28] Fox, Anthony Mark. *Quantum optics: An introduction*. Vol. 15. Oxford University Press (2006).
[29] Hadfield, Robert H. "Single-photon detectors for optical quantum information applications." *Nature Photonics* 3.12 (2009): 696–705.
[30] Kingston, Robert Hildreth. "Detection of optical and infrared radiation." 10. Springer (2013).
[31] Hirano, Takuya, et al. "Quantum cryptography using pulsed homodyne detection." *Physical Review A* 68.4 (2003): 042331.
[32] Yuen, H. P., and Shapiro, J. H., *Quantum Statistics of Homodyne and Heterodyne Detection in Coherence and Quantum Optics IV.* ed. L. Mandel and E. Wolf. Plenum (1978).
[34] Oulouda, Youssef, Mohammed El Falaki, and Mohamed Daoud. "Quantum key distribution using a single-photon added? Subtracted squeezed coherent state." *Journal of Russian Laser Research* 44.1 (2023): 13–24.
[35] Mogos, Gabriela. "Intercept-resend attack on quantum key distribution protocols with two, three and four-state systems: Comparative analysis." *2015 2nd International Conference on Information Science and Security (ICISS)*. IEEE (2015).
[36] Borelli, Luis F. M., et al. "Quantum key distribution using continuous-variable non-Gaussian states." *Quantum Information Processing* 15 (2016): 893–904.
[37] Plaga, Rainer. "A fundamental threat to quantum cryptography: Gravitational attacks." *The European Physical Journal D – Atomic, Molecular, Optical and Plasma Physics* 38 (2006): 409–413.
[33] DeLange, O. E. "Optical heterodyne detection." *IEEE Spectrum* 5.10 (1968): 77–85.
[38] Mitra, Saptarshi, et al. "Quantum cryptography: Overview, security issues and future challenges." *2017 4th International Conference on Opto-Electronics and Applied Optics (Optronix)*. IEEE, (2017).

Index

Note: Page locators in **bold** and *italics* represents tables and figures, respectively.

J

For Product Safety Concerns and Information please contact our
EU representative GPSR@taylorandfrancis.com Taylor & Francis
Verlag GmbH, Kaufingerstraße 24, 80331 München, Germany